AF812324

International Series in Operations Research & Management Science

Founding Editor
Frederick S. Hillier, Stanford University, Stanford, CA, USA

Volume 354

Series Editor

Camille C. Price, Department of Computer Science, Stephen F. Austin State University, Nacogdoches, TX, USA

Associate Editor

Joe Zhu, Business School, Worcester Polytechnic Institute, Worcester, MA, USA

Editorial Board Members

Emanuele Borgonovo, Department of Decision Sciences, Bocconi University, Milan, Italy

Barry L. Nelson, Department of Industrial Engineering and Management Sciences, Northwestern University, Evanston, IL, USA

Bruce W. Patty, Veritec Solutions, Mill Valley, CA, USA

Michael Pinedo, Stern School of Business, New York University, New York, NY, USA

Robert J. Vanderbei, Princeton University, Princeton, NJ, USA

The book series **International Series in Operations Research and Management Science** encompasses the various areas of operations research and management science. Both theoretical and applied books are included. It describes current advances anywhere in the world that are at the cutting edge of the field. The series is aimed especially at researchers, advanced graduate students, and sophisticated practitioners.

The series features three types of books:

• Advanced expository books that extend and unify our understanding of particular areas.

• Research monographs that make substantial contributions to knowledge.

• Handbooks that define the new state of the art in particular areas. Each handbook will be edited by a leading authority in the area who will organize a team of experts on various aspects of the topic to write individual chapters. A handbook may emphasize expository surveys or completely new advances (either research or applications) or a combination of both.

The series emphasizes the following four areas:

Mathematical Programming : Including linear programming, integer programming, nonlinear programming, interior point methods, game theory, network optimization models, combinatorics, equilibrium programming, complementarity theory, multiobjective optimization, dynamic programming, stochastic programming, complexity theory, etc.

Applied Probability: Including queuing theory, simulation, renewal theory, Brownian motion and diffusion processes, decision analysis, Markov decision processes, reliability theory, forecasting, other stochastic processes motivated by applications, etc.

Production and Operations Management: Including inventory theory, production scheduling, capacity planning, facility location, supply chain management, distribution systems, materials requirements planning, just-in-time systems, flexible manufacturing systems, design of production lines, logistical planning, strategic issues, etc.

Applications of Operations Research and Management Science: Including telecommunications, health care, capital budgeting and finance, economics, marketing, public policy, military operations research, humanitarian relief and disaster mitigation, service operations, transportation systems, etc.

This book series is indexed in Scopus.

Chandrasekar Vuppalapati

Assessing Policy Effectiveness using AI and Language Models

Applications for Economic and Social Sustainability

Springer

Chandrasekar Vuppalapati
Computer Engineering
San Jose State University
San Jose, CA, USA

ISSN 0884-8289 ISSN 2214-7934 (electronic)
International Series in Operations Research & Management Science
ISBN 978-3-031-56096-5 ISBN 978-3-031-56097-2 (eBook)
https://doi.org/10.1007/978-3-031-56097-2

© The Editor(s) (if applicable) and The Author(s), under exclusive license to Springer Nature Switzerland AG 2024
This work is subject to copyright. All rights are solely and exclusively licensed by the Publisher, whether the whole or part of the material is concerned, specifically the rights of translation, reprinting, reuse of illustrations, recitation, broadcasting, reproduction on microfilms or in any other physical way, and transmission or information storage and retrieval, electronic adaptation, computer software, or by similar or dissimilar methodology now known or hereafter developed.
The use of general descriptive names, registered names, trademarks, service marks, etc. in this publication does not imply, even in the absence of a specific statement, that such names are exempt from the relevant protective laws and regulations and therefore free for general use.
The publisher, the authors, and the editors are safe to assume that the advice and information in this book are believed to be true and accurate at the date of publication. Neither the publisher nor the authors or the editors give a warranty, expressed or implied, with respect to the material contained herein or for any errors or omissions that may have been made. The publisher remains neutral with regard to jurisdictional claims in published maps and institutional affiliations.

This Springer imprint is published by the registered company Springer Nature Switzerland AG
The registered company address is: Gewerbestrasse 11, 6330 Cham, Switzerland

Paper in this product is recyclable.

The author dedicates his works, book, and efforts of this book to his late father Murahari Rao Vuppalapati, who had inspired him to write the book and guided the author's pursuits of life, and to his loving mother Hanumayamma Vuppalapati, who was the inspiration to develop company and dairy products for the betterment of humanity and future generations.

Preface

A country's resilience reflects the strength and courage of its people. The true test of resilience is the spirit that overcame challenges, hardships, realities, and economic fluctuations along a country's journey: such as India. Since the partition of the country in 1947, India has faced many of these challenges and economic changes; the annual budget presentation, a genuine democratic process, by the finance minister to the republic of India has recorded these shifting dynamics and has written patterns and macro-economic indicators to be learned from the budget speeches. Studying these historical events data and modeling the impact of policy measures offer a wealth of useful insights and recommendations to achieve economic and social sustainability of the poor and marginalized.

To understand the journey of modern independent India and its historical patterns, the book uses data from 1990 onward, after economic liberalization, to February 2023. The book meticulously crafted models on policy effectiveness by combining the budget policies with the statistical models that analyze economic events, macroeconomic trends, and ratings and commerce terms from international institutions and reliable data sources such as the World Bank, International Monetary Fund, the Reserve Bank of India's *Handbook of Statistics on Indian Economy*, Asian Development Bank, U.S. Bureau of Labor Statistics, and others. By linking policies to economic outcome patterns, the book aims to provide a rich source of successful policy insights on how effective they are in bringing development to the poor and underserved communities and ensuring the distribution of wealth, social wellbeing, and standard of living to the common people rather than a selected group, even though that is the reality I have seen since 1980s.

For this book, we have used natural language processing (NLP), generative pre-trained transformer (GPT) models, autoregressive language models, and autoregressive predictive mathematical models to analyze more than 33 years of budget speech and key economic signals. This helps us to show the connection between macroeconomics and real economy. We aim to evaluate the policy effectiveness of the Indian government in terms of social welfare, poverty alleviation, and food security policy in a world that faces extreme weather, climate change, and economic shocks. The book's mission is to bring together software engineers, data scientists, social scientists, economists, and agriculture practitioners to create applications for marginalized communities and provide economic sustainability in the times of increasing austerity measures, speculative markets, climate change, and international trade tensions.

San Jose, CA, USA Chandrasekar Vuppalapati

Acknowledgment

The author is deeply indebted to the love, support, and encouragement of his wife Anitha Ilapakurti and the active support of his daughters Sriya Vuppalapati and Shruti Vuppalapati who helped draft the first version of the manuscript.

The author is deeply thankful to the support of his elder brother Rajasekar Vuppalapati and the joyful support of his younger brother Jaya Shankar Vuppalapati. Additionally, the author is also very thankful to his elder sisters Padmavati Vuppalapati and Sridevi Vuppalapati.

The author sincerely thanks Santosh Kedari and his team in India for conducting field studies in India for Agriculture, Specialty Crops, Dairy analytics products and bringing the market and small farmers' feedback for the betterment of services and for the book. The author is deeply thankful to Professor Vincent Soundaram for providing policy implications on education.

The author is deeply indebted to Sharat Kedari, who helped in setting GPT models and train budget speech data using GPT models on Google. In addition, the author is thankful for the support, hard work, and dedication of Vanaja Satish who helped to develop ML forecasting models for predicting poverty rates in India. Finally, the author is sincerely and deeply thankful to Sandhya Vissapragada for her contributions in data science and in creating machine learning models on Microsoft Azure and Google Cloud.

Contents

1 Introduction . . . 1
What Is AI? . . . 2
Brief History . . . 3
Machine Learning . . . 5
Deep Learning . . . 5
Time Series . . . 6
Multivariant Time Series . . . 11
Autocorrelation . . . 11
Partial Autocorrelation Function (PACF) . . . 13
Stationarity Check: Augmented Dickey-Fuller (ADF) and Kwiatkowski-Phillips-Schmidt-Shin (KPSS) Tests . . . 13
Non-stationary Stochastic Time Series . . . 14
Forecasting Models . . . 21
Supervised Learning . . . 21
Regression Models . . . 23
Metrics for Regression Models . . . 23
Ordinary Least Squares Linear Regression . . . 24
Logistic Regression . . . 27
Random Forest Trees . . . 29
Gradient Boosting Regressor . . . 30
Polynomial Regression . . . 30
References . . . 32

2 Natural Language Models . . . 35
Potential Use Cases . . . 36
NLP Techniques . . . 37
Tokenization . . . 38
Part-of-Speech (POS) Detection . . . 39
Stemming and Lemmatization . . . 41

Cosine Similarity Search 46
GFS Analytics Framework 48
Deep Learning and NLP 50
Word Embeddings 51
Term Frequency-Inverse Document Frequency (TF-IDF) 51
Bag of Words (BOW) 53
Economic and Social Sustainability Data Mining Using NLP and Budget Speeches 53
Development of Budget Speech Embeddings with Classical NLP 55
Building BOW Embedding 56
Economic Sustainability BOW Embedding Model 60
Data Sources 60
Step 1: Load Libraries for Performing NLP 61
Step 2: Initialize Memory Variables and NLP Stop Word Processor 61
Step 3: Set Tokenizer 62
Step 4: Lemmatize and Stem Each Word 62
Step 5: Process String 63
Step 6: Convert Budget Speech Statements Into Embeddings 63
Step 7: Entry Function 63
Step 8: Output Budget Speech Statement Word Embeddings 65
Step 9: Vector Embedding 65
Adaptive Machine Learning Model: Automated Budget Statement Labeler 66
Chapter Machine Learning Experiments 67
References 68

3 Large Language Models 71
What Is AI? 72
Brief History 73
Machine Learning 75
Deep Learning 76
Generative AI 77
Large Language Foundation Models 77
BERT (Bidirectional Encoder Representations from Transformers) 79
Large Language Model Meta AI (Llama) 83
The Transformer 83
Self-Attention in Detail 86
Large Language Model: Generate Social Sustainability Strategies with Llama 2 Model 90
Step 1: Access 91
Step 2: Install the Model and Libraries 92
Step 3: Initialize the Model 93

Step 4: Initialize Tokenizer . . . 94
Step 5: Initialize Tokenizer . . . 95
Step 6: Prepare Multiclass Multilabel Prompt Template . . . 95
Step 7: Classify the Model . . . 96
GPT Models (Generative Pre-trained Transformer) . . . 98
Prompt Engineering . . . 99
Strategies for Getting Better Results . . . 102
How GPT Model Works . . . 103
Embeddings . . . 103
Large Language Model: Generate Social Sustainability Strategies with GPT-2 Model . . . 106
Step 1: Install Transformers . . . 106
Step 2: Set Output Maximum Length and Seed Value . . . 107
Step 3: Generate Social Sustainability Strategies . . . 108
Step 4: Choose GPT-2 Large Model . . . 108
Step 5: Greedy Search Decoding Method . . . 109
Step 6: Beam Search with N-gram Penalties . . . 109
Step 6: Basic Sampling . . . 111
Step 7: Top-K Sampling . . . 113
Step 8: Top-P Sampling . . . 115
Step 9: Top-K and Top-P Sampling . . . 118
Chain of Thought . . . 119
Large Language Model: Economic and Social Sustainability of Marginalized Communities Chat with Indian Union Budget Speeches . . . 120
Step 1: Install Transformers . . . 120
Step 2: Bring Your Own Budget Speech Files . . . 121
Step 3: Recursive Character Text Splitter . . . 122
Step 4: Create Local Index Stores to Store Embeddings . . . 122
Step 5: Create Chat QA Interface . . . 123
Step 6: Prompt Engineering to Retrieve Macroeconomic Stability Statements . . . 124
Step 7: Prompt Engineering to Retrieve Macroeconomic Stability Statements . . . 125
Step 8: Prompt Engineering to Retrieve Economic Sustainability Budget Statements . . . 125
Step 9: Prompt Engineering with Few Shots to Classify Economic Sustainability Results . . . 126
Step 10: Prompt Engineering to Retrieve Social Sustainability Results . . . 127
Step 11: Prompt Engineering with Few Shots to Classify Social Sustainability Results . . . 128
References . . . 129

4 Macroeconomic Indicators, Aggregates, and Framework 133
Fiscal and Macroeconomic Stabilization 137
Fiscal Discipline 139
Fiscal Correction 142
Enhancing Credit Delivery 142
Machine Learning Model: Money Supply and Economic Growth—SLR, CRR, and BR 144
Step 1: Load the Libraries 147
Step 2: Load Statutory Liquidity Ratio (SLR) Data 147
Step 3: Load Bank Rate Data 148
Step 4: Load Bank Credit Data 149
Step 5: Load Cash Reserve Data 150
Step 6: Merge CRR, SLR, Bank Rate, and Bank Credit 151
Step 7: Correlation Analysis 152
Step 8: Linear Regressor Model 153
Step 8: Evaluate Linear Regressor Model Metrics 154
Step 8: Model Explainability 154
Step 9: Gradient Boosting Regressor Model 154
Step 10: Evaluate Linear Regressor Model Metrics 155
Step 11: Model Explainability 156
Step 12: Random Forest Regressor 156
Conclusions and Policy Implications 157
Machine Learning Model: Budget Deficit, Money Supply, and Inflation Linkage and Policy Implications 158
Data Sources 159
Step 1: Load Libraries for performing Time Series and ML Modeling 160
Step 2: Load Broad Money Growth (Annual %), Money Stock, Inflation, GDP, and Deficit Data for India 161
Step 3: Check for Nulls in the Data 162
Step 4: Correlation Analysis 162
Step 5: Plot Dataset 164
Step 6: Transformation 166
Step 7: Normality Test 166
Step 8: Check for Autocorrelation 168
Step 10: Train and Test Data 169
Step 11: Transformation 169
Step 12: Stationarity Check 170
Step 12: Plot Visualization to See Variables Stationarity 171
Step 13: Granger's Causality Test 171
Step 14: VAR Model 173
Step 15: Residual Plot 173
Conclusion 174

Impacts of Financial Development and Other Factors on Inequality and Poverty 175
Share Prices 177
Inflation 178
Imports 179
Food Grain Stocks 179
Exchange Rate 179
Removal of Bureaucratic Interference 180
Manufacturing 181
Modeling Key Macroeconomic Key Indicators 188
UNSD Statistical Yearbook 188
OCED Main Economic Indicators 189
Statistical Methodological Information for International Comparisons: Dissemination Model 190
The World Bank Open Data 193
Poverty and Inequality 194
Economic Indicators 195
Chapter Machine Learning Experiments 198
References 199

5 **Economic Sustainability** 203
Industrial Production 205
Foreign Investment 206
Exports 206
The Public Distribution System (PDS) 206
Machine Learning Linkage Model: Public Distribution System and Inflation—Protect the Poorer Sections of the Population from High Prices and Shortages 209
Step 1: Load the Libraries 210
Step 2: Load Public Distribution System Procurement, Off-Take, and Stocks Data 211
Step 3: Load Inflation, Consumer Prices (Annual Percentage) 214
Step 4: Load Population Growth (Annual Percentage) Data 214
Step 5: Load Poverty Headcount (<$1.9) Data 215
Step 6: Load Economic and Drought Years Binary Data 215
Step 7: Merge Public Distribution System (PDS), Inflation, Poverty Rates, Population Growth, and Recession/Drought Data 216
Step 8: Linear Regressive Model 217
Step 9: Model Metrics 217
Step 10: Model Equation ($Y = MX + C$) 218
Step 11: Model Explainability 219
Step 12: Gradient Boosting Regressor Model 219
Step 13: Gradient Boosting Regressor Model Metrics 220
Step 14: Model Explainability 221
Step 15: Random Forest Model 221

Step 16: Random Forest Regressor Model Metrics . . . 222
Step 17: Model Explainability . . . 223
Model Comparison and Summary . . . 223
Policy Recommendations and Future Consideration . . . 225
Machine Learning Modeling—Public Distribution System—Off-Take, a True Indicator of Economic and Social Sustainability (1990:2021) . . . 226
Coincidence of Recessions with Credit Crunch and Bursts . . . 231
Data Sources . . . 232
Step 1: Load Libraries for Performing Time Series and ML Modeling . . . 234
Step 2: Load Money-Banking, Financial Markets, Public Finances, Trade Balance of Payments Linkage Data for India . . . 234
Step 3: Check for Nulls in the Data . . . 235
Step 4: Correlation Analysis . . . 236
Step 5: Plot Dataset . . . 238
Step 6: Transformation . . . 240
Step 7: Normality Test . . . 240
Step 8: Normality Probability Plot . . . 243
Step 9: Check for Autocorrelation . . . 244
Step 10: Train and Test Data . . . 245
Step 11: Transformation . . . 245
Step 12: Stationarity Check . . . 246
Step 12: Plot Visualization to See Variables Stationarity . . . 247
Step 13: Granger's Causality Test . . . 247
Step 14: VAR Model . . . 249
Step 15: Residual Plot . . . 251
Step 16: Durbin-Watson Statistics . . . 251
Step 17: Prediction or Forecast VAR . . . 252
Step 18: Invert the Transformation . . . 253
Step 19: Evaluation . . . 254
Policy Implication . . . 254
Data Science Infused Policy Implications: Large Language Models and Public Services Forecasting Patterns . . . 256
Step 1: Prepare Time Series Prophet Model . . . 256
Step 2: Prepare Prophet Time Series Variables . . . 257
Step 3: Check Any Nulls . . . 258
Step 4: Split Data to Model Time Series . . . 258
Step 5: Prepare Prophet Time Series Model . . . 259
Step: Forecast Model . . . 260
Step: Evaluate the Model . . . 262
Agricultural Productivity and Prosperity for Farmers . . . 265
Agricultural Credit . . . 265
Sustainable Farming Practices . . . 268

Machine Learning Model: Sustained Agricultural Production and Productivity Enhancement Needs in the Era of Climate Change 269
India and Wheat Production 273
Indian Wheat Growing Zones 276
Breadbasket to the World 285
Production of 2021 Wheat Crop Officially Forecast at Record Level 286
Maximum, Minimum, and Average Temperature and Precipitation (1901–2020) 286
Machine Learning Model: Heat Waves and Wheat Yield Production Linkage in India 288
Data Sources 290
Step 1: Load Libraries 291
Step 2: Load Wheat Macroeconomic Finance and Monetary Agriculture Data Frame (1961–2020) 291
Step 3: Check Any Column Data Is Null 292
Step 4: Perform Correlation Analysis 293
Step 5: Train Regressive Model 295
Step 8: Evaluate the Model 296
Step 8: Model Explainability 297
Step 9: Model Equation ($Y = MX + C$) 297
Step 10: Conclusion 298
Chapter Machine Learning Experiments 300
References 300

6 **Social Sustainability** 305
Poverty 310
Two Approaches to Measuring Poverty and Inequality 314
Approach 1: CPHS 314
Approach 2: IMF Working Paper and Poverty Estimates 317
Effectiveness of Public Distribution System (PDS) 319
Credit to Agriculture 320
Macroeconomic Model: Employment and GDP 322
Growth and Poverty 330
Social Sustainability Measures 331
Agricultural Credit 332
Pro-poor and Marginalized Inclusive Growth Machine Learning Model: Poverty and Macroeconomic Linkage Model 333
Step 1: Load Python Libraries to Run the Experiment 339
Step 2: GDP Growth (Annual Percentage) and Time Series Plot 339
Step 3: Employment in Agricultures 341
Step 4: Employment in Services 342
Step 5: Current Account Balance 342
Step 6: Gini Index 343
Step 7: Consumer Prices (Annual Percentage) 344

Step 8: Gross Savings (% of GDP) 344
Step 9: Gross National Expenditure (% of GDP) 345
Step 10: General Election Data 346
Step 11: Droughts in India 347
Step 12: Economic Boom and Bust Years in India 347
Step 13: Public Distribution System: Procurement, Off-Take, and Stocks from the Reserve Bank of India *Handbook of Statistics on Indian Economy* 348
Step 14: Life Expectancy at the Birth Data 349
Step 15: The Minimum Support Prices (MSPs) for Food Grains 350
Step 16: Population Growth (Annual Percentage) 352
Step 17: Deployment of Bank Credit by Major Sectors 352
Step 18: Institutional Credit for Agriculture and Allied Activities 354
Step 19: Price to Earnings Ratio of the Bombay Stock Exchange 355
Step 20: Linear Regressor Model 356
Step 21: Construct Linear Regression Model Metrics 357
Step 22: Construct Model Explainability 358
Step 22: Gradient Boosting Regression Model 359
Step 23: Construct Random Forest Regression Model 361
Model Comparison and Summary 362
Conclusion and Policy Measures 365
References 367

7 Conclusion 373

Appendix 375

Index 463

Abbreviations

AAY	Antyodaya Anna Yojana
AI	Artificial Intelligence
ARIMA	Autoregressive Integrated Moving Average
BEA	The U.S. Bureau of Economic Analysis
BERT	Bidirectional Encoder Representations from Transformers
BLS	Bureau of Labor Statistics
BOP	Balance of Payments
BPL	Below the Poverty Line
CBLD	IMF's Central Bank Legislation Database
CDF	Cumulative Distribution Function
CES	Consumer Expenditure Survey
CPHS	The Consumer Pyramid Household Survey
CVT	Computer Vision
DPI	Digital Public Infrastructure
eNAM	National Agriculture Market
FAISS	Facebook AI Similarity Search
FAO	The Food and Agriculture Organization
FCI	Food Corporation of India
FPS	Fair Price Shops
FRED	Federal Reserve Economic Data
GDP	The Gross Domestic Product
GEOS	Goddard Earth Observing System
GFCF	Gross Fixed Capital Formation
GLUE	General Language Understanding Evaluation
GNI	Gross National Income
HCR	Head Count Ratio
ICT	Information Communication Technology
IFAD	The International Fund for Agricultural Development

IFS	International Financial Statistics
IIP	The International Investment Position (The IIP—https://data.imf.org/?sk=7A51304B-6426-40C0-83DD-CA473CA1FD52)
ILO	Institute of Labour Organization
IMF	The International Monetary Fund
KL Divergence	Kullback-Leibler divergence
LAC	The Latin America and Caribbean
MEI	Main Economic Indicators
MFCC	The Mel frequency Cepstrum Coefficient
MGP	Minimum Guarantee Prices
ML	Machine Learning
MLP	Multi-Layer Perceptron
MPI	Multidimensional Poverty Index
MPO	The Macro Poverty Outlook
MSP	Minimum Support Price
NABARD	National Bank for Agriculture and Rural Development
NEBR	National Bureau of Economic Research
NEER	Nominal effective exchange rate
NFSA	The National Food Security Act
NLP	Natural Language Processing techniques
NLTK	Natural Language Tool Kit
NSS	The National Sample Survey
OECD	Organisation for Economic Co-operation and Development
OLS	Ordinary Least Squares
P.L.480 (USAID & PL-480—https//history.state.gov/milestones/1961-1968/pl-480) USAID and PL–480, 1961–1969	The administrations of John F. Kennedy and Lyndon B. Johnson marked a revitalization of the U.S. foreign assistance program, signified a growing awareness of the importance of humanitarian aid as a form of diplomacy, and reinforced the belief that American security was linked to the economic progress and stability of other nations
PCA	Principal Component Analysis
PCE	Personal Consumption Expenditures
PDF	Probability Distribution Function
PDS	Public Distribution System
PLB	Poverty Line Basket
PLFS	Periodic Labor Force Survey
PPP	Purchasing Power Parity
PRS	Poverty Reduction Strategy

Public Debt Act, (Public Debt Act, 1944—https//www.indiacode.nic.in/handle/123456789/2405?view_type=browse&sam_handle=123456789/1362) 1944	The Public Debt Act, 1944 was an act of the Parliament of India which provided a legal framework for the issuance and servicing of government securities in India. The Act oversees government securities and their management by the Reserve Bank of India
RBI	Reserve Bank of India
RLHF	Reinforcement Learning with Human Feedback
RMS	Root Mean Square
S&P 500 Index	The Standard and Poor's Financial Index
SDG	Sustainable Development Goals (SDGs)
SDR (SDR Valuation—https//www.imf.org/external/np/fin/data/rms_sdrv.aspx)	Special Drawing Right (SDR—https://www.imf.org/en/About/Factsheets/Sheets/2016/08/01/14/51/Special-Drawing-Right-SDR)—The SDR is an international reserve asset, created by the IMF in 1969 to supplement its member countries' official reserves. The value of the SDR is based on a basket of five currencies—the US dollar, the euro, the Chinese renminbi, the Japanese yen, and the British pound sterling
SFT	Supervised Fine-Tuning
SLR	Statutory Liquidity Ratio
SNA	System of National Accounts
SQuAD	Stanford Question Answering Dataset
SVM	Support Vector Machines
The HUNT Study	The Trøndelag Health Study
TS	Time Series
UN	The United Nations
UNDP	The United Nation Development Programme
UNFPA	United Nations Population Fund
UNICEF	The United Nations International Children's Emergency Fund
UNSD	The United Nations Statistical Division
USDA	The U.S. DEPARTMENT OF AGRICULTURE
VAR	Vector Autoregression
WDI	World Development Indicators
WFP	World Food Programme
WHO	The World Health Organization
WPS	The World Bank Policy Research Working Paper

Chapter 1
Introduction

This chapter covers the following:

- What is artificial intelligence (AI)?
- Machine learning (ML)
- Supervised, unsupervised, and deep learning
- Metrics to evaluate classification and regression models
- Time series

Artificial intelligence is the topic of this chapter, and it covers methods that make up AI and how AI relates to traditional machine learning (ML) methods. Then, it explains time series. The chapter also covers how to process regression methods. Lastly, it presents metrics to evaluate regressive models.

Artificial intelligence (AI) stands out as a transformational technology of our digital age—and its practical application throughout the economy is growing apace [1]. One of the chief reasons why AI applications are getting prominent and industry acceptance is in its software's ability to learn, albeit continuously, from real-world use and experience and its capability to improve its performance. It is no wonder that the applications of AI span from complex high-technology equipment manufacturing to personalized exclusive recommendations. AI enabled newer business models by understanding the data patterns and creating products that touch the customer's needs.

Artificial intelligence is changing how we interact with everything, from food to healthcare, travel, and also religion [2]. Artificial intelligence applications enable agriculture, one of the oldest and most important professions in the world, to be more productive and enable to achieve sustainability, especially for small farmers. At the same time, application of AI would enable us to understand how human civilizations evolved or how architects of religion, faith, construction, and arts have interwoven

© The Author(s), under exclusive license to Springer Nature Switzerland AG 2024
C. Vuppalapati, *Assessing Policy Effectiveness using AI and Language Models*,
International Series in Operations Research & Management Science 354,
https://doi.org/10.1007/978-3-031-56097-2_1

the needs of people using engineering and technologies available thousands of years ago.

Using machine learning and artificial intelligence techniques, we can understand time reckoning, application of acoustics in historical artifacts, physical perimeter security alert systems using acoustics, interaction with nature, energy principles, commerce, and many others. Application of scientific view using AI, in my view, opens new understandings of historical relics that have influence in worldwide cultures today.

Application of AI on pre-historic etiquettes that have extensive ties to constructions, arts, architecture, music, religion, and time would provide new understanding of our human history and of relics that have shaped our modern lives!

What Is AI?

Artificial intelligence (AI) has been defined and understood in many ways, as it encompasses a vast field within its realm [3]. Perspectives of AI also vary based upon many factors, use cases, and actors involved. Therefore, no definition of AI can be so comprehensive that it will cover all the dimensions of AI. AI is a very broad and diverse field. For more details on AI, please refer to my other books.

Alan Turing generally is credited with the origin of the concept when he speculated in 1950 about "thinking machines" that could reason at the level of a human being[1] [4]. Turing was followed up a few years later by John McCarthy, who first used the term "artificial intelligence" to denote machines that could think autonomously. He described the threshold as "getting a computer to do things which, when done by people, are said to involve intelligence."

Stuart Russell and Peter Norvig in defining AI[2] have used the following taxonomy about systems that ([5, 6]):

- Think like humans (e.g., cognitive architectures or neural network);
- Think rationally like humans (e.g., inferences optimization);
- Act like humans (e.g., language processing, knowledge representation, automated reasoning and learning); and.
- Act rationally like humans (e.g., intelligent software agents, embodied robots that achieve goals via perception, planning, reasoning, learning etc.).

[1] What is Artificial Intelligence?—https://www.brookings.edu/research/what-is-artificial-intelligence/

[2] What is Artificial Intelligence? An Informed Definition—https://emerj.com/ai-glossary-terms/what-is-artificial-intelligence-an-informed-definition/#:~:text=In%20%E2%80%9CArtificial%20Intelligence%3A%20A%20Modern,many%20different%20splintered%20fields%20%E2%80%93%20speech

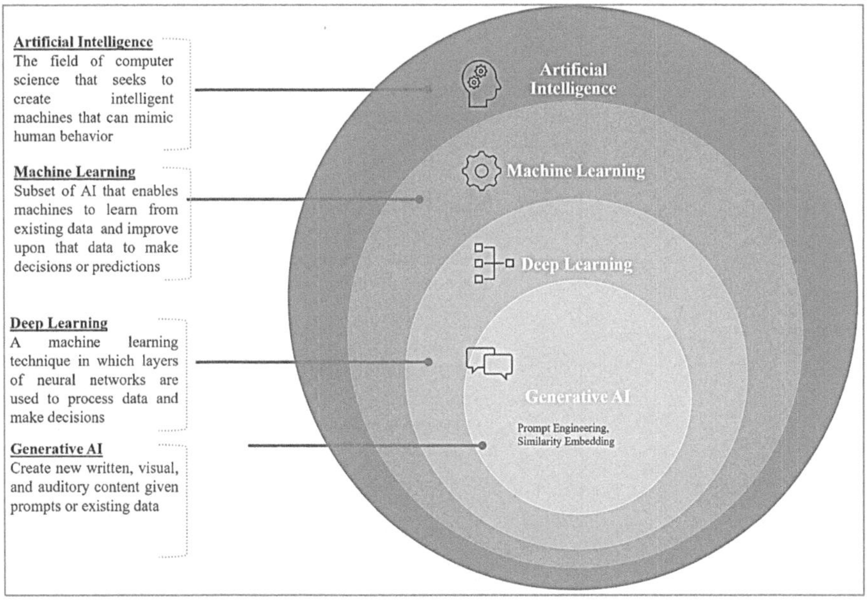

Fig. 1.1 AI framework

In computer science, artificial intelligence (AI), sometimes called machine intelligence, is intelligence demonstrated by machines, in contrast to the natural intelligence displayed by humans. According to Gartner Market Research,[3] artificial intelligence applies advanced analysis and logic-based techniques, including machine learning, to interpret events, support and automate decisions, and take actions. The typical definition of AI looks something like the composite design of AI, machine learning, and deep learning [7] (see Fig. 1.1).

Brief History

Artificial intelligence (AI) has a long and storied history that spans many decades. Some of the earliest work in AI was done in the 1950s and 1960s, when researchers first began to explore the possibility of creating machines that could "think" like humans.

In the 1950s and 1960s, AI research focused primarily on "symbolic" AI, which involved creating symbolic representations of knowledge and using them to solve problems. Researchers in this field developed early versions of what are now known

[3] Artificial Intelligence (AI)—https://www.gartner.com/en/information-technology/glossary/artificial-intelligence

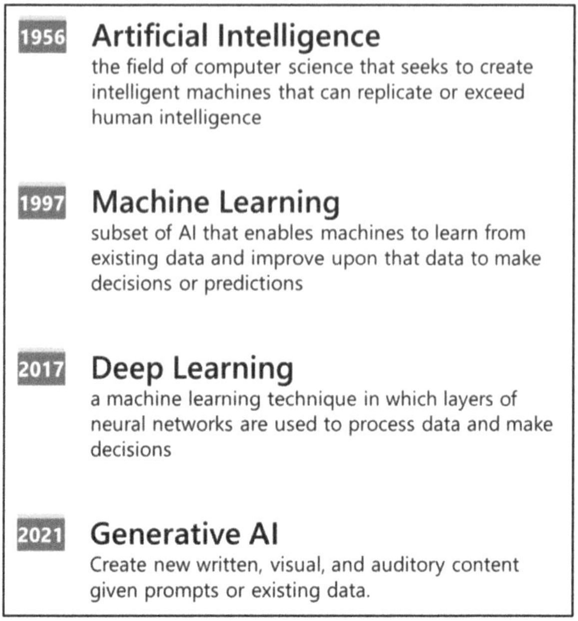

Fig. 1.2 AI history

as expert systems, which are computer programs that mimic the decision-making abilities of human experts in a specific domain. The principle behind symbolic representation of time series is to allow search time series patterns and cluster using distance (Euclidian or cosine)-based metrics [8].

In the 1970s and 1980s, AI research began to shift away from symbolic AI and towards "sub-symbolic" AI, which involves using statistical methods and mathematical models to solve problems. This shift was driven in part by the realization that symbolic AI was too complex and difficult to scale and in part by the development of new computer hardware that made it possible to process large amounts of data quickly (please see Fig. 1.2).

In the 1990s and 2000s, AI research continued to evolve, and new subfields began to emerge. One of these was "machine learning," which focuses on creating algorithms that allow computers to learn from data without being explicitly programmed. Another was "neural networks," which are mathematical models inspired by the structure and function of the human brain.

Recently, generative models like GPT-3 and DALL-E, variational autoencoders, generative adversarial networks, and many more have risen and continue to advance the field of AI. These models are able to generate new examples that are similar to examples from a given dataset. It is being used in diverse fields such as natural language processing, computer vision, speech recognition, and many more.

AI embodies techniques that enable computers to mimic and predict human behavior. Simply put, AI is the ability of a machine to perform cognitive functions that are associated with human minds, such as perceiving, reasoning, learning, and problem-solving. A full end-to-end AI solution, what we sometimes call a system of intelligence or a smart system, can ingest human-level knowledge (e.g., via machine reading and computer vision) and use this information to automate and accelerate tasks that were previously performed only by humans [7–9].

Overall, AI has come a long way since its inception in the 1950s, and it continues to evolve and improve at a rapid pace today [7, 8]. Now, we can fairly say we've entered AI age with full availability of AI across human pursuits that improve human ingenuity for the betterment of the future generations and appreciate the efforts of our forefathers that survived us through the thick and thin of times.

Using AI, we can identify the role of nature in pre-historic times and methodization of time-based principles as part of human evolution. Novel technologies such as AI and machine learning enable us to revisit historical events and factors through the scientific lens that wasn't available till the late nineteenth century. It helps us appreciate the line of thinking of architects and scientific thinkers of pre-historic Hinduism; fully understand the principles of life, science, and religion that are perceived revolutionary in those times; and understand the scientific underpinnings of such principles that are closely related to nature which is only possible through novel scientific view, with the help of data science technologies that are available to us in AI age.

Machine Learning

Machine learning (ML) is the subset of AI that uses statistical techniques to enable machines to improve with experience. Machine learning algorithms detect patterns and learn how to make predictions and recommendations by processing data and experiences rather than by receiving explicit programming instruction [10]. The algorithms also adapt in response to new data and experiences to improve efficiency and efficacy over time. ML, importantly, enables to autocalibrate, correct, or rebuild itself if it sees that the data is drifting, of course, from the original trained model. The availability of such features heralds a new era of computing that can adapt itself to course-correct or hyper-scale efficiently as it encounters new challenges. There are three types of learning that AI uses: supervised learning, unsupervised learning, and semi-supervised learning. For more on these topics, please refer to my book. In this book I will confine myself to supervised learning and apply statistical techniques to develop ML for historical time series patterns.

Deep Learning

Deep learning[4] is a subset of machine learning, which is essentially a neural network with three or more layers. These neural networks attempt to simulate the behavior of the human brain—albeit far from matching its ability—allowing it to "learn" from

[4]What is deep learning?—https://www.ibm.com/topics/deep-learning

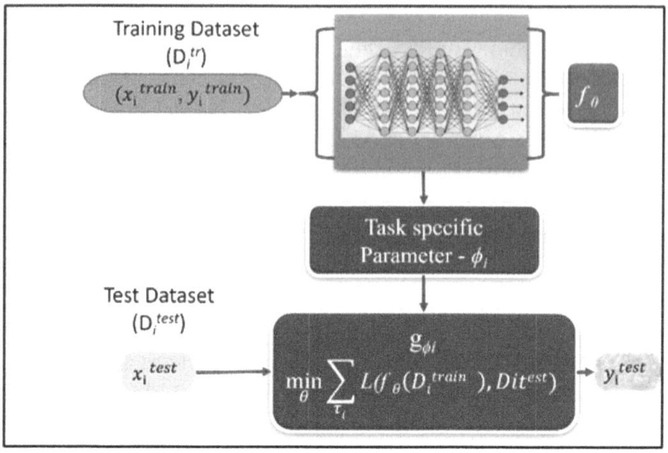

Fig. 1.3 Deep learning framework

large amounts of data. While a neural network with a single layer can still make approximate predictions, additional hidden layers can help to optimize and refine for accuracy (please see Fig. 1.3).

Deep learning drives many artificial intelligence (AI) applications and services that improve automation, performing analytical and physical tasks without human intervention. Deep learning technology lies behind everyday products and services (such as digital assistants, voice-enabled TV remotes, and credit card fraud detection) as well as emerging technologies (such as self-driving cars).

In the above the neural network then employs these parameters to make predictions for new outcomes. In this process, we're going to use both the training set (D_i^{train}) and the test set (D_i^{test}). The neural networks are trained using standard supervised learning across all tasks. The overall goal is to measure the model's performance, with parameters ϕ, in predicting outcome variable, Y, for new data.

Time Series

The time series is the universal data signature. A hierarchy of all the various time series representations in the literature is presented in Fig. 1.4. The leaf nodes refer to the actual representation, and the internal nodes refer to the classification of the approach. Symbolic representation is presented as the lower bounding symbolic approach [8].

As with most problems in computer science, the suitable choice of representation greatly affects the ease and efficiency of time series data mining [8]. With this in mind, a great number of time series representations have been introduced, including the discrete Fourier transform (DFT) [11], discrete wavelet transform (DWT) [12],

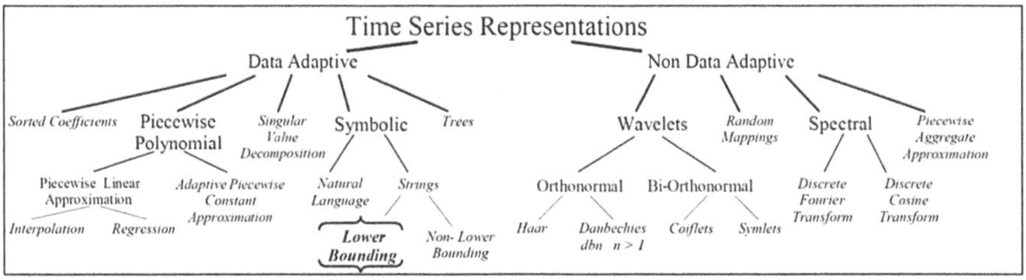

Fig. 1.4 Time series representations

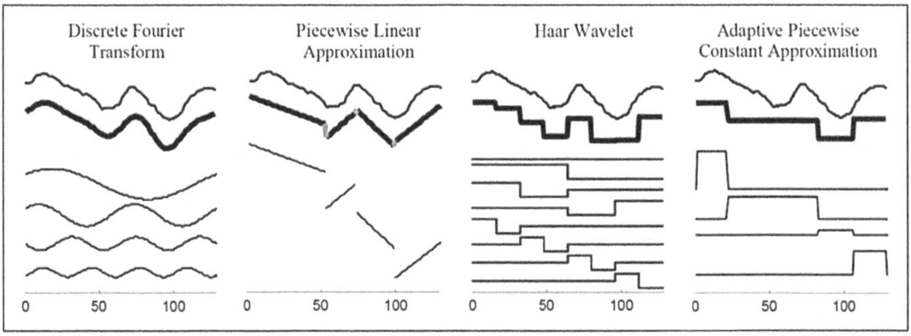

Fig. 1.5 Time series and types

piecewise linear and piecewise constant models (PAA) [13], Adaptive Piecewise Constant Approximation (APCA) [13, 14], and singular value decomposition (SVD) [14]. Figure 1.5 illustrates the most commonly used representations [8].

In addition to the importance of time series algorithms, other important aspects that enable accurate time series analyses and forecasts include history, legacy, and horizon (HLH). The history of data is very important as it embodies signals that correlate with changing dynamics. It is necessary to have sufficient data for time series modeling to capture important attributes such as trend, seasonality, and randomness.

The formula for the size of the training data is dependent on the number of variables, the number of observations, and the number of time steps. The formula for calculating the size of the training data is (please see Eq. 1.1):

$$\text{Size of training data} = \text{Number of variables}^{*}\ \text{Number of observations}^{*}\ \text{Number of time steps} \tag{1.1}$$

Please note that this formula is only for calculating the size of the training data and not the size of the entire model.

Among the many modern datasets that have considerable historical data, the most prominent one is the Producer Price Index. The Producer Price Index (PPI) is the nation's primary measure of price changes in the domestic supply chain, allowing us to monitor how price increases or decreases are passed through from producers to consumers. This Principal Federal Economic Indicator marked its 125th anniversary on March 3, 2016[5]—one of the oldest economic time series compiled by the federal government. The index, known as the Wholesale Price Index (WPI) until 1978, was established as part of a US Senate resolution on March 3, 1891, the last day of the last session of the 51st US Congress. This Congress was famously known as the "Billion-Dollar Congress," because of its expensive initiatives, such as naval expansion and creating pensions for the families of military members who served in the Civil War. The longer the history of data, the greater the model development. As you may know, "no history, no future."

The next one is legacy or Lag. Lag should be used when the relationship between the independent and dependent variables does not match up or correlate by default. For example, when trying to forecast the demand for a product, the demand in any month may depend on the price of specific commodities 3 months prior. In this example, you may want to lag the target (demand) negatively by 3 months so that the model is training on the correct relationship.

Horizon is another important characteristic of data. The forecast horizon is how many periods forward you would like to forecast. The horizon is in units of the time series frequency (e.g., daily, weekly).

To train a forecasting model, you must have a sufficient amount of historical data. This threshold quantity varies with the training configuration. If you've provided validation data, the minimum number of training observations required[6] per time series is given by (please see Eq. 1.2)

$$T \text{ user validation} = H + \max\ (L_{\text{max}}, S_{\text{window}}) + 1 \tag{1.2}$$

where H is the forecast horizon, L_{max} is the maximum lag order, and S_{window} is the window size for rolling aggregation features.

Almost all meaningful measured quantities have time associated with it. Time series (TS) represents the value of a particular characteristic with respect to time. A generalized version of time series consists of three components: trend, seasonality, and randomness. Mathematically time series is represented as (please see Eq. 1.3)

$$Y_t = m_t + s_t + \varepsilon_t \tag{1.3}$$

where m_t is trend, s_t stands for seasonality, and ε_t stands for random variable.[7]

---

[5] PPI—https://www.bls.gov/opub/mlr/2016/article/the-first-50-years-of-the-producer-price-index.htm

[6] Data Length requirements—https://learn.microsoft.com/en-us/azure/machine-learning/concept-automl-forecasting-methods?view=azureml-api-2

[7] Handle Missing Values in Time Series For Beginners—https://www.kaggle.com/code/juejuewang/handle-missing-values-in-time-series-for-beginners/report

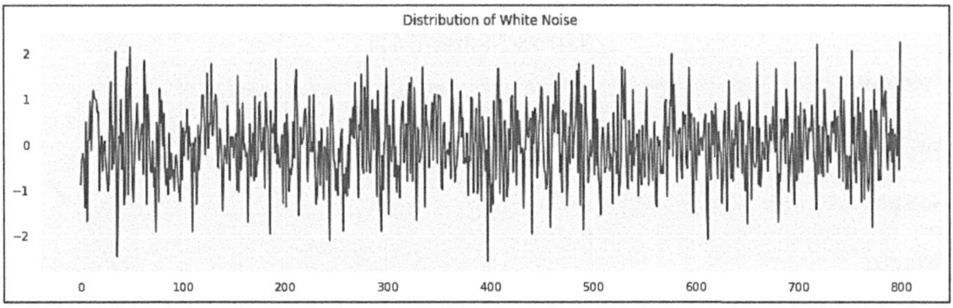

Fig. 1.6 White noise

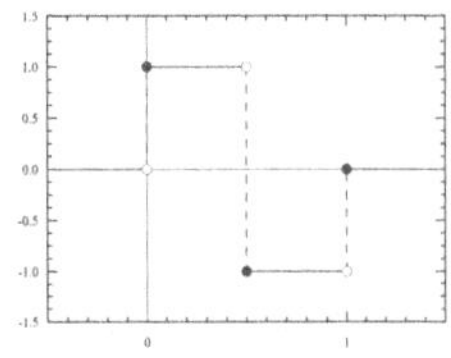

Haar wavelet
In mathematics, the Haar wavelet is a sequence of rescaled "square-shaped" functions which together form a wavelet family or basis. Wavelet analysis is similar to Fourier analysis in that it allows a target function over an interval to be represented in terms of an orthonormal basis. The Haar sequence is now recognized as the first known wavelet basis and is extensively used as a teaching example.

Three characteristics lead to the following time series behavior.

No trend or seasonality (white noise) (please see Fig. 1.6).
Has trend but no seasonality (a good example is M1 money supply (Money Supply—https://rbi.org.in/Scripts/PublicationsView.aspx?id=21846) in India from the Central Bank—Please see Fig. 1.7). M1 comprises currency outside the banking system, demand deposits with the banking system, and other deposits with the RBI. Currency outside the banking system comprises currency notes issued by the RBI and coins issued by the Government of India, with less currency notes and coin holdings of the RBI and the banking system (M1 for India—https://fred.stlouisfed.org/series/MYAGM1INM189N).
Has seasonality, but no trend (see Fig. 1.8).
A common characteristic of time series that has seasonality without trend would be repetitive patterns that are performed periodically. A good example is no noise (smoothed) Normalized Difference Vegetation Index (Vegetation Index—https://www.star.nesdis.noaa.gov/smcd/emb/vci/VH/VH-Syst_10ap30.php) (SMN) that is captured by weather satellites at the resolution of Global, 4 km, 7-day composite. The data in the above figure is from STAR-Global Vegetation Health Products (STAR—Global Vegetation Health Products: Province-Averaged VH—https://www.star.nesdis.noaa.gov/smcd/emb/vci/VH/vh_adminMean.php?type=Province_Weekly_MeanPlot): Province-Averaged VH—for California Crop land (please see Fig. 1.8).
Has both seasonality and trend (please see Fig. 1.9).
Kansas state wheat yield per area harvested (1982–2020) exhibits both trend and seasonality, particularly during periods of drought.

The data points of a time series can be stationary or non-stationary—the data points are often non-stationary or have statistical means, variances, and covariances that change over time. Non-stationary behaviors can be *trends*, *cycles*, *random walks*, or combinations of the three. To extract the maximum information from

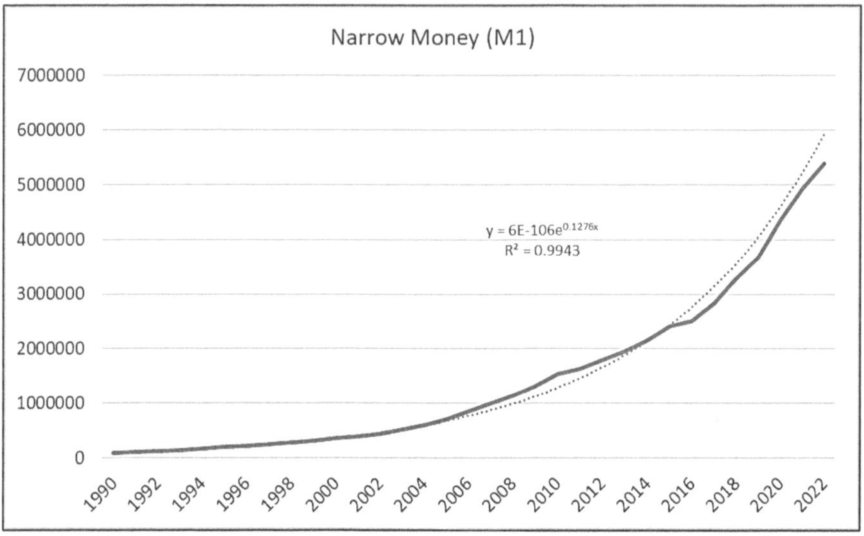

Fig. 1.7 RBI money supply

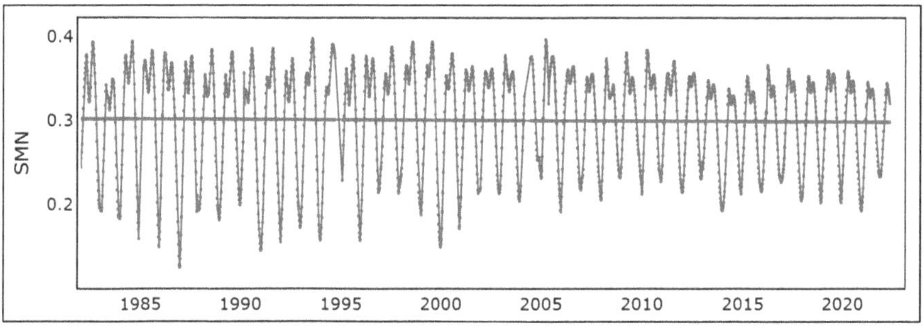

Fig. 1.8 SMN

time series and to forecast it, the data points need to be stationary. Non-stationary data, as a rule, is *unpredictable* and cannot be modeled or forecasted. The results obtained by using non-stationary time series may be spurious in that they may indicate a relationship between two variables where one does not exist. In essence, we cannot construct a meaningful predictive model using non-stationary datasets. We need to transform a non-stationary data point into stationary before applying time series models. To receive consistent, reliable results, the non-stationary data needs to be transformed into stationary data. The following diagram shows the processing of time series. The exploratory data analysis (EDA) follows standard data processing checks (please see Fig. 1.10): inspect data, check for nulls, impute strategy, and create a dataset for processing time series [7, 15].

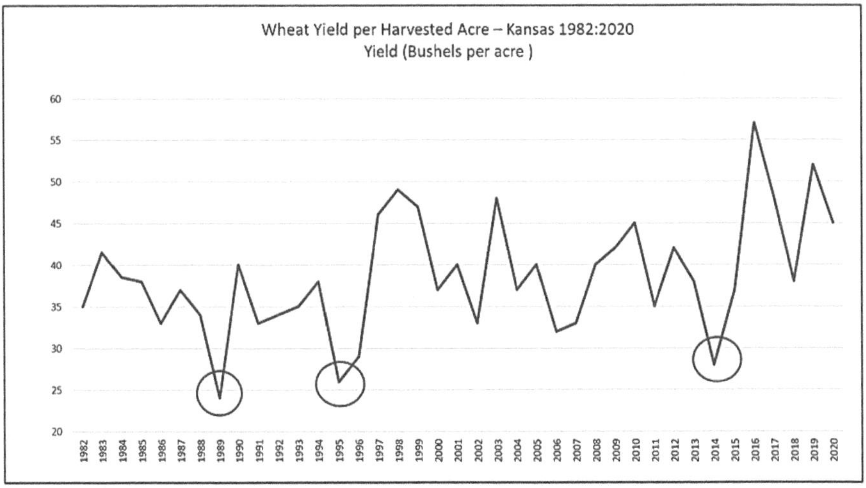

Fig. 1.9 Kansas wheat yield

Multivariant Time Series

Let's understand how a multivariate time series is formulated. Below are the simple K-equations of multivariate time series where each equation is a lag of the other series. X is the exogenous series here (please see Eqs. 1.4 and 1.5). The objective is to see if the series is affected by its own past and the past of the other series.[8]

$$y_{1,t} = f\left(y_{1,t-1} \ldots, y_{k,t-1}, \ldots y_{1,t-p} \ldots . y_{k,t-p}, \ldots X_{t-1}\ X_{t-2} \ldots \ldots \ldots\right) \tag{1.4}$$

$$y_{k,t} = f\left(y_{1,t-1} \ldots, y_{k,t-1}, \ldots y_{1,t-p} \ldots . y_{k,t-p}, \ldots X_{t-1}\ X_{t-2} \ldots \ldots \ldots\right) \tag{1.5}$$

This kind of series allows us to model the dynamics of the series itself and the interdependence of other series [1].

Autocorrelation

The term autocorrelation refers to the degree of similarity between (A) a given time series and (B) a lagged version of itself, over (C) successive time intervals. In other words, autocorrelation is intended to measure the relationship between a variable's

[8]Forecasting using Granger's Causality and Vector Auto-regressive Model—https://towardsdatascience.com/granger-causality-and-vector-auto-regressive-model-for-time-series-forecasting-3226a64889a6

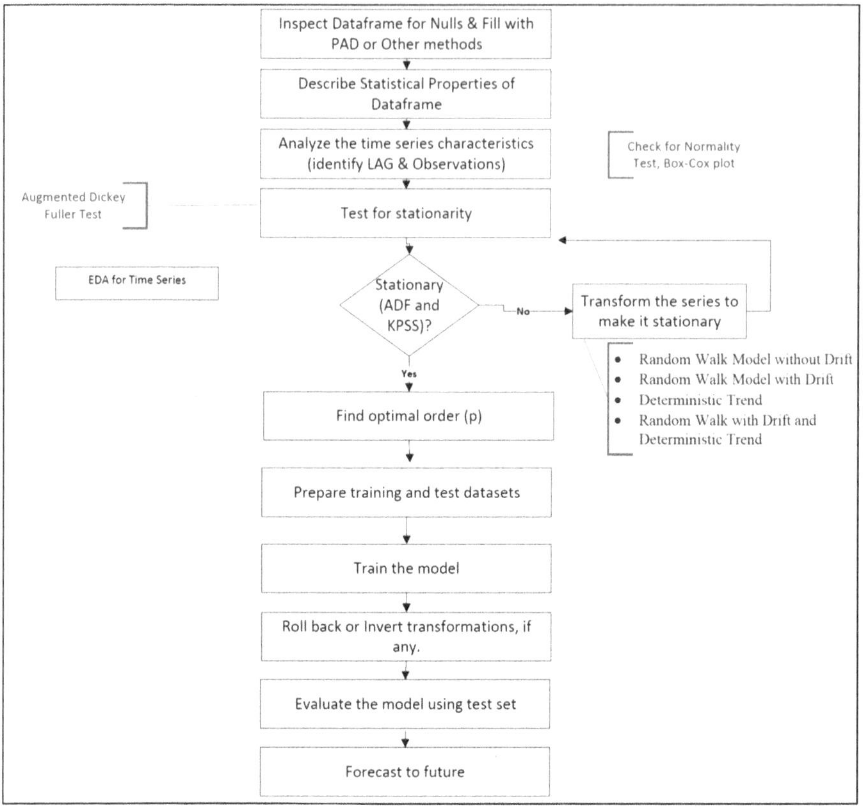

Fig. 1.10 Time series EDA

present value and any past values that you may have access to. We can calculate the correlation for time series observations with observations at previous time steps, called lags. Because the correlation of the time series observations is calculated with values of the same series at previous times, this is called a serial correlation[9] or an autocorrelation [16]. Autocorrelation or serial correlation can be a significant problem in analyzing historical data if we do not know how to look out for it [17].

To analyze autocorrelation, one of the important items to be considered is "LAG." A time series (y) with kth lag is its version that is "$t - k$" periods behind in time. A time series with lag ($k = 1$) is a version of the original time series that is 1 period behind in time, i.e., $y(t - 1)$.

[9] A Gentle Introduction to Autocorrelation and Partial Autocorrelation—https://machinelearningmastery.com/gentle-introduction-autocorrelation-partial-autocorrelation/

"plot_acf()" provides autocorrelation with option of specifying lag.

You can compute the autocorrelation by using the following (please see Eq. 1.6):

$$rk = \frac{\sum_{t=k+1}^{T} (Yt - \bar{Y})\,(Yt - k - \bar{Y})}{\sum_{t=1}^{T} (Yt - \bar{Y})2} \tag{1.6}$$

Partial Autocorrelation Function (PACF)

A partial autocorrelation is a summary of the relationship between observations in a time series and observations at prior time steps with the relationships of intervening observations removed [16]. The partial autocorrelation at lag k is the correlation that results after removing the effect of any correlations due to the terms at shorter lags [17]. The autocorrelation for an observation and an observation at a prior time step is comprised of both direct and indirect correlations. These indirect correlations are linear.

Stationarity Check: Augmented Dickey-Fuller (ADF) and Kwiatkowski-Phillips-Schmidt-Shin (KPSS) Tests

A time series is said to be stationary if it has the following three properties:

- It has a constant statistical meaning, i.e., mean remains constant over time.
- It has a constant variance, i.e., variance remains constant over time.
- It has a constant covariance. The value of the covariance between the two time periods depends on *the lag between the two time periods* and not on the time at which the covariance is computed. This means that the covariance between series at time $t = 2$ and $t = 4$ should be roughly the same as the covariance between series at time $t = 7$ and time $t = 9$.

A stationary time series will have the same statistical properties, i.e., same mean, variance, and covariance. No matter at what point we measure them, these properties are therefore time invariant. The series will have no predictable patterns in the long term. Time plot will show the series to be roughly horizontal with constant mean and variance.

Why do we need stationary time series? Because if the time series is not stationary, we can study its behavior only for that period. Each period of the time series will have its own distinct behavior, and it is not possible to predict or generalize for future time periods if the series is not stationary.

A stationary time series will tend to return to its mean value, and the fluctuations around this mean will have a constant magnitude. Thus, a stationary time series will not drift too much from its mean value because of the finite variance [18].

Non-stationary Stochastic Time Series

The presence of a trend component, in most observed time series, results in the series being non-stationary.[10] Furthermore, the trend can be either deterministic or stochastic, depending on which appropriate transformations must be applied to obtain a stationary series. For example, a stochastic trend, or commonly known as a unit root, is eliminated by differencing the series. However, differencing a series that in fact contains a deterministic trend results in a unit root in the moving-average process. Similarly, subtracting a deterministic trend from a series that in fact contains a stochastic trend does not render a stationary series [17]. Hence, it is important to identify whether non-stationarity is due to a deterministic or a stochastic trend before applying the proper transformations. There are two kinds of non-stationary stochastic time series that are observed [1]:

- Random walk model without drift
- Random walk model with drift
- Deterministic trend
- Random walk with drift and deterministic trend

Random Walk Without Drift

In random walk[11] without drift, it assumes that, at each point in time, the series merely takes a random step away from its last recorded position, with steps whose mean value is zero. The mean value of the series is constant and is equal to its initial or starting value, but the variance increases with time. Thus, the series is non-stationary.

The value of the time series Y at time t (please see Eq. 1.7) is equal to its value at time $t - 1$ plus a random shock, i.e.,

$$Y_{(t)} = Y_{(t-1)} + \varepsilon_t \tag{1.7}$$

Random walk predicts that the value at time "t" will be equal to the last period value plus a stochastic (non-systematic) component that is a white noise, which means ε_t is independent and identically distributed with mean "0" and variance "σ^2."

[10] Unit-root tests in Stata—https://blog.stata.com/2016/06/21/unit-root-tests-in-stata/

[11] Random Walk Model—https://people.duke.edu/~rnau/411rand.htm

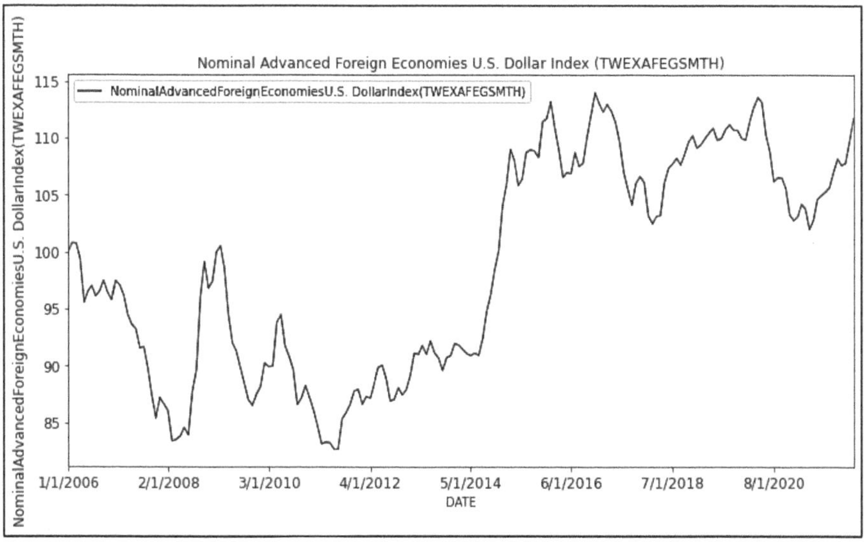

Fig. 1.11 US Dollar Index

Usually, the stock prices and exchange rates are following random walk. The data exhibiting random walk has long periods of up or down trend and sudden and unpredictable changes in direction. Consider, for example, the data distribution of Nominal Advanced Foreign Economies US Dollar Index[12] (TWEXAFEGSMTH). One can see the random change in the time series (please see Fig. 1.11). The Augmented Dickey-Fuller Test would result in a p-value greater than 0.05, and hence the series is non-stationary.

Nominal Advanced Foreign Economies U.S. Dollar Index (TWEXAFEGSMTH)
ADF Statistic: −1.059046
p-value: 0.731125
Critical Values:

1%: −3.466
5%: −2.877
10%: −2.575

[12]Nominal Advanced Foreign Economies U.S. Dollar Index (TWEXAFEGSMTH)—https://fred.stlouisfed.org/series/TWEXAFEGSMTH

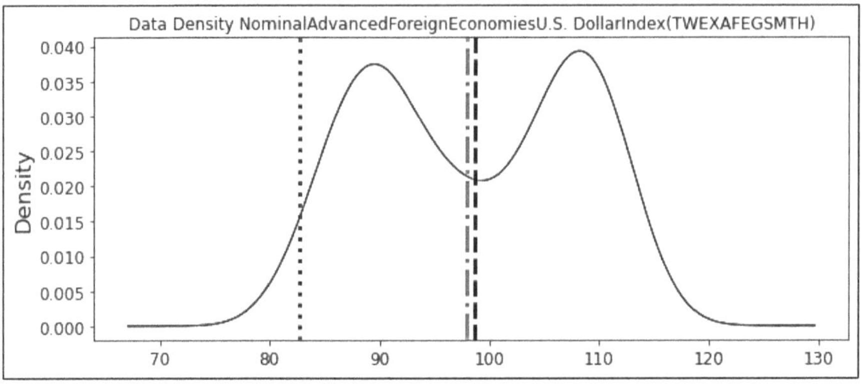

Fig. 1.12 Distribution US Dollar Index (2006–2020)

The data distribution has bi-modal distribution indicating two modes of data peaks or distribution.

The distribution of the data includes the following: minimum 82.68, mean 98.66, median 97.97, mode 82.68, and maximum 114.01; range 31.34; variance 90.65; and standard deviation 9.52 (please see Fig. 1.12).

Random Walk with Drift

$$Y_t = \alpha + Y_{(t-1)} + \varepsilon_t \tag{1.8}$$

If the random walk model predicts that the value at time "t" will equal the last period's value plus a constant, $Y_{(t-1)}$, or drift (α), and a white noise term (ε_t), then the process is random walk with a drift (please see Eq. 1.8). It also does not revert to a long-run mean and has a variance dependent on time.[13]

For example, consider the yearly cheddar cheese price from USDA. As the average cheddar cheese price from 1995 is increasing with random drift: $Y_t = \alpha + Y_{(t-1)} + \varepsilon_t$.

This is obviously non-stationary (please see Fig. 1.13). To prove it, let's apply the Augmented Dickey-Fuller Test.

[13]Time Series—https://www.investopedia.com/articles/trading/07/stationary.asp

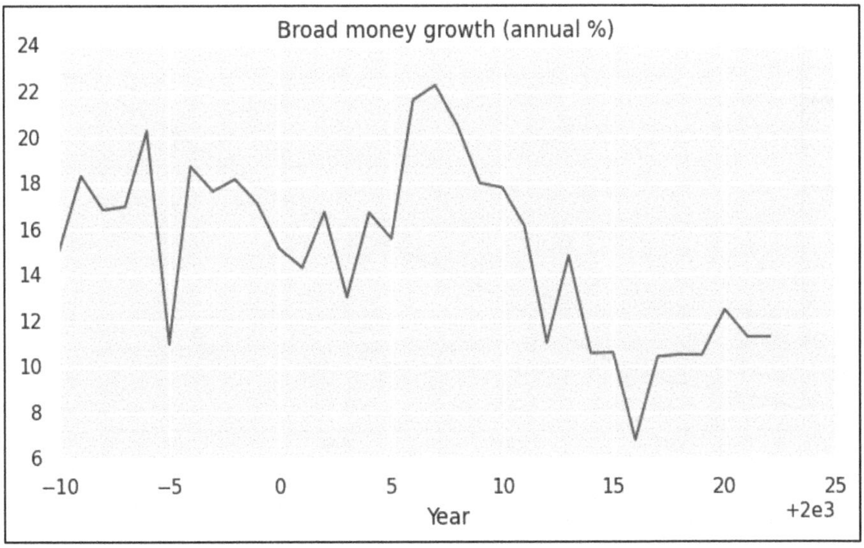

Fig. 1.13 Cheddar cheese prices (1995–2020)

```
import statsmodels.tsa.stattools as sm
def augmented_dickey_fuller_statistics(time_series):
    result = sm.adfuller(time_series.values, autolag='AIC')
    print('ADF Statistic: %f' % result[0])
    print('p-value: %f' % result[1])
    print('Critical Values:')
    for key, value in result[4].items():
        print('\t%s: %.3f' % (key, value))
```

Output:

Augmented dickey-Fuller test: Broad money growth (annual %)
ADF statistic: **−9.394242**
p-value: 0.000000
Critical values:

1%: −3.689
5%: −2.972
10%: −2.625

Since the p-value is less than 0.05, it rejects the null hypothesis. That is, the time series is stationary.

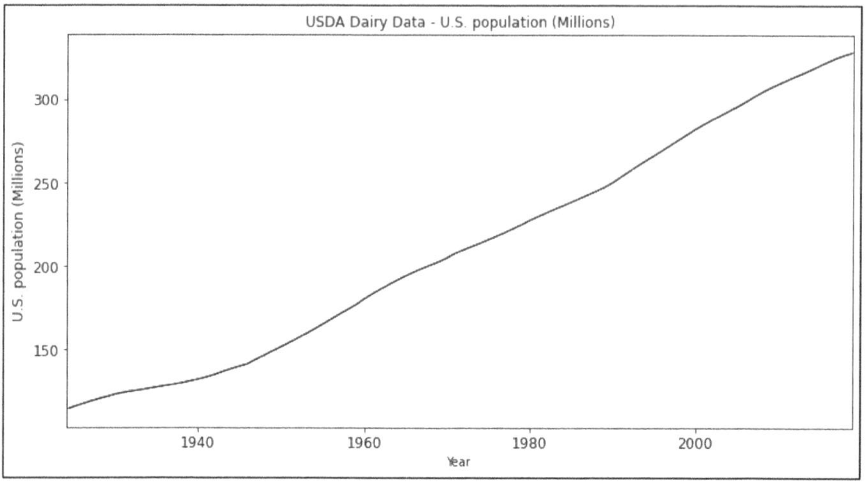

Fig. 1.14 US population (millions) 1936–2020

Deterministic Trend

Often a random walk with a drift is confusing for a deterministic trend. Both include a drift and a white noise component, but the value at time "*t*" in the case of a random walk is regressed on the last period's value (Y_{t-1}), while in the case of a *deterministic trend it is regressed on a time trend* (βt). A non-stationary process with a deterministic trend has mean that grows around a fixed trend, which is constant and independent of time (please see Eq. 1.9).

$$Y_t = \alpha + \beta t + \varepsilon_t \tag{1.9}$$

For example, consider the US population growth since the 1940s. As it can be seen, the population growth is pretty deterministic and trended upside (please see Fig. 1.14).

Trend + Cycle

Consider the US dairy shipments to the US territories (million pounds) (please see Fig. 1.15).

The US dairy shipment data has a constant increasing trend[14] plus cycle inside the data—for instance, every 15 years there is spike in shipments. So, we have trend plus cycle.

[14]Trendline options in Offic e—https://support.microsoft.com/en-us/office/trendline-options-in-office-92157920-fee4-4905-bc89-6a0f48152c52

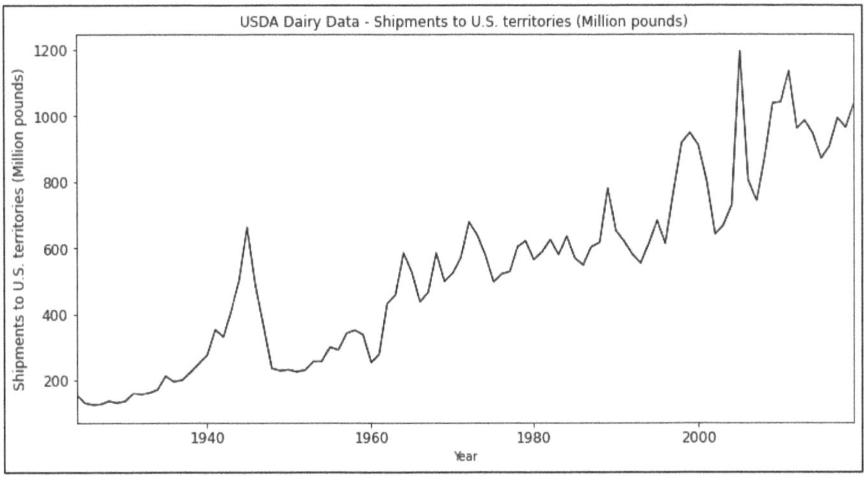

Fig. 1.15 Dairy data shipments (1936–2020)

Random Walk with Drift and Deterministic Trend

$$Y_t = \alpha + Y_{(t-1)} + \varepsilon_t + \beta t \tag{1.10}$$

Another example is a non-stationary process that combines a random walk with a drift component (α) and a deterministic trend (βt) (please see Eq. 1.10). It specifies the value at time "t" by the last period's value, a drift, a trend, and a stochastic component. A good example is the price movements of imported crude oil ($/barrel).[15] It shows a deterministic downward trend, especially during the recession periods due to a drop in demand. In the below figure, recessionary periods, gray-shaded areas,[16] marked a deterministic price[17] movement behavior. Random walk of the imported oil prices is difficult to predict (please see Fig. 1.16).

The distribution (please see Fig. 1.17) of the data includes the following: minimum 21.35, mean 62.55, median 57.01, mode 21.35, and maximum 131.52; range 110.17; variance 1006.57; and standard deviation 31.73.

Finally, the Augmented Dickey-Fuller Test confirms non-stationarity of the series.

[15] eia—https://www.eia.gov/outlooks/steo/realprices/

[16] U.S. Recessionary periods—https://fredhelp.stlouisfed.org/fred/data/understanding-the-data/recession-bars/

[17] All real and nominal price series—https://www.eia.gov/outlooks/steo/realprices/real_prices.xlsx

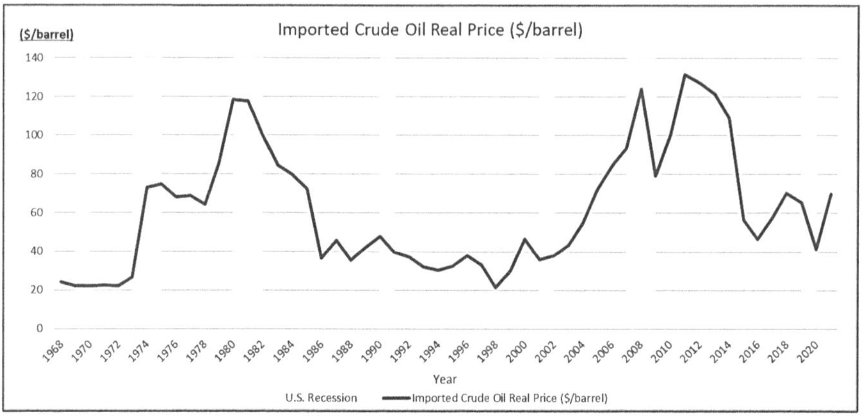

Fig. 1.16 Imported crude oil (1968–2022)

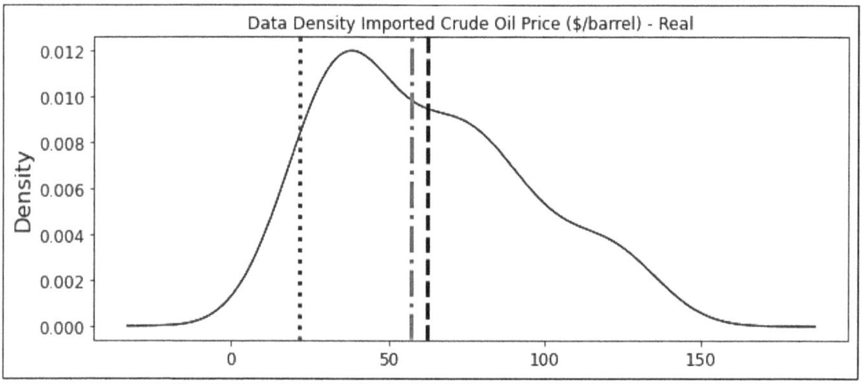

Fig. 1.17 Density crude oil

ADF statistic: −2.239740
p-value: 0.192140
Critical values:

1%: −3.555
5%: −2.916
10%: −2.596

The p-value greater than 0.05 rejects the null hypothesis, confirming non-stationary time series!

The first difference of the random walk without drift is a stationary time series. For the above equation, $Y_{(t)} = Y_{(t-1)} + \varepsilon_t$.

The error term is a white noise or stationary series with mean 0 and constant variance. Thus, the forecast for this model is the last observation as future movements are unpredictable and are equally likely to go up or down.

In contrast to the non-stationary process that has a variable variance and a mean that does not remain near or returns to a long-run mean over time, the stationary process reverts around a constant long-term mean and has a constant variance independent of time.

Forecasting Models

ARIMA stands for autoregressive integrated moving average. ARIMA can also be broken down into part autoregressive (AR), part moving average (MA), and part integrated (I). Models AR and MA can be used as separate models to produce predictions of the time series. It can however produce better predictions when combined as an ARMA model [19].

Auto ARIMA is essentially ARIMA with a grid search part to find the best metric parameters such as the Akaike Knowledge Criterion (AIC) [19].

Vector autoregression (VAR) is the statistical model used to capture multiple quantity relationships as they change over time. VAR is a model of the stochastic process [19].

The VAR models generalize the autoregressive single variable (univariate) model by allowing multivariate time series. The model of vector autoregression (VAR) adds the idea of univariate autoregression to kk time-series regressions, in which all kk series lagged values appear as regressors. Put differently, a vector of time series variables on lagged vectors of those variables is regressed in a VAR model [19].

The SVAR model is a type of vector autoregressive (VAR) model that imposes some restrictions based on economic theory. It allows for the simultaneous modeling of multiple time series, making it a good choice for studying the interrelationships between budget deficits, money supply, and inflation.

Supervised Learning

In supervised learning, an algorithm uses training data and feedback from humans to learn the relationship of given inputs (features) to a given output (labels) [6]. That is, the humans label the desired class or output so that the machine identifies pattern between given input and labeled output. For example, in the following regression model, supervised learning was employed[18] to show that the *rural population*, life expectancy, *foreign direct investment*, rate of inflation, level of exports of all goods and services, and ratio of agricultural exports to agricultural imports significantly determined the percentage contribution of agriculture to GDP in the selected Latin America and the Caribbean countries. Before use case could be introduced, here is the synopsis of regression model.

If agriculture fails to develop at a suitable pace, this could prove to be a critical constraint to the growth of the industrial sector as well as other sectors of the economy.

The learning process continues until the model achieves the desired level of accuracy on the training data. Future incoming data (without known labels) are tested on the constructed model.

The following table summarizes supervised learning algorithms:

Algorithm	Definition
Linear regression	Regression aims to model the relationship between variables that is improved gradually using a measure of error in the model's predictions. The most common regression algorithms are [17] ordinary least squares regression (OLSR), linear regression, logistic regression, stepwise regression, multivariate adaptive regression splines (MARS), and locally estimated scatterplot smoothing (LOESS). Linear regression models are highly understandable, and they model the connection between input independent variables and output variables (class or dependent variable) to create a model or classifier equation that helps to forecast future values. Highly interpretable,[19] linear regression models between input independent variables and output variables (class or dependent variable) to generate model or classifier equation that helps to predict future values
Logistic regression	Logistic regression[20] is a machine learning classification algorithm that is used to estimate the likelihood of a categorical outcome variable [21]. In logistic regression, the outcome variable is a binary variable that contains data coded as 1 (yes, success, etc.) or 0 (no, failure, etc.). In other words, the logistic regression model predicts $P(Y = 1)$ as a function of X. In simple terms, logistic regression is a version of linear regression that is used for classification tasks, meaning the output variable is discrete rather than continuous [17]. Examples of logistic regression include recognizing digits

(continued)

[18]Factors Influencing Agriculture's Contribution to GDP—https://ageconsearch.umn.edu/record/187331/files/Factors%20Influencing%20Agriculture_s%20Contribution%20to%20GDP%20Latin%20America%20and%20the%20Caribbean.pdf

[19]Model interpretability in Azure Machine Learning service—https://docs.microsoft.com/en-us/azure/machine-learning/service/how-to-machine-learning-interpretability

[20]Building A Logistic Regression in Python, Step by Step—https://towardsdatascience.com/building-a-logistic-regression-in-python-step-by-step-becd4d56c9c8

Algorithm	Definition
Random forest	Random forest is based on a simple but powerful idea—The wisdom of crowds. A large group of relatively uncorrelated models (trees) working as a committee will perform better than any of the individual models alone. In other words, random forest is a classification or regression model that improves the accuracy of a simple decision tree by creating multiple decision trees and taking a majority vote of them to predict the output, which is a continuous variable (e.g., age) for a regression problem and a discrete variable (e.g., either black, white, or red) for classification [17]
Gradient boosting trees	A method for classification or regression that creates decision trees one by one, where each tree tries to fix the mistakes made by the previous tree model. The final output is a mix of the outcomes from all trees [17]

In agricultural use cases, the most useful algorithms include decision tree and regression (linear, logistic).

Regression Models

In statistics, linear regression is an approach for modeling the relationship between a scalar-dependent variable y and one or more explanatory variables denoted X (please see Eq. 1.11). The case of one explanatory variable is called simple linear regression. When the outcome, or class, is numeric, and all the attributes are numeric, linear regression is a natural technique to consider. This is a staple method in statistics. The idea is to express the class as a linear combination of the attributes, with predetermined weights [21]:

$$x = w_0 + w_1 a_1 + w_2 a_2 + \ldots + w_k a_k \tag{1.11}$$

where x is the class; a_1, $a_2 \ldots$, a_k are the attribute values; and w_0, $w_1 \ldots$, w_k are weights.

The weights are calculated from the training data. Here, the notation gets a little heavy, because we need a way of expressing the attribute values for each training instance. The first instance will have a class, say $x(1)$, and attribute values, ${a_1}^{(1)}$, ${a_2}^{(1)} \ldots$, ${a_k}^{(1)}$, where the superscript denotes that it is the first example. Moreover, it is convenient to assume an extra attribute a_0, whose value is always 1.

Metrics for Regression Models

The metrics returned for regression models[21] are generally designed to estimate the amount of error. A model is considered to fit the data well if the difference between

[21] Scikit Regressors—https://scikit-learn.org/stable/modules/classes.html

observed and predicted values is small. However, looking at the pattern of the residuals (the difference between any one predicted point and their corresponding actual value) can tell you a lot about potential bias in the model.

The following metrics are generally used to report regression models [18, 21]:

- *Negative log likelihood* measures the loss function; *a lower score is better*. Note that this metric is only calculated for *Bayesian linear regression and decision forest regression*; for other algorithms, the value is *infinity which means for nothing*.
- *Mean absolute error (MAE)* measures how close the predictions are to the actual outcomes; thus, a lower score is better.
- *Root mean squared error (RMSE)* creates a single value that summarizes the error in the model. By squaring the difference, the metric disregards the difference between overprediction and underprediction.
- *Relative absolute error (RAE)* is the relative absolute difference between expected and actual values, relative because the mean difference is divided by the arithmetic mean.
- *Relative squared error (RSE)* similarly normalizes the total squared error of the predicted values by dividing by the total squared error of the actual values.
- *Mean zero one error (MZOE)* indicates whether the prediction was correct or not. In other words: ZeroOneLoss(x,y) = 1 when $x! = y$; otherwise, 0.
- *Coefficient of determination, often referred to as* R^2, represents the predictive power of the model as a value between 0 and 1. Zero means the model is random (explains nothing); 1 means there is a perfect fit. However, caution should be used in interpreting R^2 values, as low values can be entirely normal and high values can be suspect [22].

Ordinary Least Squares Linear Regression

Linear regression fits a linear model with coefficients $w = (w_1, \ldots, w_p)$ to minimize the residual sum of squares between the observed targets in the dataset and the targets predicted by the linear approximation. Linear regression is a quiet and the simplest statistical regression method used for predictive analysis in machine learning. Linear regression shows the linear relationship between the independent (predictor) variable, i.e., X-axis, and the dependent (output) variable, i.e., Y-axis. If there is a single input variable X (independent variable), such linear regression is called simple linear regression (please see Eq. 1.12).

$$\text{The simplest theoretical model}: Y = \beta_0 + \beta_1\, X + \varepsilon \qquad (1.12)$$

where β_0 and β_1 are parameters to be estimated and ε represents the influence of random factors (noise) on Y. Assumptions implicit in the model are as follows: the only "systematic" determinant of Y is X, the relationship between Y and X is linear,

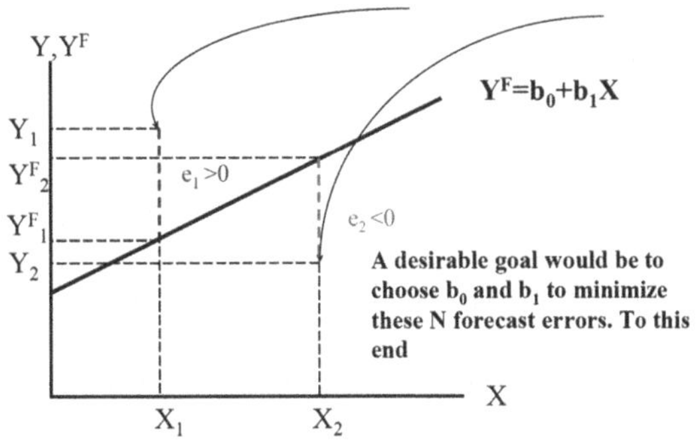

Fig. 1.18 Linear regression

and ε has three very special properties. To put it in another way, there are no missing variables except "noise."

$$Y^F = b_0 + b_1 X, \text{the forecast of } Y \text{ for a given } X$$

since $Y^F = b_0 + b_1 X$

$$b_1 = dY^F/dX, \text{the sensitivity of forecasted } Y \text{ to changes in } X$$

Plot the first two data points: ($Y1$,$X1$) and ($Y2$,$X2$) (please see Fig. 1.18).

A desirable goal would be to choose b_0 and b_1 to minimize these N forecast errors. To this end, define: $Yi - YF_i = e_i$ or $Y - \tilde{Y} = e$ the forecast error or residual in period i.

The following metrics would help: the cost function helps to work out the optimal values for b_0 and b_1, which provides the best fit line for the data points. In the following section, regression metrics are described.

Mean Squared Errors (MSE)

Mean squared error gives the mean of squared difference between model prediction and target value. It can be used as the measure of the quality of an estimator. Default thresholds: upper limit = 80%. In statistics, the mean squared error (MSE) of an estimator (of a procedure for estimating an unobserved quantity) measures the average of the squares of the errors,[22] that is, the average squared difference between the estimated values and what is estimated. MSE is a risk function, corresponding to the expected value of the squared error loss. The fact that MSE (please see Eq. 1.13) is almost always strictly positive (and not zero) is because of randomness or because

[22]MSE—https://medium.com/free-code-camp/machine-learning-mean-squared-error-regression-line-c7dde9a26b93

the estimator does not account for information that could produce a more accurate estimate [23].

$$\mathrm{MSE} = \frac{1}{n}\sum_{i=1}^{n}\left(Y-\tilde{Y}\right)^2 \tag{1.13}$$

For each point, we take the y true and the y' predicted. We subtract the y value from the y' value and calculate the square of the result. Our goal is to minimize this mean, which will provide us with the best line that goes through all the points.

Root Mean Squared Error (RMSE)

The root mean squared error (RMSE) view shows the difference between the predicted and observed values in your model. The root mean squared error (RMSE) is a standard way to measure[23] the error of a model in predicting quantitative data. Formally it is defined as follows (please see Eq. 1.14):

$$\mathrm{RMSE} = \sqrt{\sum_{i=1}^{n}\frac{\left(\tilde{Y}-Y\right)^2}{n}} \tag{1.14}$$

where $\tilde{Y}$ is predicted value, Y is observed value, and n is the number of observations.

Another important metric is R-squared which is a number that explains the amount of variation that is explained/captured by the developed model (please see Fig. 1.19). It always ranges between 0 and 1. Overall, the higher the value of R-squared, the better the model fits the data. The significance of R-squared is shown by the following figures:

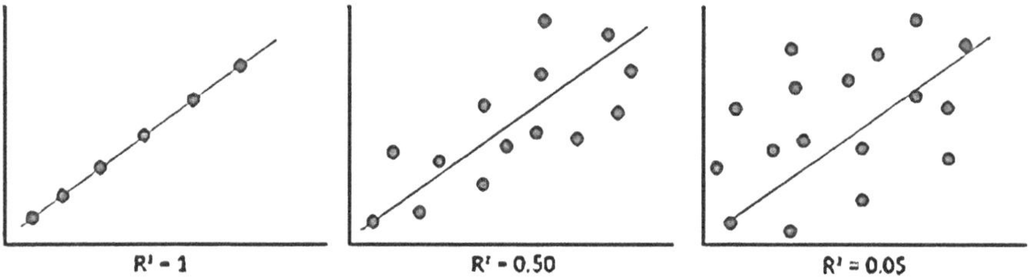

Fig. 1.19 R^2

[23] What does RMSE mean?—https://towardsdatascience.com/what-does-rmse-really-mean-806b65f2e48e

For linear regression, the following five assumptions need to be valid:

1. The residuals are normally distributed.
2. The residuals are independent of one another (serial independence); this is achieved through the DW Test—The Durbin-Watson Statistic.
3. The residuals have a constant variance (homoskedasticity).
4. *Y* is linear in the *X*'s.
5. No "systematic" influence on *Y* is missing from the model.

For more on these tests, please refer to statistical textbooks.

Let's build a basic linear regression model that includes an intercept and two independent variables.[24] This model is a very simple construct of scikit-learn [20].

$$Y = 1^* X_0 + 2^* X_1 + 3$$

Logistic Regression

Logistic regression is a linear model that is used for classification instead of regression, despite what its name suggests. It is also called logit regression, MaxEnt, or the log-linear classifier in literature. In this model, a logistic function is used to model the probabilities of the different outcomes of a single trial [24].

We applied linear regression in the previous step. Linear regression and logistic regression are common models in data science, and tools like Python and R make them easy and fast to compute. Linear regression models are used to find the relationship between a continuous outcome variable and one or more predictor variables. When there is only one predictor and one outcome, it is called simple linear regression, but when there are more predictors, it is called multiple linear regression. For both types of linear regression, it tries to draw a line that fits the data points best, which is usually done using the *least squares method* [25].

Logistic regression is similar to linear regression in that it also uses one or more independent variables to estimate their relationship with a dependent variable, but it is used to predict a variable that has categories instead of a continuous value. A variable that has categories can be true or false, yes or no, 1 or 0, and so on. The way of measuring the outcome also differs from linear regression as it gives a probability, but the logit function makes the *S-curve into a straight line*. Logistic regression uses the logistic sigmoid function to apply weights to the input values and produce a prediction of the data class. Logistic regression calculates the probability of an event happening, such as voting or not voting, based on a given dataset of independent variables. Since the result is a probability, the dependent variable is limited to between 0 and 1. In logistic regression, a logit transformation is done on the odds,

[24] Scikit-learn: Machine Learning in Python, Pedregosa et al., JMLR 12, pp. 2825–2830, 2011.

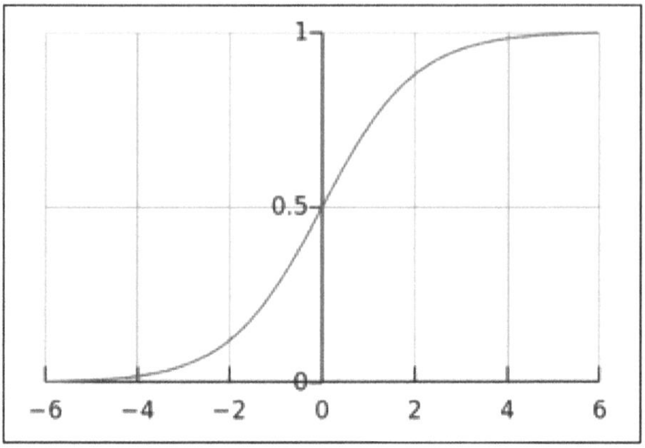

Fig. 1.20 *S*-curve

that is, the probability of success divided by the probability of failure. This is also often called the log odds, or the natural logarithm of odds, and this logistic function is shown by the following formulas (please see Fig. 1.20):

$$\text{The logistic sigmoid function}\, f(x) = \frac{1}{1 + e^{-x}} \tag{1.15}$$

A generalized linear regression can work with any distribution of the data input. It doesn't require the normal distribution. Logistic regression is a type of generalized linear regression where the logit function follows the response variable. The logit function takes a probability p, which is between 0 and 1, as its input. The odds ratio for probability p is $p/(1 - p)$, and the logit function is the log odds or the logarithm of the odds ratio.

Linear regression and logistic regression are both models for predicting future outcomes based on regression analysis, but linear regression is usually simpler to comprehend. Linear regression also needs a smaller sample size than logistic regression, which needs a big enough sample to cover all the values in the response categories. If the sample is not large or representative enough, the model may not be able to find a meaningful effect (please see Eqs. 1.16 and 1.17).

$$\begin{gathered} \text{Logit(pi)} = 1/(1 + \exp(-\text{pi})) \\ \ln(\text{pi}/(1 - \text{pi})) = \text{Beta\_0} + \text{Beta\_1}^{*}X\_1 + \ldots + B\_k^{*}K\_k \end{gathered} \tag{1.16}$$

In this logistic regression equation, logit(pi) is the dependent or response variable, and x is the independent variable. The beta parameter, or coefficient, in this model is commonly estimated via maximum likelihood estimation (MLE). This method tests different values of beta through multiple iterations to optimize for the best fit of log odds.

Let $y_i \in 1, \ldots, K$ be the label (ordinal) encoded target variable for observation i. Instead of a single coefficient vector, we now have a matrix of coefficients W where each row vector W_k corresponds to class k. We aim at predicting the class probabilities $P(y_i = k|X_i)$ via predict_proba[25] as (please see Eq. 1.17):

$$\hat{p}_k(X_i) = \frac{\exp(X_i W_k + W_{0,k})}{\sum_{l=0}^{K-1} exp(X_i W_l + W_{0,l})}. \tag{1.17}$$

The objective for the optimization becomes L (please see Eq. 1.18):

$$\min_W -C\sum_{i=1}^{n}\sum_{k=0}^{K-1}[y_i = k]\log(\hat{p}_k(X_i)) + r(W). \tag{1.18}$$

where $[P]$ represents the Iverson bracket which evaluates to 0 if P is false, otherwise it evaluates to 1.

The quality of a logistic regression model is determined by measures of fit and predictive power. R-squared is a measure of how well the independent variable in the logistic function can be predicted from the dependent variables and ranges from 0 to 1. Many different ways exist to calculate R-squared including the Cox-Snell R^2 and the McFadden R^2. Goodness of fit, on the other hand, can be measured using tests such as the Pearson chi-square, Hosmer-Lemeshow, and Stukel tests. The right type of test to use will depend on factors such as the distribution of p-values, interactions, and quadratic effects and grouping of data [25, 26].

Random Forest Trees

Random forest is a supervised learning algorithm that uses the ensemble learning method and many decision trees. Random forest is a bagging technique, so each calculation is done in parallel, and there is no influence between the decision trees when constructing them. RF can be used for both classification and regression tasks. The name "random forest" comes from the bagging idea of using random data (random) and creating multiple decision trees (forest). Overall, it is a strong ML algorithm that avoids the drawbacks of a decision tree model (we will explain that later on). Moreover, random forest is quite popular as shown by many Kaggle competitions, academic papers, and technical posts [25].

The sklearn.ensemble module contains two averaging algorithms that use random decision trees: the RandomForest algorithm and the Extra-Trees method. These algorithms are both techniques that introduce randomness in the tree building

[25] Predic_prob—https://scikit-learn.org/stable/modules/generated/sklearn.linear_model.LogisticRegression.html#sklearn.linear_model.LogisticRegression.predict_proba

process to create a diverse set of classifiers.[26] The ensemble prediction is the average prediction of the individual classifiers. In random forests, each tree in the ensemble is built from a sample with replacement (i.e., a bootstrap sample) from the training set. Moreover, when splitting each node during the tree construction, the best split is chosen either from all input features or a random subset of size max_features.

These two sources of randomness help to make the forest estimator more stable. This is because single decision trees often have high variance and tend to fit too closely to the data. The randomness in forests makes the prediction errors of different trees less dependent on each other. By taking an average of those predictions, some errors can balance out. Random forests improve stability by combining diverse trees, sometimes with a small trade-off of a slight increase in bias. In practice, this often leads to a better model overall.

Gradient Boosting Regressor

Gradient boosting is a method that stands out for its fast and accurate predictions, especially with large and complex datasets. This algorithm has achieved the best results in many situations, from Kaggle competitions to machine learning solutions for business. We already know that errors are very important in any machine learning algorithm. There are mainly two types of error: bias error and variance error. Gradient boost algorithm helps us reduce the bias errors of the model. The main idea behind this algorithm is to build models one by one, and these later models try to lower the errors of the previous model [23].

Gradient boosting (GB) regressor is used when the target column is continuous, while gradient boosting classifier is used when it is a categorical problem. The main difference between them is the "loss function," which is the objective to minimize by adding weak learners using gradient descent. Depending on the type of problem, we can have different loss functions, such as mean squared error (MSE) for regression or log likelihood for classification. GB builds a model by adding components sequentially; it can optimize any differentiable loss function. In each step, a regression tree fits on the negative gradient of the chosen loss function.

Polynomial Regression

Polynomial regression is a type of linear regression where the relation between the independent variable x and the dependent variable y is modeled as a polynomial of

[26]Random Forest—https://scikit-learn.org/stable/modules/ensemble.html#forest

degree n. Polynomial regression fits a nonlinear relation[27] between the value of x and the expected value of y, given by $E = \left(\frac{Y}{X}\right)$. The method of least squares is usually used to fit polynomial regression models. The least squares method makes the variance of the coefficients as small as possible, according to the Gauss-Markov theorem. In statistics, the Gauss-Markov theorem (or sometimes Gauss theorem for some authors) says that the ordinary least squares (OLS) estimator has the smallest sampling variance among the class of linear unbiased estimators, if the errors in the linear regression model are not correlated and have the same variances and zero mean.

- There are some relationships that a researcher will hypothesize as curvilinear. Clearly, such types of cases will include a polynomial term.
- Inspection of residuals. If we try to fit a linear model to curved data, a scatterplot of residuals (Y-axis) on the predictor (X-axis) will have patches of many positive residuals in the middle. Hence in such a situation, it is not appropriate.
- An assumption in usual multiple linear regression analysis is that all the independent variables are independent. In polynomial regression model, this assumption is not satisfied.

Cost function is a function that *measures the performance* of a machine learning model for given data. Cost function is basically the calculation of the error between predicted values and expected values and presents it in the form of a single real number [27]. The cost function of polynomial regression is as follows (please see Eq. 1.19):

$$\mathrm{MSE} = \frac{1}{n} \sum_{i=1}^{n} \left(Y - \tilde{Y}\right)^2 \tag{1.19}$$

Advantages of using polynomial regression[28] [28]:

- A broad range of functions can be fit under it.
- Polynomial basically fits a wide range of curvatures.
- Polynomial provides the best approximation of the relationship between dependent and independent variables.

Disadvantages of using polynomial regression:

- These are too sensitive to the outliers.
- The presence of one or two outliers in the data can seriously affect the results of nonlinear analysis.
- In addition, there are unfortunately fewer model validation tools for the detection of outliers in nonlinear regression than there are for linear regression.

[27] Python | Implementation of Polynomial Regression—https://www.geeksforgeeks.org/python-implementation-of-polynomial-regression/

[28] ML—https://ggbaker.ca/data-science/content/ml.html

As we increase the degree in the model, it tends to increase the performance of the model. However, increasing the degrees of the model also increases the risk of overfitting and underfitting the data.

After reading this chapter, you should be able to answer artificial intelligence, machine learning, and time series techniques. You should be able to answer the types of learning and should be able to evaluate performance metrics of classification and regression techniques.

References

1. Lutz Goedde, Joshua Katz, Alexandre Ménard, and Julien Revellat, "Agriculture's connected future: How technology can yield new growth", October 9, 2020, https://www.mckinsey.com/industries/agriculture/our-insights/agricultures-connected-future-how-technology-can-yield-new-growth, Access Date: 09/18/2020
2. Sofia Bettiza, God and robots: Will AI transform religion?, 21 October 2021, https://www.bbc.com/news/av/technology-58983047, Access Date: June 11, 2023
3. Dr. Prabhat Kumar, Artificial Intelligence: Reshaping Life and Business, Chapter 2, BPB Publications; 1 edition (August 17, 2019), ISBN-10: 9388511077
4. Darrell M. West, "What is artificial intelligence?", Thursday, October 4, 2018, https://www.brookings.edu/research/what-is-artificial-intelligence/, Access Date: November 22, 2020
5. Daniel Faggella, "What is Artificial Intelligence? An Informed Definition", December 21, 2018, https://emerj.com/ai-glossary-terms/what-is-artificial-intelligence-an-informed-definition/#:~:text=In%20%E2%80%9CArtificial%20Intelligence%3A%20A%20Modern,many%20different%20splintered%20fields%20%E2%80%93%20speech, Access Date: October 2020
6. Russell Stuart and Peter Norvig, Artificial Intelligence: A Modern Approach (3rd edition), Publisher: Pearson; 3 edition (December 11, 2009), ISBN-10: 0136042597
7. Chandrasekar Vuppalapati (Author), Artificial Intelligence and Heuristics for Enhanced Food Security (International Series in Operations Research & Management Science, 331) 1st ed. 2022 Edition, Publisher: Springer; 1st ed. 2022 edition (September 17, 2022), ISBN-13: 978-3031087424
8. Jessica Lin Eamonn Keogh Stefano Lonardi Bill Chiu, A Symbolic Representation of Time Series, with Implications for Streaming Algorithms, June 13, 2003, https://www.cs.ucr.edu/~eamonn/SAX.pdf, Access Date:June 19, 2023
9. Matt Taddy, Business Data Science: Combining Machine Learning and Economics to Optimize, Automate, and Accelerate Business Decisions 1st Edition, Publisher: McGraw-Hill Education; 1 edition (August 21, 2019), ISBN-10: 1260452778
10. Kai Hwang and Min Chen, "Big-Data Analytics for Cloud, IoT and Cognitive Computing", Publisher: Wiley; 1 edition (August 14, 2017), ISBN-10: 9781119247029
11. Faloutsos, C., Ranganathan, M., & Manolopoulos, Y. (1994). Fast Subsequence Matching in Time-Series Databases. In proceedings of the ACM SIGMOD Int'l Conference on Management of Data. May 24–27, Minneapolis, MN. pp 419–429.
12. Chan, K. & Fu, A. W. (1999). Efficient Time Series Matching by Wavelets. In proceedings of the 15th IEEE Int'l Conference on Data Engineering. Sydney, Australia, Mar 23–26. pp 126–133.
13. Geurts, P. (2001). Pattern Extraction for Time Series Classification. In proceedings of the 5th European Conference on Principles of Data Mining and Knowledge Discovery. Sep 3-7, Freiburg, Germany. pp. 115–127.

14. Keogh, E., Chakrabarti, K., Pazzani, M. & Mehrotra, S. (2001). Locally Adaptive Dimensionality Reduction for Indexing Large Time Series Databases. In proceedings of ACM SIGMOD Conference on Management of Data. Santa Barbara, CA, May 21-24. pp 151–162.
15. Chandrasekar Vuppalapati (Author), Artificial Intelligence and Advanced Analytics for Food Security, Publisher: CRC Press; 1st edition (July 17, 2023), ISBN-13: 978-1032346182
16. Jason Brownlee, "A Tour of Machine Learning Algorithms", Publication Date: August 12, 2019, URL: https://machinelearningmastery.com/a-tour-of-machine-learning-algorithms/, Access Date: Sep 18, 2019
17. Susan Li, "Building A Logistic Regression in Python, Step by Step", Publication Date: Sep 28, 2017, URL: https://towardsdatascience.com/building-a-logistic-regression-in-python-step-by-step-becd4d56c9c8, Access Date: Nov 22, 2019
18. Chandrasekar Vuppalapati, Machine Learning and Artificial Intelligence for Agricultural Economics Prognostic Data Analytics to Serve Small Scale Farmers Worldwide, Edition, ISBN-13: 978-3-030-77484-4, https://link.springer.com/book/10.1007/978-3-030-77485-1, Access Date: December 15, 2022
19. H. P. Suresha, Mutturaj Uppaladinni, and Krishna Kumar Tiwari, Indian Commodity Market Price Comparative Study of Forecasting Methods—A Case Study on Onion, Potato and Tomato, 22 December 2021, https://www.researchgate.net/profile/Suresh-Hp-2/publication/358037089_Indian_Commodity_Market_Price_Comparative_Study_of_Forecasting_Methods_-_A_Case_Study_on_Onion_Potato_and_Tomato/links/622a067484ce8e5b4d163012/Indian-Commodity-Market-Price-Comparative-Study-of-Forecasting-Methods-A-Case-Study-on-Onion-Potato-and-Tomato.pdf, Access Date: June 06, 2023
20. A. Ramalingam, S. Kedari and C. Vuppalapati, "IEEE FEMH Voice Data Challenge 2018," 2018 IEEE International Conference on Big Data (Big Data), Seattle, WA, USA, 2018, pp. 5272–5276, https://doi.org/10.1109/BigData.2018.8622164.
21. Vishal Morde, "XGBoost Algorithm: Long May She Reign", Publication Date: Apr 7, 2019, URL: https://towardsdatascience.com/https-medium-com-vishalmorde-xgboost-algorithm-long-she-may-rein-edd9f99be63d , Access Date: Sep 18, 2019
22. Chandrasekar Vuppalapati, Artificial Intelligence and Heuristics for Enhanced Food Security, Springer; 1st ed. 2022 edition (September 17, 2022), ISBN-13: 978-3031087424, Access Date: December 03, 2022
23. Moshe Binieli, Machine learning: an introduction to mean squared error and regression lines, examples and documentation, https://medium.com/free-code-camp/machine-learning-mean-squared-error-regression-line-c7dde9a26b93, Access Date: December 07, 2022
24. Fabian Pedregosa, Gaël Varoquaux, Alexandre Gramfort, Vincent Michel, Bertrand Thirion, Olivier Grisel, Mathieu Blondel, Peter Prettenhofer, Ron Weiss, Vincent Dubourg, Jake Vanderplas, Alexandre Passos, David Cournapeau, Matthieu Brucher, Matthieu Perrot, Édouard Duchesnay; 12(85):2825−2830, 2011, Journal of Machine Learning Research 12 (2011) 2825–2830, Access Date: December 05, 2022
25. What is logistic regression?, https://www.ibm.com/topics/logistic-regression, Access Date: December 05, 2022 What is logistic regression?—https://www.ibm.com/topics/logistic-regression
26. What is Regression?, https://www.nvidia.com/en-us/glossary/data-science/linear-regression-logistic-regression/, Access Date: December 05, 2022
27. H. Ngo and G. Owens, THE PROFITABILITY OF MANGOES IN THE TOP END, July 2002, https://industry.nt.gov.au/__data/assets/pdf_file/0013/233500/tb301.pdf, Access Date: Decemner 28 2022
28. Jennifer, An Indic Palm-leaf Manuscript: Please Help Identify It!, 1999, http://www.columbia.edu/cu/lweb/data/indiv/southasia/cuvl/indicmss/palm.html, Access Date: June 13, 2023

Chapter 2
Natural Language Models

This chapter covers the following:

- Natural language processing (NLP) models:
  - Tokenization
  - Part-of-speech identification
  - Stemming, lemmatization
- Deep learning models
- Embeddings:
  - Bag of words
- NLP model to analyze economic and social sustainability models

This chapter covers natural language processing (NLP) techniques such as tokenization, part-of-speech identification, stemming, lemmatization, and creation of complete NLP models. Then, the chapter explains cosine similarity to measure how similar documents are. The chapter also presents deep learning framework and bag-of-words embedding models. The chapter shows how NLP can be used for social and economic sustainability models. Lastly, the chapter helps to use NLP techniques for other language models.

Natural language processing (NLP) has many uses. It is used for tasks such as sentiment analysis, topic detection, language detection, key phrase extraction, and document categorization.

Specifically, you can use NLP to:

- Classify documents. For instance, you can label documents as sensitive or spam.
- Do subsequent processing or searches. You can use NLP output for these purposes.

© The Author(s), under exclusive license to Springer Nature Switzerland AG 2024

C. Vuppalapati, *Assessing Policy Effectiveness using AI and Language Models*,
International Series in Operations Research & Management Science 354,
https://doi.org/10.1007/978-3-031-56097-2_2

- Summarize text by identifying the entities that are present in the document.
- Tag documents with keywords. For the keywords, NLP can use identified entities.
- Do content-based search and retrieval. Tagging makes this functionality possible.
- Summarize a document's important topics. NLP can combine identified entities into topics.
- Categorize documents for navigation. For this purpose, NLP uses detected topics.
- Enumerate related documents based on a selected topic. For this purpose, NLP uses detected topics.
- Score text for sentiment. By using this functionality, you can assess the positive or negative tone of a document.

Potential Use Cases

Social and economic sustainability scenarios that can benefit from custom NLP include:

- Document intelligence for policy identification and effectiveness of policies for public welfare. For example, identification of tone and action is required for such. An example is included in the 1991–1992 Indian Budget Speech:[1] there has been inadequate emphasis on the reduction of costs, upgradation of technology, and improvement of living standards [1].
- Industry-agnostic NLP tasks for text processing, such as name entity recognition (NER), classification, summarization, and relation extraction. These tasks automate the process of retrieving, identifying, and analyzing document information like text and unstructured data. Examples of these tasks include progress development, risk stratification models, ontology classification, and summarizations. For instance, even after 52 years of independence,[2] the provision of basic services in rural areas remains very unsatisfactory. These large gaps in basic services in rural areas are not acceptable, and the government is committed to removing them rapidly [2].
- Information retrieval and knowledge graph creation for semantic search. This functionality makes it possible to make economic and social knowledge graphs that support sustainability initiatives.
- Text translation for conversational AI systems in governmental customer-facing applications department and other industries.

[1] Budget Speech 1992–1993 https://www.indiabudget.gov.in/doc/bspeech/bs199293.pdf

[2] Budget Speech 2000–2001 https://www.indiabudget.gov.in/doc/bspeech/bs200001.pdf

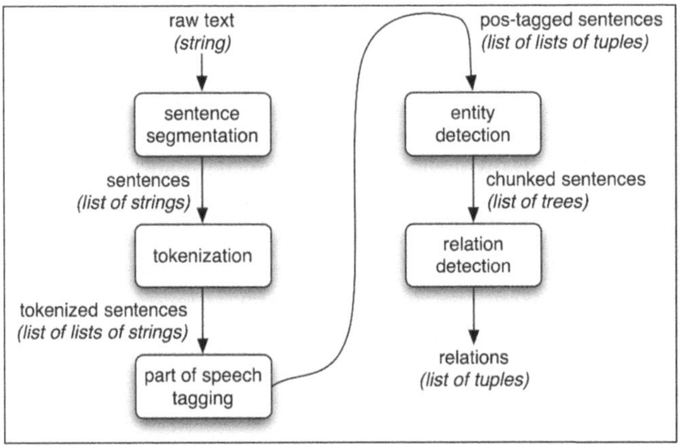

Fig. 2.1 NLP process

NLP Techniques

- Tokenizer. Splitting the text into words or phrases
- Stemming and lemmatization. Normalizing words so that the different forms map to the canonical word with the same meaning. For example, "running" and "ran" map to "run"
- Entity extraction. Identifying subjects in the text
- Part-of-speech detection. Identifying text as a verb, noun, participle, verb phrase, and so on
- Sentence boundary detection. Detecting complete sentences within paragraphs of text

In order to run the code in this chapter, please install Natural Language Tool Kit (NLTK) from the web: https://www.nltk.org/data.html. Detailed instructions can be found at https://www.nltk.org/install.html.

Figure 2.1 shows the architecture for a simple information extraction system. It begins by processing a document using several procedures: First, the raw text of the document is split into sentences using sentence segmentation, and each sentence is further subdivided into words using a tokenizer. Next, each sentence is tagged with part-of-speech tags, which will prove very helpful in the next step, named entity detection. In this step, we search for mentions of potentially interesting entities in each sentence. Finally, we use relation detection to search for likely relations between different entities in the text [3].

Tokenization

Splitting the text into words or phrases. Given a character sequence and a defined document unit, tokenization is the task of chopping it up into pieces, called tokens, perhaps at the same time throwing away certain characters, such as punctuation. These tokens are often loosely referred to as terms or words, but it is sometimes important to make a type/token distinction. A token is an instance of a sequence of characters in some particular document that are grouped together as a useful semantic unit for processing. A type is the class of all tokens containing the same character sequence. Large language models also employ tokens as a billing factor to charge the models [4].

budgetspeechstatement_scioeconomic_tokenizer.py

In the following code, tokenization technique was performed to split a budget speech document.

```
# -*- coding: utf-8 -*-
"""
Created on Thu Nov  8 02:47:21 2018

@author: cvuppalapati
"""
import nltk
from nltk.tokenize import word_tokenize

s = '''The headline inflation remained high for most part of the year. It was only in December 2011 that
it moderated to 8.3 per cent followed by 6.6 percent in January 2012.
print(word_tokenize(s))

tagged = nltk.pos_tag(word_tokenize(s))
print(tagged)



from nltk.tokenize import wordpunct_tokenize
print(wordpunct_tokenize(s))


from nltk.tokenize import sent_tokenize, word_tokenize
print(sent_tokenize(s))
```

The word_tokenize returns a tokenized copy of *text*, using NLTK's recommended word tokenizer (currently an improved :class:`.TreebankWordTokenizer` along with : class:`.PunktSentenceTokenizer` for the specified language). The Treebank tokenizer[3] uses regular expressions to tokenize text as in Penn Treebank.

[3]TreebankWordTokenizer—https://www.nltk.org/api/nltk.tokenize.TreebankWordTokenizer.html

The wordpunct_tokenize tokenizes a text into a sequence of alphabetic and non-alphabetic characters, using the regexp ``\w+|[^\w\s]+``..

Finally, sent_tokenize returns a sentence-tokenized copy of *text*, using NLTK's recommended sentence tokenizer (currently :class:`. PunktSentenceTokenizer` for the specified language).

Output

```
PS C:\Hanumayamma\Clustering> python .\budgetspeechstatement_scioeconomic_tokenizer.py
['The', 'headline', 'inflation', 'remained', 'high', 'for', 'most', 'part', 'of', 'the', 'year', '.', 'It', 'was', 'only', 'in', 'December', '2011', 'that', 'it', 'moderated', 'to', '8.3', 'per', 'cent', 'followed', 'by', '6.6', 'percent', 'in', 'January', '2012', '.']
[('The', 'DT'), ('headline', 'NN'), ('inflation', 'NN'), ('remained', 'VBD'), ('high', 'JJ'), ('for', 'IN'), ('most', 'JJS'), ('part', 'NN'), ('of', 'IN'), ('the', 'DT'), ('year', 'NN'), ('.', '.'), ('It', 'PRP'), ('was', 'VBD'), ('only', 'RB'), ('in', 'IN'), ('December', 'NNP'), ('2011', 'CD'), ('that', 'IN'), ('it', 'PRP'), ('moderated', 'VBD'), ('to', 'TO'), ('8.3', 'CD'), ('per', 'IN'), ('cent', 'NN'), ('followed', 'VBN'), ('by', 'IN'), ('6.6', 'CD'), ('percent', 'NN'), ('in', 'IN'), ('January', 'NNP'), ('2012', 'CD'), ('.', '.')]
['The', 'headline', 'inflation', 'remained', 'high', 'for', 'most', 'part', 'of', 'the', 'year', '.', 'It', 'was', 'only', 'in', 'December', '2011', 'that', 'it', 'moderated', 'to', '8', '.', '3', 'per', 'cent', 'followed', 'by', '6', '.', '6', 'percent', 'in', 'January', '2012', '.']
['The headline inflation remained high for most part of the year.', 'It was only in December 2011 that it moderated to 8.3 per cent followed by 6.6 percent in January 2012.']
```

Part-of-Speech (POS) Detection

Part-of-speech detection uses NLTK's currently recommended part-of-speech tagger to tag the given list of tokens and identifies text as a verb, noun, participle, verb phrase, and so on.

Here is the list of POS tags:

CC coordinating conjunction	RB adverb very, silently,
CD cardinal digit	RBR adverb, comparative better
DT determiner	RBS adverb, superlative best
EX existential there (like: "there is" … think of it like "there exists")	RP particle give up
FW foreign word	TO to go 'to' the store.
IN preposition/subordinating conjunction	UH interjection errrrrrrm
JJ adjective 'big'	VB verb, base form take
JJR adjective, comparative 'bigger'	VBD verb, past tense took
JJS adjective, superlative 'biggest'	VBG verb, gerund/present participle taking
LS list marker 1)	VBN verb, past participle taken
MD modal could, will	VBP verb, sing. present, non-3d take
NN noun, singular 'desk'	VBZ verb, 3rd person sing. present takes
NNS noun plural 'desks'	WDT wh-determiner which
NNP proper noun, singular 'Harrison'	WP wh-pronoun who, what
NNPS proper noun, plural 'Americans'	WP$ possessive wh-pronoun whose
PDT predeterminer 'all the kids'	WRB wh-abverb where, when
POS possessive ending parent's	
PRP personal pronoun I, he, she	
PRP$ possessive pronoun my, his, hers	

budgetspeechstatement_scioeconomic_postagging.py

```
# -*- coding: utf-8 -*-
"""
Created on Mon Aug 20 14:29:52 2018

@author: cvuppalapati
"""


import nltk
from nltk.tokenize import word_tokenize
from nltk.tokenize import wordpunct_tokenize
text = '''Even after 52 years of Independence the provision of basic services in rural areas remains very unsatisfactory. These large gaps in basic services in rural areas are not acceptable and Government is committed to removing them rapidly. '''

print(wordpunct_tokenize(text))



# split into words by white space
words = text.split()
print("\n Words \n")
print(words)

print("\n Part of Speech \n")
#  Using a Tagger. Which is part-of-speech
# tagger or POS-tagger.
tagged = nltk.pos_tag(words)
print(tagged)
```

Output

```
PS C:\Hanumayamma\Clustering> python .\budgetspeechstatement_scioeconomic_postagging.py
['Even', 'after', '52', 'years', 'of', 'Independence', 'the', 'provision', 'of', 'basic', 'services', 'in', 'rural', 'areas', 'remains', 'very', 'unsatisfactory', '.', 'These', 'large', 'gaps', 'in', 'basic', 'services', 'in', 'rural', 'areas', 'are', 'not', 'acceptable', 'and', 'Government', 'is', 'committed', 'to', 'removing', 'them', 'rapidly', '.']

 Words

['Even', 'after', '52', 'years', 'of', 'Independence', 'the', 'provision', 'of', 'basic', 'services', 'in', 'rural', 'areas', 'remains', 'very', 'unsatisfactory.', 'These', 'large', 'gaps', 'in', 'basic', 'services', 'in', 'rural', 'areas', 'are', 'not', 'acceptable', 'and', 'Government', 'is', 'committed', 'to', 'removing', 'them', 'rapidly.']

 Part of Speech

[('Even', 'RB'), ('after', 'IN'), ('52', 'CD'), ('years', 'NNS'), ('of', 'IN'), ('Independence', 'NNP'), ('the', 'DT'), ('provision', 'NN'), ('of', 'IN'), ('basic', 'JJ'), ('services', 'NNS'), ('in', 'IN'), ('rural', 'JJ'), ('areas', 'NNS'), ('remains', 'VBZ'), ('very', 'RB'), ('unsatisfactory.', 'JJ'), ('These', 'DT'), ('large', 'JJ'), ('gaps', 'NNS'), ('in', 'IN'), ('basic', 'JJ'), ('services', 'NNS'), ('in', 'IN'), ('rural', 'JJ'), ('areas', 'NNS'), ('are', 'VBP'), ('not', 'RB'), ('acceptable', 'JJ'), ('and', 'CC'), ('Government', 'NNP'), ('is', 'VBZ'), ('committed', 'VBN'), ('to', 'TO'), ('removing', 'VBG'), ('them', 'PRP'), ('rapidly.', 'VB')]
PS C:\Hanumayamma\Clustering>
```

Stemming and Lemmatization

Normalizing words so that the different forms map to the canonical word with the same meaning. For example, "running" and "ran" map to "run."

	budgetspeechstatement_scioeconomic_textProcwithStemmer.py

In the below code, we have performed tokenization and stemmed the tokens using porter stemmer. The word stemmer is based on the Porter stemming algorithm.[4] The Porter stemming algorithm (or "Porter stemmer") is a process for removing the commoner morphological and inflexional endings from words in English. Its main use is as part of a term normalization process that is usually done when setting up information retrieval systems [5].

```
"""
Created on Mon Aug 20 14:29:52 2018

@author: cvuppalapati
"""


import nltk
from nltk.tokenize import word_tokenize
from nltk.tokenize import wordpunct_tokenize

text = '''Even after 52 years of Independence the provision of basic services in rural areas remains very
unsatisfactory. These large gaps in basic services in rural areas are not acceptable and Government is
committed to removing them rapidly. '''

print(wordpunct_tokenize(text))

from nltk.tokenize import sent_tokenize
print(sent_tokenize(text))

# split into words by white space
words = text.split()
print("\n Words \n")
print(words)
# remove punctuation from each word
import string
print("\n string.punctuation \n")
print(string.punctuation)
table = str.maketrans('', '', string.punctuation)
stripped = [w.translate(table) for w in words]
```

[4]Porter, M. "An algorithm for suffix stripping." Program 14.3 (1980): 130–137. http://www.tartarus.org/~martin/PorterStemmer/

```
print("\n stripped \n")
print(stripped)

# split into sentences
from nltk import sent_tokenize
sentences = sent_tokenize(text)

print("\n sentences \n")
print(sentences[0])

# split into words
from nltk.tokenize import word_tokenize
tokens = word_tokenize(text)
print("\n NLTK SPLIT WORDS \n")

# remove all tokens that are not alphabetic
words = [word for word in tokens if word.isalpha()]

print("\n remove all tokens that are not alphabetic \n")

print(tokens[:100])


print("\n PORTER STEMMER  \n")

# Stemming of words
from nltk.stem.porter import PorterStemmer
porter = PorterStemmer()
stemmed = [porter.stem(word) for word in tokens]
print(stemmed[:100])


print("\n Part of Speech \n")
#  Using a Tagger. Which is part-of-speech
# tagger or POS-tagger.
tagged = nltk.pos_tag(words)
print(tagged)
```

Output

```
PS C:\Hanumayamma\Clustering> python
.\budgetspeechstatement_scioeconomic_textProcwithStemmer.py
['Even', 'after', '52', 'years', 'of', 'Independence', 'the', 'provision', 'of', 'basic', 'services', 'in', 'rural', 'areas',
'remains', 'very', 'unsatisfactory', '.', 'These', 'large', 'gaps', 'in', 'basic', 'services', 'in', 'rural', 'areas', 'are',
'not', 'acceptable', 'and', 'Government', 'is', 'committed', 'to', 'removing', 'them', 'rapidly', '.']
['Even after 52 years of Independence the provision of basic services in rural areas remains very
unsatisfactory.', 'These large gaps in basic services in rural areas are not acceptable and Government is
committed to removing them rapidly.']

PORTER STEMMER

['even', 'after', '52', 'year', 'of', 'independ', 'the', 'provis', 'of', 'basic', 'servic', 'in', 'rural', 'area', 'remain',
'veri', 'unsatisfactori', '.', 'these', 'larg', 'gap', 'in', 'basic', 'servic', 'in', 'rural', 'area', 'are', 'not', 'accept',
'and', 'govern', 'is', 'commit', 'to', 'remov', 'them', 'rapidli', '.']
```

Entity Extraction

Entity extraction is the process of identifying subjects in the text. Named entity recognition (NER) is probably the first step towards information extraction that seeks to locate and classify named entities in text into pre-defined categories such

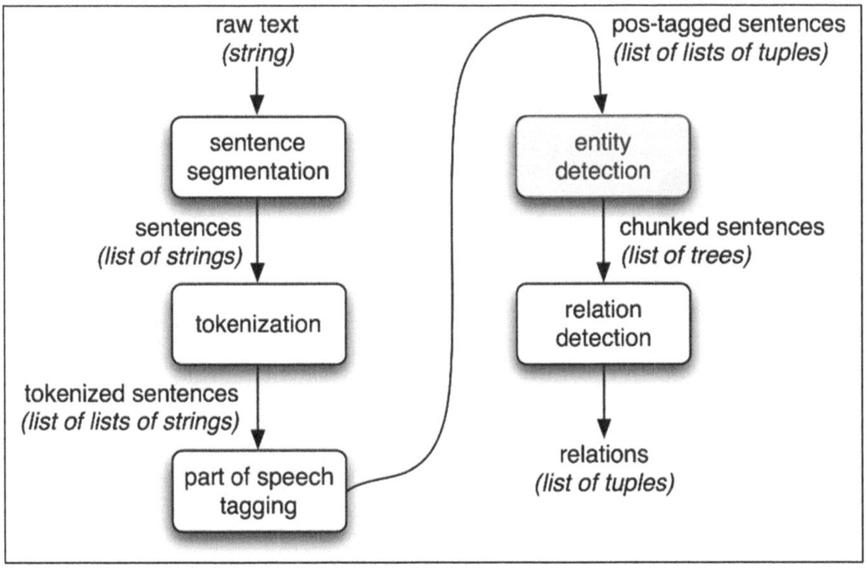

Fig. 2.2 NLP entity detection

as the names of persons (please see Fig. 2.2), organizations, locations, expressions of times, quantities, monetary values, percentages, etc. [6].

Please find commonly used types of[5] named entity [3]:

NE Type	Examples
ORGANIZATION	Georgia-Pacific Corp., WHO
PERSON	Eddy Bonte, President Obama
LOCATION	Murray River, Mount Everest
DATE	June, 2008-06-29
TIME	two fifty a m, 1:30 p.m.
MONEY	175 million Canadian Dollars, GBP 10.40
PERCENT	twenty pct, 18.75 %
FACILITY	Washington Monument, Stonehenge
GPE	South East Asia, Midlothian

I took a sentence from the Bangladesh 2019–2020[6] Budget Speech: "Madam Speaker, in one word, Bangladesh is known today to the whole world by the Prime Minister's name. She is now a world leader. A survey conducted by Fortune, the magazine of the Time Corporation in the United States, has placed Hon'ble Prime Minister among the top ten world leaders. Time, the New York-based famous weekly, has included our Hon'ble Prime Minister Sheikh Hasina in its 2018 list of

[5] Information retrieval—https://www.nltk.org/book/ch07.html

[6] Budget Speech 2019–20—https://nbr.gov.bd/uploads/budget/Budget_Speech_2019-20_English_Final_Print.pdf

100 most influential world leaders, along with US President Donald Trump and Japan's Prime Minister Shinzo Abe. Hon'ble Prime Minister, we the countrymen are grateful to you for taking Bangladesh to such a great height."

Let's extract entities from the statements.

budgetspeechstatement_entityextraction.py

```
# -*- coding: utf-8 -*-
"""
Created on Sun Aug 19 19:16:12 2018

@author: cvuppalapati
"""
# https://spacy.io/
# pip install spacy
# python -m spacy download en_core_web_sm
import string
import spacy
import nltk
from nltk.corpus import stopwords

nltk.download('stopwords')

# Load English tokenizer, tagger, parser, NER and word vectors
nlp = spacy.load('en_core_web_sm')

stopwords = list(set(stopwords.words('english')))

print("\n Stopwords \n")
print(len(stopwords))
print(stopwords)
print("\n Done Stopwords \n")


punct = [i for i in string.punctuation]
print(punct)


# Process whole documents
text = (u"Madam Speaker, in one word, Bangladesh is known today to the whole world by the Prime
Minister's name. She is now a world leader. A survey conducted by Fortune, the magazine of the Time
Corporation in the United States, has placed Hon'ble Prime Minister among the top ten world leaders.
Time, the New York-based famous weekly, has included our Hon'ble Prime Minister Sheikh
Hasina in its 2018 list of 100 most influential world leaders, along with US President Donald
Trump and Japan's Prime Minister Shinzo Abe. Hon'ble Prime Minister, we the countrymen are
grateful to you for taking Bangladesh to such a great height. ")
doc = nlp(text)
print(text)
# Find named entities, phrases and concepts
for entity in doc.ents:
  print(entity.text, entity.label_)
```

doc.ents performs the entity extraction.

Output

```
PS C:\Hanumayamma\Clustering> python .\budgetspeechstatement_entityextraction.py
[nltk_data] Downloading package stopwords to C:\Users\chvuppal.NORTHAM
[nltk_data]     ERICA\AppData\Roaming\nltk_data...
[nltk_data]   Package stopwords is already up-to-date!

 Stopwords

179
['it', 'me', 'such', 'don', 'ma', 'themselves', 'her', 'until', 'yourselves', 'through', 'here', 'few', 'couldn', 'same',
'weren', 'whom', 'our', 'shouldn', "shan't", 'in', 'wouldn', "weren't", 'm', 'his', 'have', 'some', 'mustn',
"shouldn't", 'you', 'won', 'am', 'ours', "mustn't", 'the', 'my', 'yours', 'what', 'off', 'wasn', 'after', 'but', 'other',
'has', "that'll", 'll', 'before', 'had', 'their', 'shan', 'hadn', 'both', "mightn't", "you've", 'so', 'now', 'your', 'd',
'an', 'those', 'over', 'below', 'at', 'with', 'out', 'doesn', 'didn', "isn't", 're', 'than', 'been', "don't", "hadn't",
'who', 'most', 'y', 'as', "aren't", 'a', 'why', 'them', 's', 'he', 'to', 'further', "wouldn't", 'himself', 'only', 'just',
'there', 'into', 'do', "couldn't", 'theirs', 'herself', 'that', 'own', 'again', 't', 'because', 'not', 'mightn', "doesn't",
'she', 'where', 'for', 'i', 'will', "needn't", 'having', "she's", 'then', 'hers', "hasn't", 'should', 'be', 'down', 'no',
'when', 'nor', 'each', "it's", 'ain', 'which', 'on', 'all', 'him', 'myself', 'of', 'ourselves', 'are', 'we', 'needn',
'does', 'o', 'its', "won't", 'yourself', 'doing', "didn't", 'aren', 'up', 'under', 'while', 'between', "you'd", 'being',
'and', 'against', 'above', 'from', 'or', 'about', 'itself', 'was', 'any', 'more', 'were', 'can', 'isn', "wasn't", 'is',
'this', 'they', 'did', 'haven', 'how', 'hasn', 've', 'by', "haven't", 'these', 'very', 'during', "you're", 'if', 'too',
"should've", "you'll", 'once']

 Done Stopwords

['!', '"', '#', '$', '%', '&', "'", '(', ')', '*', '+', ',', '-', '.', '/', ':', ';', '<', '=', '>', '?', '@', '[', '\\', ']', '^', '_', '`', '{', '|', '}',
'~']
Madam Speaker, in one word, Bangladesh is known today to the whole world by the Prime Minister's
name. She is now a world leader. A survey conducted by Fortune, the magazine of the Time
Corporation in the United States, has placed Hon'ble Prime Minister among the top ten world leaders.
Time, the New York-based famous weekly, has included our Hon'ble Prime Minister Sheikh
Hasina in its 2018 list of 100 most influential world leaders, along with US President Donald
Trump and Japan's Prime Minister Shinzo Abe. Hon'ble Prime Minister, we the countrymen are
grateful to you for taking Bangladesh to such a great height.
one CARDINAL
Bangladesh ORG
today DATE
Fortune ORG
the Time Corporation ORG
the United States GPE
Hon'ble NORP
ten CARDINAL
Time ORG
York GPE
Hasina GPE
2018 DATE
100 CARDINAL
US GPE
Donald Trump PERSON
Japan GPE
Shinzo Abe PERSON
Bangladesh GPE
```

Cosine Similarity Search

The cosine distance makes sense in spaces that have dimensions, including Euclidean spaces and discrete versions of Euclidean spaces, such as spaces where points are vectors with integer components or Boolean (0 or 1) components. In such a space, points may be thought of as directions. We do not distinguish between a vector and a multiple of that vector. Then the cosine distance between two points is the angle that the vectors to those points make. This angle will be in the range from 0° to 180°, regardless of how many dimensions the space has.

We can calculate the cosine distance by first computing the cosine of the angle and then applying the arc-cosine function to translate to an angle in the 0°–180° range. Given two vectors x and y, the cosine of the angle between them is the dot product $x.y$ divided by the L2-norms of x and y (i.e., their Euclidean distances from the origin). Please note that the dot product is the result of multiplying two vectors together.

The cosine similarity between two vectors (or two documents on the vector space) is a measure that calculates the cosine of the angle between them (please see Fig. 2.3). This metric is a measurement of orientation and not magnitude; it can be seen as a comparison between documents on a normalized space because we're not taking into consideration only the magnitude of each word count (text feature extraction (tf-idf)) of each document, but the angle between the documents. What we have to do to build the cosine similarity equation is to solve the equation of the dot product for the

$$\vec{a} \cdot \vec{b} = \|\vec{a}\|\|\vec{b}\| \cos\theta$$

$$\cos\theta = \frac{\vec{a} \cdot \vec{b}}{\|\vec{a}\|\|\vec{b}\|}$$

Let's compare poverty-reducing strategies from two different decades.

Budget Speech of Shri Manmohan Singh—Poverty Reducing Strategy 1991–1992:[7] "Investing in education, particularly elementary education, to develop human resources and provide quality education to all, especially children from marginalized groups."

Fig. 2.3 Cosine

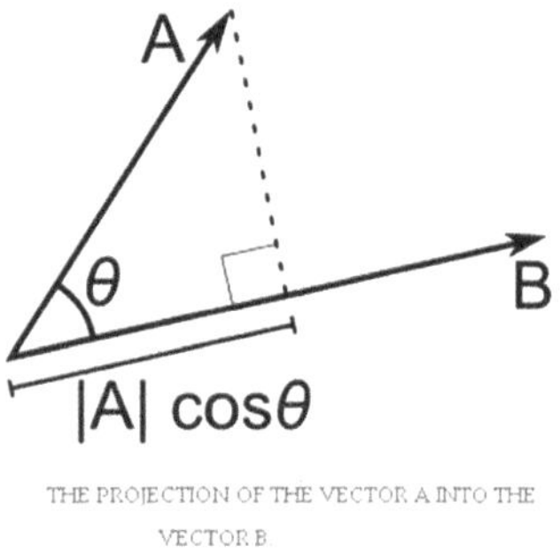

[7] Budget Speech 1991–1992 https://www.indiabudget.gov.in/doc/bspeech/bs199192.pdf

Budget Speech of Shri Yashwant Sinha—Poverty Reducing Strategy 2000–2001:[8] "Promoting credit flow to agriculture and strengthening the rural economy." Let's compare both of semantic relevance:

budgetspeechstatement_docsimilarity.py

```
# -*- coding: utf-8 -*-
"""
Created on Sun Aug 19 19:16:12 2018

@author: cvuppalapati
"""
# https://spacy.io/
# pip install spacy
# python -m spacy download en_core_web_sm
import string
import spacy
import nltk
from nltk.corpus import stopwords

nltk.download('stopwords')

# Load English tokenizer, tagger, parser, NER and word vectors
nlp = spacy.load('en_core_web_sm')

# Determine semantic similarities
print("\n Determine semantic similarities\n")
budgetstatement_socialsustanability_200001_YSinha = nlp(u"Even after 52 years of Independence the
provision of basic services in rural areas remains very unsatisfactory. These large gaps in basic services in
rural areas are not acceptable and Government is committed to removing them rapidly")
budgetstatement_socialsustanability_200607_PChidambaram = nlp(u"2. The government will actively
promote the establishment of new towns focused on specific industries or themes, such as Information
Technology, education, or health. Projects are being planned in West Bengal and Karnataka.")
similarity =
budgetstatement_socialsustanability_200001_YSinha.similarity(budgetstatement_socialsustanability_200
607_PChidambaram)
print(budgetstatement_socialsustanability_200001_YSinha.text,
budgetstatement_socialsustanability_200607_PChidambaram.text, similarity)



# Determine semantic similarities
print("\n Poverty Reducing Strategies \n")
# Budget Speech 1991-1992  - Shri Manmohan Singh

budgetstatement_199192_MSingh = nlp(u"Investing in education, particularly elementary education, to
develop human resources and provide quality education to all, especially children from marginalized
groups")
#Budget Speech 2000-2001  - Shri Yashwant Sinha

budgetstatement_200001_YSinha = nlp(u"Promoting credit flow to agriculture and strengthening the
rural economy.")
similarity = budgetstatement_199192_MSingh.similarity(budgetstatement_200001_YSinha)
print(budgetstatement_199192_MSingh.text, budgetstatement_200001_YSinha.text, similarity)
```

extraction.

[8] Budget Speech 2000–2001 https://www.indiabudget.gov.in/doc/bspeech/bs200001.pdf

Output

```
PS C:\Hanumayamma\Clustering> python .\budgetspeechstatement_docsimilarity.py
[nltk_data] Downloading package stopwords to C:\Users\chvuppal.NORTHAM
[nltk_data]    ERICA\AppData\Roaming\nltk_data...
[nltk_data]   Package stopwords is already up-to-date!

 Determine semantic similarities

.\budgetspeechstatement_docsimilarity.py:24: UserWarning: [W007] The model you're using has no word vectors loaded, so the result of the Doc.similarity method will be based on the tagger, parser and NER, which may not give useful similarity judgements. This may happen if you're using one of the small models, e.g. `en_core_web_sm`, which don't ship with word vectors and only use context-sensitive tensors. You can always add your own word vectors, or use one of the larger models instead if available.
  similarity =
budgetstatement_socialsustanability_200001_YSinha.similarity(budgetstatement_socialsustanability_200607_PChidambaram)
Even after 52 years of Independence the provision of basic services in rural areas remains very unsatisfactory. These large gaps in basic services in rural areas are not acceptable and Government is committed to removing them rapidly 2.   The government will actively promote the establishment of new towns focused on specific industries or themes, such as Information Technology, education, or health. Projects are being planned in West Bengal and Karnataka. 0.7108443256062732

 Poverty Reducing Strategies

.\budgetspeechstatement_docsimilarity.py:37: UserWarning: [W007] The model you're using has no word vectors loaded, so the result of the Doc.similarity method will be based on the tagger, parser and NER, which may not give useful similarity judgements. This may happen if you're using one of the small models, e.g. `en_core_web_sm`, which don't ship with word vectors and only use context-sensitive tensors. You can always add your own word vectors, or use one of the larger models instead if available.
  similarity = budgetstatement_199192_MSingh.similarity(budgetstatement_200001_YSinha)
Investing in education, particularly elementary education, to develop human resources and provide quality education to all, especially children from marginalized groups Promoting credit flow to agriculture and strengthening the rural economy. 0.6005312654725209
PS C:\Hanumayamma\Clustering>
```

Please note that UserWarning: [W007] may happen if you're using one of the small models, e.g., "en_core_web_sm", which don't ship with word vectors and only use context-sensitive tensors.

GFS Analytics Framework

The set of four financial statements is the basis for the GFS analytics framework.[9] The financial statements include:

- Statements of operations
- Statements of other economic flows
- Balance sheet
- Statements of sources and uses of cash

[9] GFSM 2014—https://www.imf.org/external/Pubs/FT/GFS/Manual/2014/gfsfinal.pdf

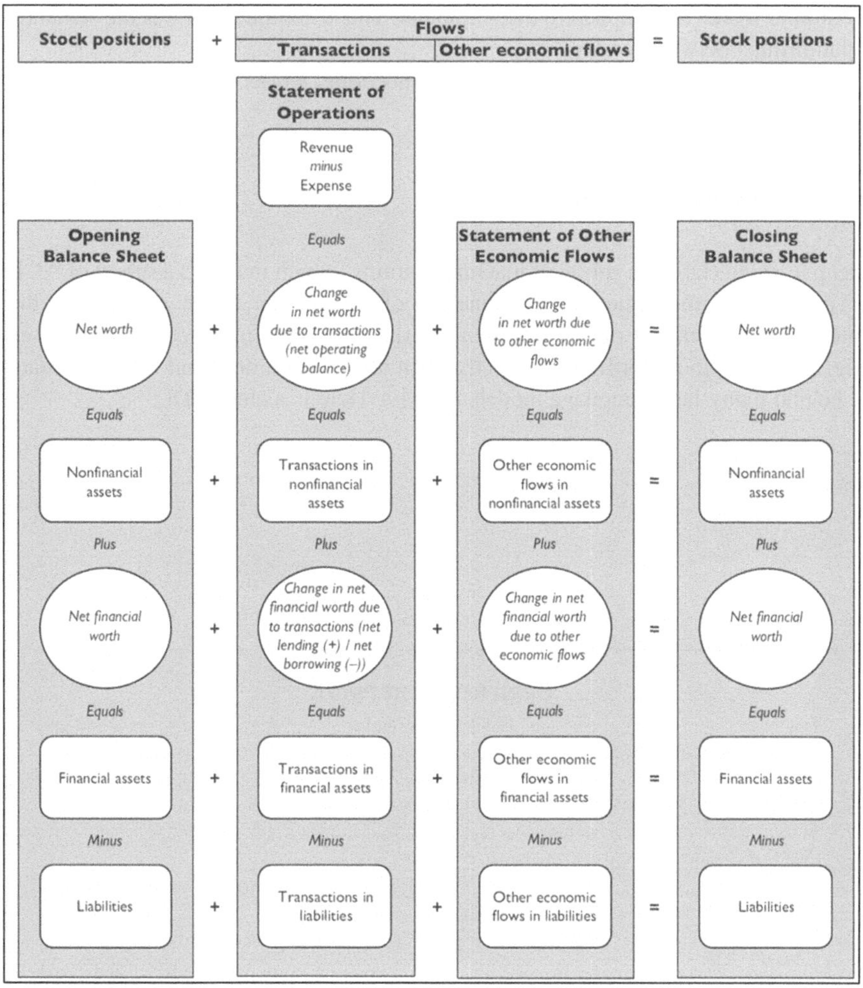

Fig. 2.4 GFS analytics

GFS Analytics Framework 2014 (please see Fig. 2.4) builds on top of the GSM 1986 framework and extends it by incorporating additional elements that are useful in assessing fiscal policy. Some of the notable additions are:

- The definitions of individual statistical variables are closely aligned with economics.
- Concepts are harmonized with the 2008 System of National Accounts (SNA).

Word embedding is a technique in NLP where words or phrases from the vocabulary are mapped to vectors of real numbers. It involves representing words in a dense vector space where similar words are mapped to nearby points and dissimilar words are mapped to distant points. This technique captures the semantic relationships between words.

Here is a simple diagram of how word embedding works:

Deep Learning and NLP

Deep learning (DL) is a subset of machine learning, which in turn is a subset of AI. It is inspired by the structure and function of the human brain, specifically the interconnecting nodes of the neural network. Deep learning models (please see Fig. 2.5) learn to perform tasks directly from text, image, or sound. Deep learning is behind many large language models (LLMs). Here is a simple DL:

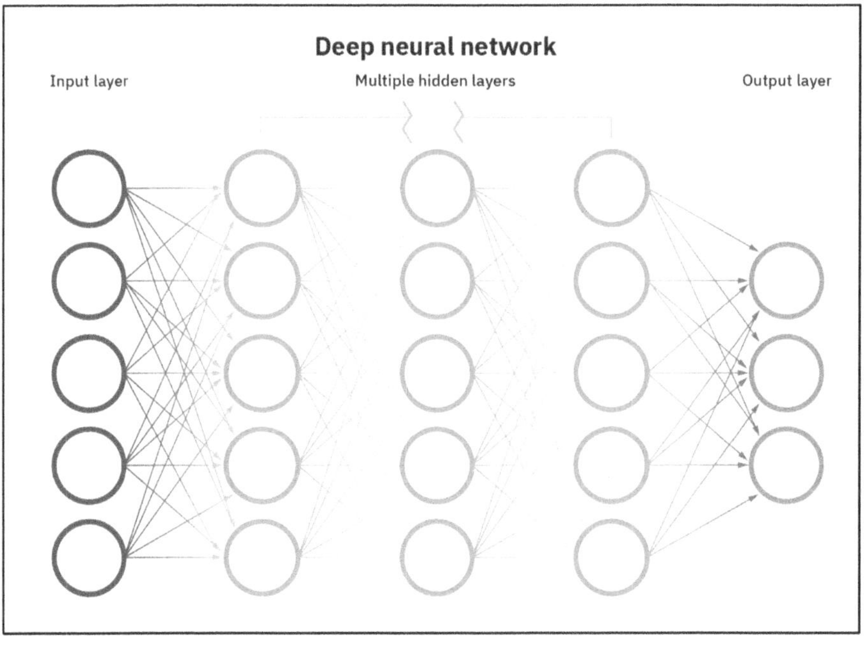

Fig. 2.5 Deep learning

Word Embeddings

Word embedding in NLP is an important term that is used for representing words for text analysis in the form of real-valued vectors. It is an advancement in NLP that has improved the ability of computers to understand text-based content in a better way. It is considered one of the most significant breakthroughs of deep learning for solving challenging natural language processing problems. In this approach, words and documents are represented in the form of numeric vectors allowing similar words to have similar vector representations. The extracted features are fed into a machine learning model so as to work with text data and preserve the semantic and syntactic information. This information once received in its converted form is used by NLP algorithms that easily digest these learned representations and process textual information.

Word embedding or word vector is an approach with which we represent documents and words. It is defined as a numeric vector input that allows words with similar meanings to have the same representation. It can have approximate meaning and represent a word in a lower-dimensional space. These can be trained much faster than the hand-built models that use graph embeddings like WordNet.

Term frequency-inverse document frequency is the machine learning algorithm that is used for word embedding for text. It comprises two metrics, namely, term frequency (TF) and inverse document frequency (IDF).[10]

This algorithm works on a statistical measure of finding word relevance in the text that can be in the form of a single document or various documents that are referred to as corpus.

Term Frequency-Inverse Document Frequency (TF-IDF)

The term frequency (TF) score measures the frequency of words in a particular document. In simple words, it means that the occurrence of words is counted in the documents. The inverse document frequency or the IDF score measures the rarity of the words in the text. It is given more importance over the term frequency score because even though the TF score gives more weightage to frequently occurring words, the IDF score focuses on rarely used words in the corpus that may hold significant information. TF-IDF algorithm finds application in solving simpler natural language processing and machine learning problems for tasks like information retrieval, stop words removal, keyword extraction, and basic text analysis. However, it does not capture the semantic meaning of words efficiently in a sequence [7].

[10] A Guide on Word Embeddings in NLP—https://www.turing.com/kb/guide-on-word-embeddings-in-nlp

Now let's understand it further with an example. We will see how vectorization is done in TF-IDF.

budgetspeechstatement_tfidf.py

In the below code, TF-IDF was used to generate word embeddings.

```
# -*- coding: utf-8 -*-
"""
Created on Sat Dec  9 12:59:03 2023

@author: chvuppal
"""
# https://scikit-learn.org/stable/modules/generated/sklearn.feature_extraction.text.TfidfVectorizer.html
from sklearn.feature_extraction.text import TfidfVectorizer


corpus = [
    'This is the first document.',
    'This document is the second document.',
    'hello',
    'Is this the first document?',
    ]

vectorizer = TfidfVectorizer()
X = vectorizer.fit_transform(corpus)
vectorizer.get_feature_names_out()

print(vectorizer.get_feature_names_out())
print(X)
print(X.shape)
```

Output

```
PS C:\Hanumayamma\Clustering> python .\budgetspeechstatement_tfidf.py
['document' 'first' 'hello' 'is' 'second' 'the' 'this']
  (0, 0)	0.42540804913721164
  (0, 1)	0.5254635733493682
  (0, 5)	0.42540804913721164
  (0, 3)	0.42540804913721164
  (0, 6)	0.42540804913721164
  (1, 4)	0.5095233013298407
  (1, 0)	0.6504442461675999
  (1, 5)	0.32522212308379994
  (1, 3)	0.32522212308379994
  (1, 6)	0.32522212308379994
  (2, 2)	1.0
  (3, 0)	0.42540804913721164
  (3, 1)	0.5254635733493682
  (3, 5)	0.42540804913721164
  (3, 3)	0.42540804913721164
  (3, 6)	0.42540804913721164
```

The rows represent each document, the columns represent the vocabulary, and the values of tf-idf(i,j) are obtained through the above formula. This matrix obtained can be used along with the target variable to train a machine learning/deep learning model.

Bag of Words (BOW)

A bag of words is one of the popular word embedding techniques of text where each value in the vector would represent the count of words in a document/sentence. In other words, it extracts features from the text. We also refer to it as vectorization.

To get you started, here's how you can proceed to create BOW.

- In the first step, you have to tokenize the text into sentences.
- Next, the sentences tokenized in the first step have further tokenized words.
- Eliminate any stop words or punctuation.
- Then, convert all the words to lowercase.
- Finally, move to create a frequency distribution chart of the words.

Economic and Social Sustainability Data Mining Using NLP and Budget Speeches

The budget speech is a vital part of the democratic process. The budget speech contains information on yearly financial reports or plans of the government, showing the details of income and spending in the past, and the expected spending and forecasts for the next year. The budget speech follows a standard template and, for many countries, is based on the International Monetary Fund (IMF) *Government Finance Statistics Manual 2014*[11] template. Using a common template to present government finance statistics, importantly, helps the world of international finance and capital markets. Having internationally comparable data as recent financial crisis events have revealed enables early identification of sources of risk and timely actions to correct them [8].

The National Budget Speech affects the profitability of small to medium enterprises (SMEs) and influences long-term business decisions. The budget speech affects many sectors of the economy, especially agriculture and social programs. For example, subsidies to farmers, food, and exports are some of the crucial items that affect the whole market. Also, the budget speech discusses the new development to address shortages on the supply side of items, for instance, increasing the

[11] Government Finance Statistics Manual 2014—https://www.imf.org/external/Pubs/FT/GFS/Manual/2014/gfsfinal.pdf

production of oil seeds. More importantly, the budget speech and the budget policies after that have a direct impact on the common man's wallet and checkbook. Some may directly reject budget speech with little to no relief for the common man. Considering the highly uncertain business environment, it is very important for SMEs to pay close attention to the National Budget Speech, as key decisions made by the Minister will affect their growth and survival [9, 10].

Budget speeches in times of economic downturn and major recessions, on the other hand, describe the economic situation, the challenges, and the bleak outlook that the country faces and offer fiscal and monetary policies to overcome financial difficulties and austerity measures. For example, Hong Kong Financial Secretary Paul Chan revealed in 2021 a budget of over $15 billion for measures to boost the economy out of recession. He added as part of budget speech that, "Hong Kong experienced hardships in the past 2 years and with the epidemic still ongoing, our economy has not recovered from recession" [11].

Budget speeches give a comprehensive overview of the national economic activities and show a realistic picture of the state of the economy. They have many benefits and offer a unique perspective on fiscal and trade policy. For example, by analyzing the 2018–2019 Sierra Leone[12] Budget Speech, we can get an idea of how the global market affects Sierra Leone's national economy. As stated in the 2018–2019 Sierra Leone Budget Speech,[13] "the medium term outlook for global growth is expected to be weaker. The factors that could undermine global growth are policy uncertainty due to trade conflicts between the US and its trading partners; the anticipated tightening of global financial conditions, the reversal of the US fiscal stimulus in 2020; the projected slow growth of the Chinese economy; increasing debt levels in developing countries and currency volatility" [8].

Furthermore, by analyzing budget speech, we can get a sense of the sectors of the economies that are vulnerable to global changes and depend on the sectors that sustain the economy. Policymakers need this information about the sectors to support the weaker parts of the economies to achieve economic and social stability. For example, as stated in the 2018–2019 Sierra Leone Budget Speech, "the non-iron ore economy is expected to grow by 5.7%, despite major sectors being affected by global crisis, driven by increased activities in agriculture and fisheries, recovery in diamond and other mining activities as well as improvements in the services sector." Therefore, having a global economic perspective would be very helpful to design national economic policies and protect local economies [8].

[12] Sierra Leone—Budget Framework—2019—https://www.cabri-sbo.org/en/documents/government-budget-and-statement-of-economic-and-financial-policies-for-2019

[13] Sierra Leone—Budget Framework—2019—https://www.cabri-sbo.org/en/documents/government-budget-and-statement-of-economic-and-financial-policies-for-2019

Development of Budget Speech Embeddings with Classical NLP

We have taken budget speech statements from 1990 to 2020. The first step in building the model is to manually code the budget speech statements into relevant classification. To annotate, we have defined agricultural vocabulary dictionary and label the speech statement (please see Fig. 2.6).

As an example, consider the following budget statement from 1991 to 1992 budget speech:[14] "I would like to draw the attention of the House to the fact that there has been no increase in *fertilizer prices* since July 1981. In these 10 years, there has been a continuous increase in the *procurement prices* of paddy and wheat, as also in the market prices of other crops, received by the *agricultural sector*."

In the above statement, the key terms are bolded, as can be seen from the statement, state of the economy referring to fertilizer prices and commodity prices within the agriculture sector.

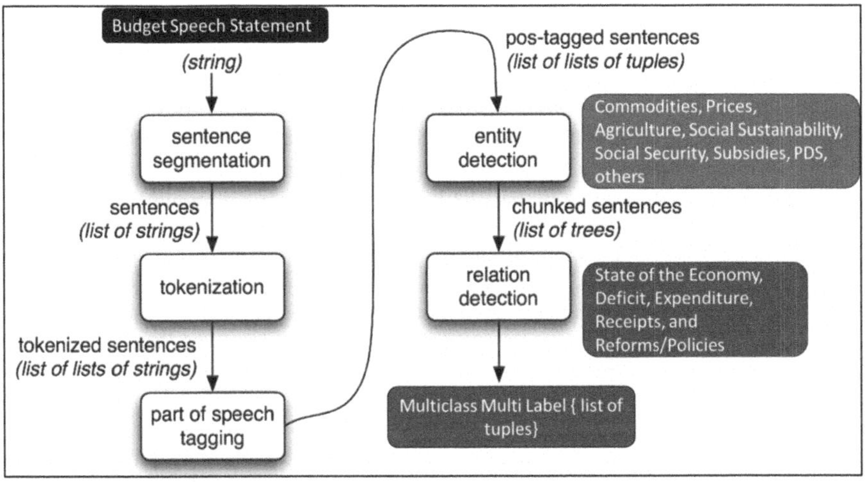

Fig. 2.6 NLP overall

[14] Budget 1991–92 Speech of Shri Manmohan Singh Minister of Finance 24th July, 1991—https://www.indiabudget.gov.in/doc/bspeech/bs199192.pdf

Year	SS #	BudgetSpeech	Main Classification	Sub Classificatio	Third	Net Impact To Agriculture	Score
1991-92	2	The **balance of payments** situation is precarious	State of the Economy	Crisis	payments	Negative	-800
1991-92	2	There has been a sharp decline in capital inflows through commercial borrowing and non-resident deposits.	State of the Economy	capital inflows	decrease	Negative	-800
1991-92	2	There has been a sharp decline in capital inflows through commercial borrowing and non-resident deposits. As a result, despite large borrowings from the International Monetary Fund in July 1990 and January 1991, there was a **sharp reduction** in our **foreign exchange reserves.**	State of the Economy	foreign exchange reserves	decrease	Negative	-800
1991-92	2	The foreign exchange crisis constitutes a serious threat to the sustainability of growth processes and orderly implementation of our development programmes. Due to the combination of unfavourable internal and external factors, the **inflationary pressures** on the price level have increased very substantially since mid-1990. The people of India have to face **double digit inflation** which hurts most the poorer sections of our society. In sum, the crisis in the economy is both acute and deep. We have not experienced anything similar in the history of independent India.	State of the Economy	inflationary pressures & double digit inflation	crisis	Negative	-600
1991-92	3	The increasing difference between the income and expenditure of the Government has led to a widening of the gap between the income and expenditure of the economy as a whole. This is reflected in growing current account **deficits** in the **balance of payments.**	Deficit	Fiscal Deficit	Increase	Negative	-500

Building BOW Embedding

The following BOW model extracts budget speech embeddings from 1990 to 2022 budget speeches. The model works with supervised data that has budget specific primary and secondary classifications. The manual annotation was something we have started in 2018, prior to the availability of large language models.

> Strengthening macroeconomic fundamentals: This involves striking a balance between fiscal consolidation and strengthening macroeconomic fundamentals to create adequate headroom to deal with future shocks in the global environment [8].
>
> Shri Pranab Mukherjee
> Minister of Finance, 2012–2013[15]

Prior to the availability of large language models, we have devised macroeconomic structure to analyze and categorize each budget statement. We have stated the following primary and secondary categories to bucket each budget statement.

Primary category
Deficit
Expenditure
Receipts
Reforms/policies
State of the economy

[15] Budget Speech Budget 2012–2013 https://www.indiabudget.gov.in/doc/bspeech/bs201213.pdf

Deficit The secondary categories include budget deficit, revenue deficit, fiscal deficit, reduced budget deficit, and reduced budget deficit. While presenting the budget for 2009–2010, I had expressed my concern about the high level of fiscal deficit. The first step in codification is to understand the statement. Second, classify the primary category. Next, classify the secondary category. Finally, score the policy in terms of effectiveness.

Budget Statement for 2009–2010

I had also stated that the Government will address this issue in right earnest to come back to the path of fiscal consolidation at the earliest. I am happy to report that against a *fiscal deficit* of 7.8% in 2008–09, inclusive of *oil and fertiliser bonds*, the comparable fiscal deficit is 6.9% as per the Revised Estimates for 2009–10. Both these *deficit* figures are based on the revised GDP numbers published by the Central Statistical Organization and include what were earlier referred to as below the line items. This marks an improvement of about 1% in fiscal deficit during the current year. I have made a conscious effort to avoid issuing bonds to oil and fertiliser companies. I would like to continue with this practice of extending Government subsidy in cash, thereby bringing all subsidy related liabilities into our fiscal accounting.

We relied on manual processes until we developed our own Artificial Intelligence (AI) engine to classify the data. The automated data labeling helped to rapidly scale budget statements (please see Fig. 2.7).

State of the Economy Another category would provide more underpinning of our model design. Take the state of the economy as a label. When the economy is going through stress, many of the macroeconomic variables are listed in budget speeches to resolve or provide a sustainable economic and social policy. Codifying all

Fig. 2.7 Deficit model

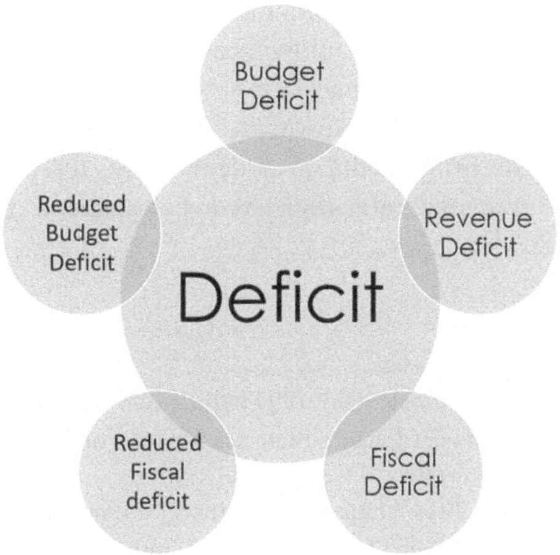

subcategories to state of the economy could be impossible. So, we have focused on economic and social sustainability from a point of view.

Budget Speech 1992–1993[16]

Our first and immediate *challenge* was to *arrest* the slide and restore *India's credibility* both domestically and in the eyes of the world. To achieve this objective, we had to take immediate measures to *avert a default on international payments* and also take steps to restore macro-economic balance in the economy in the short run, with a view to *controlling inflation and reducing the balance of payments deficit* to a manageable level. Our medium term objective was to place the economy back on the path of high and sustainable growth.

Shri Manmohan Singh
Finance Minister

We used the IMF ontology and the government finance statistics (GFS) framework, which are designed to support fiscal analysis.[17] The government finance statistics (GFS) has data for all countries that report in the framework of the *Government Finance Statistics Manual 2014* (GFSM 2014).[18] The GFSM 2014 classifications are mostly the same as the GFSM 2001 presentations. However, some tables have more GFS items than before, while others have fewer GFS items. Figure 2.8 contains the revenue for selected countries from GFS Query.[19]

Receipts Category

Receipts components include taxes on income, profits, and gains; taxes on payroll and workforce; taxes on property; taxes on goods and services; taxes on international trade and transactions; other taxes; social contributions; grants; and other revenues.

Budget Speech 1999–1999

Finally, I have something to say on behalf of my Honorable colleague, the Minister of Communications. Postal service is highly *employment-intensive*, and salary and allowances constitute a major part of the operating expenses of the postal department. A revision of tariff for postal services has become unavoidable. However, in the *interest of the common man and* the role of the print media in a democracy, there will be no change in the *tariff for postcard* and registered newspapers. However, the rate of competition

(continued)

[16]Budget Speech 1992–1993 https://www.indiabudget.gov.in/doc/bspeech/bs199293.pdf

[17]Government Finance Statistics Manuals and Guides—https://www.imf.org/external/pubs/ft/gfs/manual/gfs.htm

[18]GFSM 2014—https://www.imf.org/external/Pubs/FT/GFS/Manual/2014/gfsfinal.pdf

[19]GFS Revenue Query—https://data.imf.org/?sk=FA66D646-6438-4A65-85E5-C6C4116C4416

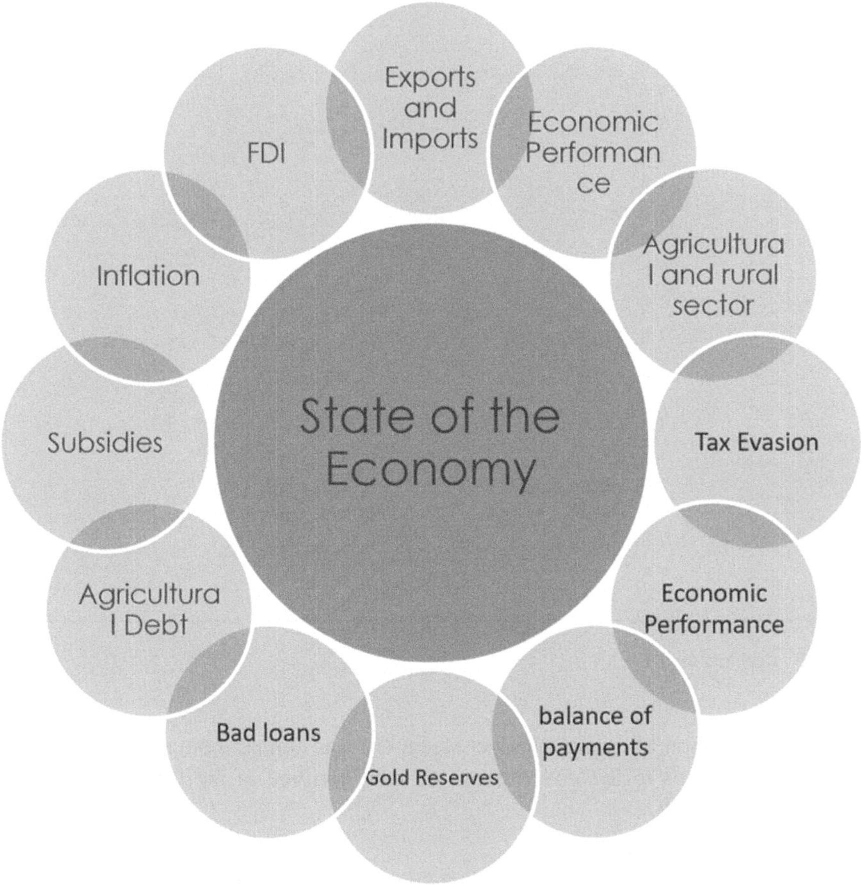

Fig. 2.8 State of the economy

postcards is being raised from Rs. 2 to Rs. 3, Inland letter from Rs. 1 to Rs. 1.50, letter from Rs. 2 to Rs. 3 for every 20 g or part thereof and parcels from Rs. 8 to Rs. 10 for every 500 g or part thereof. There are also certain other changes which are explained in the Memorandum circulated along with the budget documents. The changes would take effect from a date to be notified after the Finance Bill is passed. The revisions proposed are estimated to yield an additional revenue of about Rs. 270 crore in a full year and about Rs. 180 crore in 1998–99. Even this modest increase, which is necessary for sustaining postal development, will only partially meet the cost of postal services leaving an uncovered postal deficit of about Rs. 695 crore.

Shri P. Chidambaram
Finance Minister

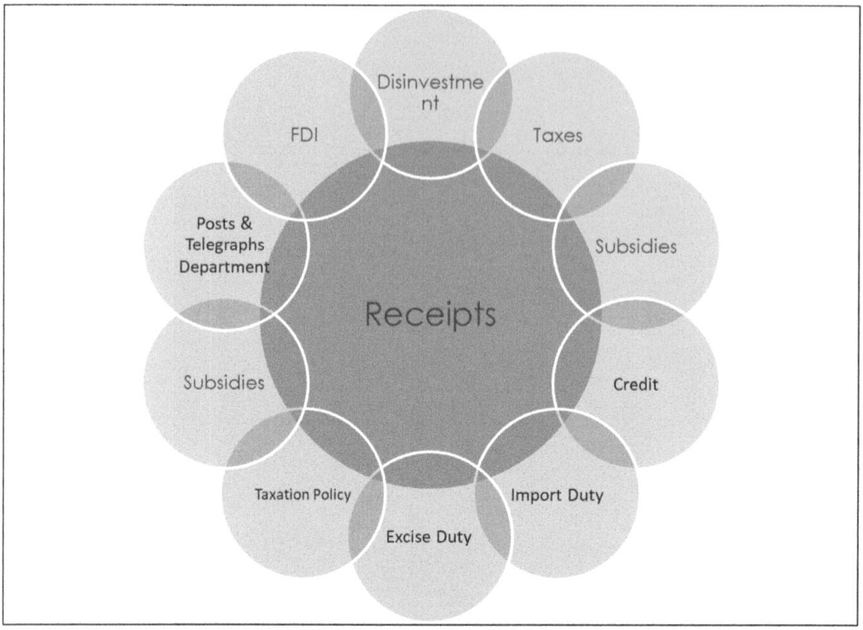

Fig. 2.9 Receipts model

The codification of receipts budget statement was simpler compare to others as the direct revenues to the government are well certified in the GSM framework (please see Fig. 2.9).

Economic Sustainability BOW Embedding Model

Let's use the bag of words and load economic and social sustainability embeddings.

	Final_BagOfWords_EconomicandSocialSustainability.ipynb (Jupyter Notebook Code)

Data Sources

Source	Files	Source
Union budgets	Budget speeches[20] 1947–2022	केन्द्रीय बजट UNION BUDGET

[20] Budget Speeches—https://www.indiabudget.gov.in/bspeech.php

Step 1: Load Libraries for Performing NLP

Load scikit-learn and Matplot libraries.

```
import csv
import re
import time
import string
import subprocess
import pandas as pd
from sklearn import preprocessing
from pylab import rcParams
import scipy
import numpy as np
import seaborn as sb
import nltk
nltk.download('stopwords')
nltk.download('wordnet')

from sklearn.model_selection import train_test_split
from nltk.corpus import stopwords
from nltk.stem import PorterStemmer, WordNetLemmatizer
from nltk.tokenize import word_tokenize
```

Output

All required libraries are loaded to run the code.

Step 2: Initialize Memory Variables and NLP Stop Word Processor

Let's initialize memory, data store, and stop word processor.

```
dataSet = []
stopWords = []
#Function that initalizes the system
def initializeSystem():
    print ("Preparing set of stop words")
    stop = stopwords.words('english') + punctuation + ['rt', 'via','i\'m','us','it']
    for x in stop:
        stopWords.append(stemmer.stem(lemmatiser.lemmatize(x, pos="v")))
```

Output: able to extract stop words from budget speech statements.

```
['i', 'me', 'my', 'myself', 'we', 'our', 'our', 'ourselv', 'you', "you'r", "you'v", "you'll", "you'd", 'your', 'your',
'yourself', 'yourselv', 'he', 'him', 'hi', 'himself', 'she', "she'", 'her', 'her', 'herself', 'it', "it'", 'it', 'itself', 'they',
'them', 'their', 'their', 'themselv', 'what', 'which', 'who', 'whom', 'thi', 'that', "that'll", 'these', 'those', 'be',
'be', 'be', 'be', 'be', 'be', 'be', 'be', 'have', 'have', 'have', 'have', 'do', 'do', 'do', 'do', 'a', 'an', 'the', 'and', 'but',
'if', 'or', 'becaus', 'as', 'until', 'while', 'of', 'at', 'by', 'for', 'with', 'about', 'against', 'between', 'into', 'through',
'dure', 'befor', 'after', 'abov', 'below', 'to', 'from', 'up', 'down', 'in', 'out', 'on', 'off', 'over', 'under', 'again',
'further', 'then', 'onc', 'here', 'there', 'when', 'where', 'whi', 'how', 'all', 'ani', 'both', 'each', 'few', 'more',
'most', 'other', 'some', 'such', 'no', 'nor', 'not', 'onli', 'own', 'same', 'so', 'than', 'too', 'veri', 's', 't', 'can', 'will',
'just', 'don', "don't", 'should', "should'v", 'now', 'd', 'll', 'm', 'o', 're', 've', 'y', 'ain', 'aren', "aren't", 'couldn',
"couldn't", 'didn', "didn't", 'doesn', "doesn't", 'hadn', "hadn't", 'hasn', "hasn't", 'haven', "haven't", 'isn',
"isn't", 'ma', 'mightn', "mightn't", 'mustn', "mustn't", 'needn', "needn't", 'shan', "shan't", 'shouldn',
"shouldn't", 'wasn', "wasn't", 'weren', "weren't", 'win', "won't", 'wouldn', "wouldn't", '!', '"', '#', '$', '%',
'&', "'", '(', ')', '*', '+', ',', '-', '.', '/', ':', ';', '<', '=', '>', '?', '@', '[', '\\', ']', '^', '_', '`', '{', '|', '}', '~', 'rt', 'via', "i'm",
'us', 'it']
```

Step 3: Set Tokenizer

Set up tokenizer to perform NLP tasks.

```
#Function to convert string into a list of tokens represented in the regular expression
def tokenize(s):
    return tokens_re.findall(s)
```

Output
  Performs budget speech tokenization.

Step 4: Lemmatize and Stem Each Word

Convert the input string into a list of words. Each word is lemmatized and stemmed after the stop words are removed.

```
#Converts the input string into a list of words
#Each word is lemmatized and stemmed after the stop words are removed
def preprocess(s, lowercase=True):
    tokens = tokenize(s)
    if lowercase:
        tokens = [stemmer.stem(lemmatiser.lemmatize(token.lower(), pos="v")) for token in tokens]
    return tokens
```

Output
  Word array with canonical form of the word.

Step 5: Process String

Initializer of NLP pipeline.

```
def processString(string):
    terms_stop = [term for term in preprocess(string) if term not in stopWords and len(str(term)) >1 and
not term.isnumeric()]
    return terms_stop
```

Output
Outputs stop words.

Step 6: Convert Budget Speech Statements Into Embeddings

Generate budget statement embeddings.

```
#Removes stop words, stems and lemmatizes each word in all the budget speech statement passed
def convertBudgetspeechstatementsintoembeddings(reviews):
    budgetspeechStatementEmbeddings=[]
    for a in reviews:
        budgetspeechStatementEmbeddings.append(processString(str(a).lower()))
    return budgetspeechStatementEmbeddings
```

Output
Outputs stop words.

Step 7: Entry Function

Entry function that performs grammar and lexicon initialization, stemmer, and lemmatizer.

```
print ("Defining grammar..")
tokens_re = re.compile(r'('+'|'.join(regex_str)+')', re.VERBOSE | re.IGNORECASE)
punctuation = list(string.punctuation)
stemmer = PorterStemmer()
lemmatiser = WordNetLemmatizer()

initializeSystem()
```

Output

```
Defining grammar.
Preparing set of stop words.
```

Load the data.

```
dataSet=pd.read_csv("financebudgetspeechdataset_India_allstates_economic_and_social_sustainability.c
sv")

#preparing the test and training set of data
train,test =train_test_split(dataSet,test_size=0.21)
dataSet
```

Output

	budgetspeech	Label	Primary\nClassification	Secondary\nClassification	Third	Net Impact To Agriculture	Score	Model\nAg
0	I would like to draw the attention of the Hous...	[state of the economy] [agricultural and rural ...	state of the economy	agricultural and rural sector	promotion	positive	-200.0	
1	We would continue to ensure that 50 per cent o...	[state of the economy] [agricultural and rural ...	state of the economy	agricultural and rural sector	assistance	positive	400.0	
2	The provision for assistance for fresh water a...	[state of the economy] [agricultural and rural ...	state of the economy	agricultural and rural sector	assistance	positive	400.0	
3	Another new scheme that would be implemented f...	[state of the economy] [agricultural and rural ...	state of the economy	agricultural and rural sector	assistance	positive	300.0	
4	In order to safeguard any possible loss...	[state of the economy] [agricultural...	state of the economy	agricultural and rural sector	assistance	positive	300.0	

```
#Parse the testing dataset read from the file
testDataFeaturesBudgetSpeechStatements =
pd.DataFrame(dataSet,columns=["budgetspeech","Label"])
#targetRating = testDataFeaturesReviews.values.reshape(-1,1)
budgetspeechStmnts = testDataFeaturesBudgetSpeechStatements['budgetspeech']
#Label = testDataFeaturesReviews['Label']
budgetspeechStmnts
```

Start parse the budget speech statements data.

Output

```
   I would like to draw the attention of the Hous...
   We would continue to ensure that 50 per cent o...
   The provision for assistance for fresh water a...
   Another new scheme that would be implemented f...
   In order to safeguard any possible loss in pro...
                         ...
   There  is  an  urgent  need  for  improving  t...
   For  realizing  the  potential  for  agricultu...
   Growth is not just an end in itself. It is the...
   At the end of the Second Plan the country woul...
   It is perhaps a platitude to say that in this ...
: budgetspeech, Length: 114, dtype: object
```

```
## Convert the reviews by removing stop words, lemmatize and stem each word
print ("Preprocessing the data set...")
budgetspeechStmntEmbeddings =
convertBudgetspeechstatementsintoembeddings(budgetspeechStmnts)
```

Output

Step 8: Output Budget Speech Statement Word Embeddings

Print matrix with the list of all the unique root words.

```
#final matrix with the list of all the unique root words
np.matrix(budgetspeechStmntEmbeddings)
```

Output

```
['govern', 'project', 'nation', 'economi', 'grow', 'rang', 'percent'
['we', 'deepen', 'reform', 'of', 'the', 'purchas', 'and', 'storag',
['deepen', 'reform', 'purchas', 'storag', 'system', 'import', 'agric
['the', 'year', '1998', 'particularli', 'have', 'be', 'a', 'year', '
['year', 'particularli', 'year', 'unpreced', 'global', 'turmoil', 'e
['abov', 'all', ',', 'despit', 'our', 'continu', 'concern', 'year',
['despit', 'continu', 'concern', 'year', 'year', 'make', 'limit', 'i
['water', 'and', 'credit', 'must', 'flow', 'togeth', 'for', 'maximum
['water', 'credit', 'must', 'flow', 'togeth', 'maximum', 'impact', '
['the', 'rural', 'infrastructur', 'develop', 'fund', '(', 'ridf', ')
['rural', 'infrastructur', 'develop', 'fund', 'ridf', 'emerg', 'impo
```

Step 9: Vector Embedding

Print vector with embedding.

```
from sklearn.feature_extraction.text import TfidfVectorizer

tfidf_vec = TfidfVectorizer()

# Using for loop
for i in budgetspeechStmntEmbeddings:
    tfidfv = tfidf_vec.fit(i)
    print(i)
    print(tfidfv.vocabulary_)
    # encode document
    vector = tfidfv.transform(i)
    # summarize encoded vector
    print(vector.shape)
    print(type(vector))
    print(vector.toarray())
```

Output

```
['would', 'continu', 'ensur', 'per', 'cent', 'plan', 'resourc', 'invest', 'agricultur', 'rural', 'sector']
{'would': 10, 'continu': 2, 'ensur': 3, 'per': 5, 'cent': 1, 'plan': 6, 'resourc': 7, 'invest': 4, 'agricultur': 0,
'rural': 8, 'sector': 9}
(11, 11)
<class 'scipy.sparse._csr.csr_matrix'>
[[0. 0. 0. 0. 0. 0. 0. 0. 0. 0. 1.]
 [0. 0. 1. 0. 0. 0. 0. 0. 0. 0. 0.]
 [0. 0. 0. 1. 0. 0. 0. 0. 0. 0. 0.]
 [0. 0. 0. 0. 0. 1. 0. 0. 0. 0. 0.]
 [0. 1. 0. 0. 0. 0. 0. 0. 0. 0. 0.]
 [0. 0. 0. 0. 0. 0. 1. 0. 0. 0. 0.]
 [0. 0. 0. 0. 0. 0. 0. 1. 0. 0. 0.]
 [0. 0. 0. 0. 1. 0. 0. 0. 0. 0. 0.]
 [1. 0. 0. 0. 0. 0. 0. 0. 0. 0. 0.]
 [0. 0. 0. 0. 0. 0. 0. 0. 1. 0. 0.]
 [0. 0. 0. 0. 0. 0. 0. 0. 0. 1. 0.]]
```

```
['order', 'safeguard', 'possibl', 'loss', 'product', 'increas', 'declin', 'consumpt', 'credit', 'structur', 'would',
'strengthen', 'ensur', 'adequ', 'avail', 'credit', 'particularli', 'small', 'margin', 'farmer']
{'order': 10, 'safeguard': 14, 'possibl': 12, 'loss': 8, 'product': 13, 'increas': 7, 'declin': 4, 'consumpt': 2,
'credit': 3, 'structur': 17, 'would': 18, 'strengthen': 16, 'ensur': 5, 'adequ': 0, 'avail': 1, 'particularli': 11,
'small': 15, 'margin': 9, 'farmer': 6}
(20, 19)
<class 'scipy.sparse._csr.csr_matrix'>
[[0. 0. 0. 0. 0. 0. 0. 0. 0. 0. 1. 0. 0. 0. 0. 0. 0. 0. 0.]
 [0. 0. 0. 0. 0. 0. 0. 0. 0. 0. 0. 0. 0. 0. 1. 0. 0. 0. 0.]
 [0. 0. 0. 0. 0. 0. 0. 0. 0. 0. 0. 0. 1. 0. 0. 0. 0. 0. 0.]
 [0. 0. 0. 0. 0. 0. 0. 0. 1. 0. 0. 0. 0. 0. 0. 0. 0. 0. 0.]
 [0. 0. 0. 0. 0. 0. 0. 0. 0. 0. 0. 0. 0. 1. 0. 0. 0. 0. 0.]
 [0. 0. 0. 0. 0. 0. 0. 1. 0. 0. 0. 0. 0. 0. 0. 0. 0. 0. 0.]
 [0. 0. 0. 0. 1. 0. 0. 0. 0. 0. 0. 0. 0. 0. 0. 0. 0. 0. 0.]
 [0. 0. 1. 0. 0. 0. 0. 0. 0. 0. 0. 0. 0. 0. 0. 0. 0. 0. 0.]
 [0. 0. 0. 1. 0. 0. 0. 0. 0. 0. 0. 0. 0. 0. 0. 0. 0. 0. 0.]
 [0. 0. 0. 0. 0. 0. 0. 0. 0. 0. 0. 0. 0. 0. 0. 0. 0. 1. 0.]
 [0. 0. 0. 0. 0. 0. 0. 0. 0. 0. 0. 0. 0. 0. 0. 0. 0. 0. 1.]
 [0. 0. 0. 0. 0. 0. 0. 0. 0. 0. 0. 0. 0. 0. 0. 0. 1. 0. 0.]
 [0. 0. 0. 0. 0. 1. 0. 0. 0. 0. 0. 0. 0. 0. 0. 0. 0. 0. 0.]
 [1. 0. 0. 0. 0. 0. 0. 0. 0. 0. 0. 0. 0. 0. 0. 0. 0. 0. 0.]
 [0. 1. 0. 0. 0. 0. 0. 0. 0. 0. 0. 0. 0. 0. 0. 0. 0. 0. 0.]
 [0. 0. 0. 1. 0. 0. 0. 0. 0. 0. 0. 0. 0. 0. 0. 0. 0. 0. 0.]
 [0. 0. 0. 0. 0. 0. 0. 0. 0. 0. 0. 1. 0. 0. 0. 0. 0. 0. 0.]
 [0. 0. 0. 0. 0. 0. 0. 0. 0. 0. 0. 0. 0. 0. 0. 1. 0. 0. 0.]
 [0. 0. 0. 0. 0. 0. 0. 0. 0. 1. 0. 0. 0. 0. 0. 0. 0. 0. 0.]
 [0. 0. 0. 0. 0. 0. 1. 0. 0. 0. 0. 0. 0. 0. 0. 0. 0. 0. 0.]]
```

Adaptive Machine Learning Model: Automated Budget Statement Labeler

The House will, I am sure, be interested to get some information on the subject of the sterling balances, the recent agreement regarding which between ourselves and His Majesty's Government in the United Kingdom I placed on the table of the House a few days ago. The peak figure which the sterling balances reached was Rs. 1733 crores on the 5th of April 1946. Thereafter, they have

(continued)

declined very rapidly. At the end of March 1947, they stood at Rs. 1612 crores showing a reduction of Rs. 121 crores in 12 months. In the middle of July 1947, from when our new agreement became effective, they stood at about Rs. 1547 crores. We had thus drawn as much as Rs. 65 crores in a little over 6 months. These large decreases were due largely to heavy imports mainly of food grains and of consumer goods, of which the country had been starved during the period of the war. They also reflected some movement of capital from India, largely British.

Shri R.K. Shanmukham Chetty, Minister of Finance,[21] India,
Introducing the Budget for the Year 1947–1948

Chapter Machine Learning Experiments

ML Experiment 1 Perform NLP tasks on the 1991 and 2012 budget statements.

1991 Budget Speech snippets: Our economy would return to a path of high sustained growth with reasonable price stability and greater social equity. There has been inadequate emphasis on the reduction of costs, upgradation of technology, and improvement of living standards. This task has become imperative as we prepare ourselves to be an internationally competitive economy. The main objective of the National Renewal Fund will be to ensure that the cost of technical change and modernization does not devolve on the workers.

2012 Budget Speech snippets: Address supply bottlenecks in agriculture, energy, and transport sectors, particularly in coal, power, national highways, railways, and civil aviation. Intervene decisively to address the problem of malnutrition, especially in the 200 high-burden districts. Expedite coordinated implementation of decisions being taken to improve delivery systems, governance, and transparency; and address the problem of black money and corruption in public life. The government has further evolved its approach to divestment of central public sector enterprises (CPSEs). The CPSEs are being given a level playing field vis-a-vis the private sector with regard to practices like buybacks and listing at stock exchange.

ML Experiment 2 Create embeddings for the following economic and social sustainability budget speeches from 1991 to 2012, and compare how the priorities have changed during these two decades.

[21] India Budget Speech—1947–1948—https://www.indiabudget.gov.in/doc/bspeech/bs194748.pdf

1991 Budget Speech snippets: Our economy would return to a path of high sustained growth with reasonable price stability and greater social equity. There has been inadequate emphasis on the reduction of costs, upgradation of technology, and improvement of living standards. This task has become imperative as we prepare ourselves to be an internationally competitive economy. The main objective of the National Renewal Fund will be to ensure that the cost of technical change and modernization does not devolve on the workers.

2012 Budget Speech snippets: Address supply bottlenecks in agriculture, energy, and transport sectors, particularly in coal, power, national highways, railways, and civil aviation. Intervene decisively to address the problem of malnutrition, especially in the 200 high-burden districts. Expedite coordinated implementation of decisions being taken to improve delivery systems, governance, and transparency; and address the problem of black money and corruption in public life. The government has further evolved its approach to divestment of central public sector enterprises (CPSEs). The CPSEs are being given a level playing field vis-a-vis the private sector with regard to practices like buybacks and listing at stock exchange.

This chapter will teach you how to use NLP techniques and build NLP model techniques. You will also learn how to apply NKP models and create bag-of-words embeddings and Tf-Idf embeddings. At the end, you will practice using NLP techniques on budget speeches.

References

1. Shri Manmohan Singh, Budget Speech 1991-1992, 24th July, 1991, https://www.indiabudget.gov.in/doc/bspeech/bs199192.pdf, Access Date: November 26, 2023
2. Shri Yashwant Sinha, Budget 2000-2001, February 2000, https://www.indiabudget.gov.in/doc/bspeech/bs200001.pdf, Access Date: November 29, 2023
3. Steven Bird, Ewan Klein and Edward Loper, Natural Language Processing with Python, Publisher: O'Reilly Media; 1st edition (August 4, 2009), ISBN-13: 978-0596516499
4. Christopher D. Manning, Prabhakar Raghavan and Hinrich Schütze, Introduction to Information Retrieval, Cambridge University Press. 2008, ISBN-13: 978-0521865715
5. Martin Porter, The Porter Stemming Algorithm, 1980, https://tartarus.org/martin/PorterStemmer/, Access Date: December 09, 2023 https://doi.org/10.1186/s40008-023-00308-9
6. Susan Li, Named Entity Recognition with NLTK and SpaCy, Aug 16 2018, https://towardsdatascience.com/named-entity-recognition-with-nltk-and-spacy-8c4a7d88e7da, Access Date: December 09, 2023
7. A Guide on Word Embeddings in NLP, 2020, https://www.turing.com/kb/guide-on-word-embeddings-in-nlp, Access Date: December 09, 2023
8. C. Vuppalapati et al., "Application of Machine Learning and Government Finance Statistics for macroeconomic signal mining to analyze recessionary trends and score policy effectiveness,"

2021 IEEE International Conference on Big Data (Big Data), Orlando, FL, USA, 2021, pp. 3274–3283, https://doi.org/10.1109/BigData52589.2021.9672025

9. Mayank Bhardwaj and Rajendra Jadhav, "India to launch $1.48 billion plan to raise oilseed output, cut imports", August 9, 2021, https://www.yahoo.com/news/india-launch-1-48-billion-111942866.html, Access Date: 16 August 20218
10. Khalid Iqbal, "Little to no relief for common man in budget", Saturday, Jun 12, 2021, https://www.thenews.com.pk/print/848281-little-to-no-relief-for-common-man-in-budget, Access Date: 16 August 2021
11. Yen Nee Lee, "Hong Kong announces over $15 billion budget to lift economy out of recession", PUBLISHED TUE, FEB 23 2021 11:23 PM ESTUPDATED WED, FEB 24 2021 1:19 AM EST, https://www.cnbc.com/2021/02/24/hong-kong-announces-budget-to-lift-economy-out-of-covid-recession.html, Access Date: 1 August 2021

Chapter 3
Large Language Models

This chapter covers the following:

- Generative pre-trained transformer
- What is a large language model (LLM)?
- Application of LLM on economic and social sustainability
- GPT internal workings
- Prompt engineering
- Application of GPT models to bring your own data

This chapter presents large language models. It begins with LangChain model, and then it covers GPT-2 model and uses GPT model to classify social sustainability and economic sustainability classifiers from the data. The chapter also covers prompt engineering. Finally, it finishes GPT model on budget speeches as shown in Appendix A–F.

Artificial intelligence (AI) stands out as a transformational technology of our digital age—and its practical application throughout the economy is growing apace [1]. One of the chief reasons why AI applications are getting prominent and industry acceptance is in its software's ability to learn, albeit continuously, from real-world use and experience and its capability to improve its performance. It is no wonder that the applications of AI span from complex high-technology equipment manufacturing to personalized exclusive recommendations. AI enabled newer business models by understanding the data patterns and creating products that touch the customer's needs.

Artificial intelligence is changing how we interact with everything, from food to healthcare, travel, and also religion [2]. Artificial intelligence applications enable agriculture, one of the oldest and most important professions in the world, to be more productive and enable to achieve sustainability, especially for small farmers. At the

© The Author(s), under exclusive license to Springer Nature Switzerland AG 2024
C. Vuppalapati, *Assessing Policy Effectiveness using AI and Language Models*,
International Series in Operations Research & Management Science 354,
https://doi.org/10.1007/978-3-031-56097-2_3

same token, application of AI would enable us to understand how human civilizations evolved or how architects of religion, faith, construction, and arts have interwoven the needs of people using engineering and technologies available thousands of years ago.

Using machine learning and artificial intelligence techniques, we can understand time reckoning, application of acoustics in historical artifacts, physical perimeter security alert systems using acoustics, interaction with nature, energy principles, commerce, and many others. Application of scientific view using AI, in my view, opens new understandings of historical relics that have influence in worldwide cultures today.

> Application of AI on pre-historic etiquettes that have extensive ties to constructions, arts, architecture, music, religion, and time would provide new understanding of our human history and of relics that've shaped our modern lives!

What Is AI?

Artificial intelligence (AI) has been defined and understood in many ways, as it encompasses a vast field within its realm [3]. Perspectives of AI also vary based upon many factors, use cases, and actors involved. Therefore, no definition of AI can be so comprehensive that it will cover all the dimensions of AI. AI is a very broad and diverse field. For more details on AI, please refer to my other books.

Alan Turing generally is credited with the origin of the concept when he speculated in 1950 about "thinking machines" that could reason at the level of a human being[1] [4]. Turing was followed up a few years later by John McCarthy, who first used the term "artificial intelligence" to denote machines that could think autonomously. He described the threshold as "getting a computer to do things which, when done by people, are said to involve intelligence."

Stuart Russell and Peter Norvig in defining AI[2] have used the following taxonomy about systems that ([5, 6]):

- Think like humans (e.g., cognitive architectures or neural network);
- Think rationally like humans (e.g., inferences optimization);
- Act like humans (e.g., language processing, knowledge representation, automated reasoning and learning); and

---

[1] What is Artificial Intelligence?—https://www.brookings.edu/research/what-is-artificial-intelligence/

[2] What is Artificial Intelligence? An Informed Definition—https://emerj.com/ai-glossary-terms/what-is-artificial-intelligence-an-informed-definition/#:~:text=In%20%E2%80%9CArtificial%20Intelligence%3A%20A%20Modern,many%20different%20splintered%20fields%20%E2%80%93%20speech

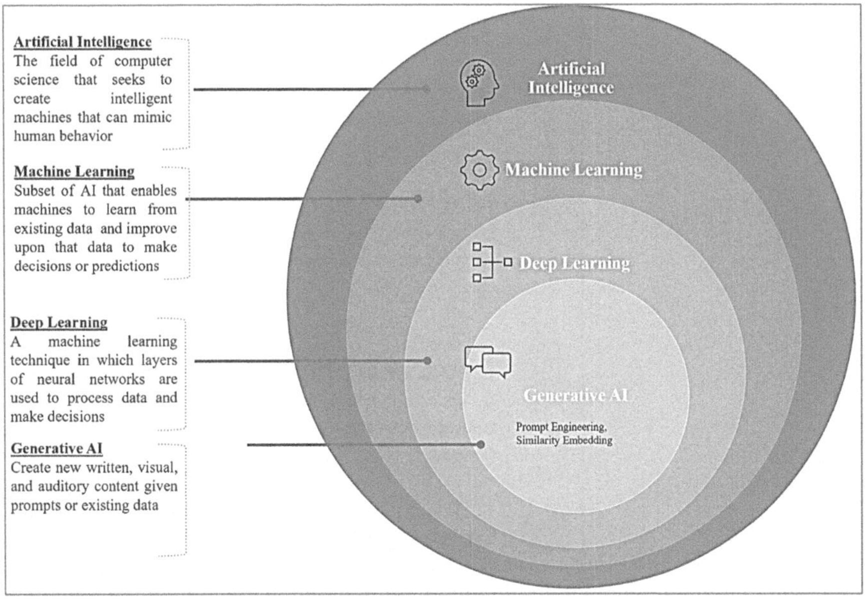

Fig. 3.1 AI framework

- Act rationally like humans (e.g., intelligent software agents, embodied robots that achieve goals via perception, planning, reasoning, learning etc.).

In computer science, artificial intelligence (AI), sometimes called machine intelligence, is intelligence demonstrated by machines, in contrast to the natural intelligence displayed by humans. According to Gartner Market Research,[3] artificial intelligence applies advanced analysis and logic-based techniques, including machine learning, to interpret events, support and automate decisions, and take actions. The typical definition of AI looks something like the composite design of AI, machine learning, and deep learning [7] (see Fig. 3.1).

Brief History

Artificial intelligence (AI) has a long and storied history that spans many decades. Some of the earliest work in AI was done in the 1950s and 1960s, when researchers first began to explore the possibility of creating machines that could "think" like humans.

[3] Artificial Intelligence (AI)—https://www.gartner.com/en/information-technology/glossary/artificial-intelligence

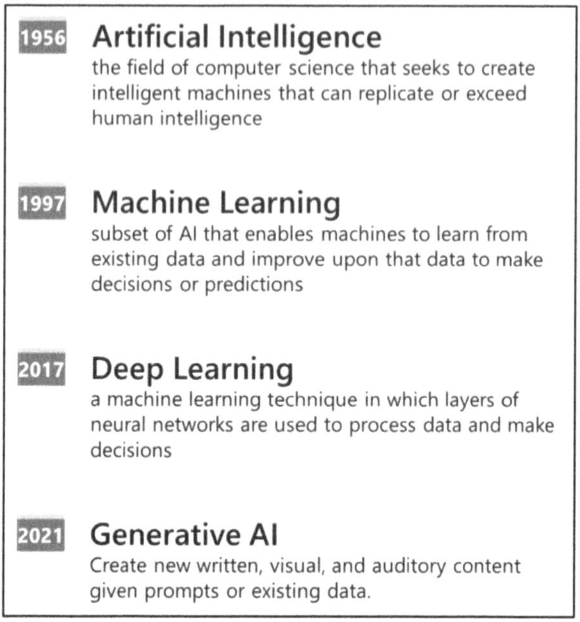

Fig. 3.2 AI history

In the 1950s and 1960s, AI research focused primarily on "symbolic" AI, which involved creating symbolic representations of knowledge and using them to solve problems. Researchers in this field developed early versions of what are now known as expert systems, which are computer programs that mimic the decision-making abilities of human experts in a specific domain. The principle behind symbolic representation of time series is to allow search time series patterns and cluster using distance (Euclidian or cosine)-based metrics [8].

In the 1970s and 1980s, AI research began to shift away from symbolic AI and towards "sub-symbolic" AI, which involves using statistical methods and mathematical models to solve problems. This shift was driven in part by the realization that symbolic AI was too complex and difficult to scale and in part by the development of new computer hardware that made it possible to process large amounts of data quickly (please see Fig. 3.2).

In the 1990s and 2000s, AI research continued to evolve, and new subfields began to emerge. One of these was "machine learning," which focuses on creating algorithms that allow computers to learn from data without being explicitly programmed. Another was "neural networks," which are mathematical models inspired by the structure and function of the human brain.

Recently, generative models like GPT-3 and DALL-E, variational autoencoders, generative adversarial networks, and many more have risen and continue to advance the field of AI. These models are able to generate new examples that are similar to examples from a given dataset. It is being used in diverse fields such as natural language processing, computer vision, speech recognition, and many more.

AI embodies techniques that enable computers to mimic and predict human behavior. Simply put, AI is the ability of a machine to perform cognitive functions that are associated with human minds, such as perceiving, reasoning, learning, and problem-solving. A full end-to-end AI solution, what we sometimes call a system of intelligence or a smart system, can ingest human-level knowledge (e.g., via machine reading and computer vision) and use this information to automate and accelerate tasks that were previously performed only by humans [7–9].

Overall, AI has come a long way since its inception in the 1950s, and it continues to evolve and improve at a rapid pace today [7, 8]. Now, we can fairly say we've entered AI age with full availability of AI across human pursuits that improve human ingenuity for the betterment of the future generations and appreciate the efforts of our forefathers that survived us through the thick and thin of times.

Using AI, we can identify the role of nature in pre-historic times and methodization of time-based principles as part of human evolution. Novel technologies such as AI and machine learning enable us to revisit historical events and factors through the scientific lens that wasn't available till the late nineteenth century. It helps us appreciate the line of thinking of architects and scientific thinkers of pre-historic Hinduism; fully understand the principles of life, science, and religion that are perceived revolutionary in those times; and understand the scientific underpinnings of such principles that are closely related to nature which is only possible through novel scientific view, with the help of data science technologies that are available to us in AI age.

Machine Learning

Machine learning (ML) is the subset of AI that uses statistical techniques to enable machines to improve with experience. Machine learning algorithms detect patterns and learn how to make predictions and recommendations by processing data and experiences rather than by receiving explicit programming instruction [10]. The algorithms also adapt in response to new data and experiences to improve efficiency and efficacy over time. ML, importantly, enables to autocalibrate, correct, or rebuild itself if it sees that the data is drifting, of course, from the original trained model. The availability of such features dawns a new era of commuting that adapts itself to course correctly or to hyper-scale efficiently as it sees new. There are three types of learning that AI uses: supervised learning, unsupervised learning, and semi-supervised learning. For more on these topics, please refer to my book. In this book I will confine myself to supervised learning and apply statistical techniques to develop ML for historical time series patterns.

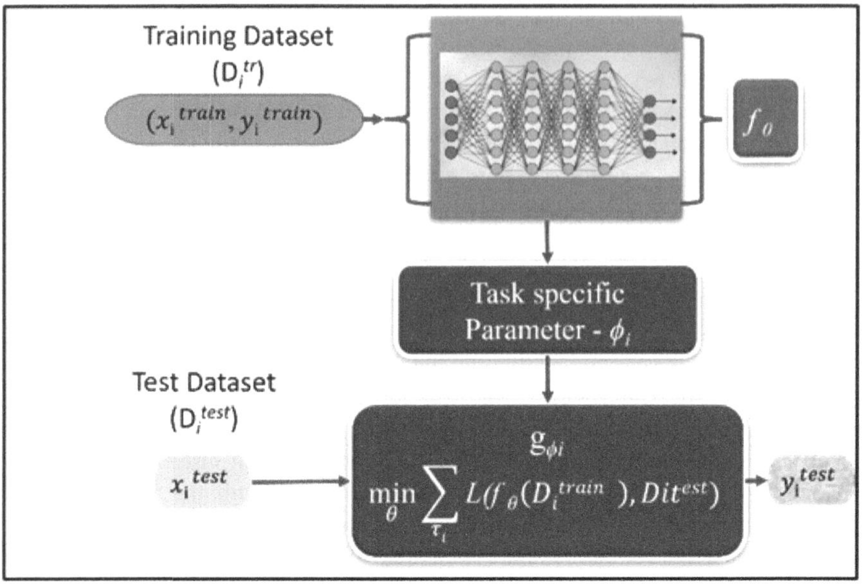

Fig. 3.3 Deep learning framework

Deep Learning

Deep learning[4] is a subset of machine learning, which is essentially a neural network with three or more layers. These neural networks attempt to simulate the behavior of the human brain—albeit far from matching its ability—allowing it to "learn" from large amounts of data. While a neural network with a single layer can still make approximate predictions, additional hidden layers can help to optimize and refine for accuracy (please see Fig. 3.3).

Deep learning drives many artificial intelligence (AI) applications and services that improve automation, performing analytical and physical tasks without human intervention. Deep learning technology lies behind everyday products and services (such as digital assistants, voice-enabled TV remotes, and credit card fraud detection) as well as emerging technologies (such as self-driving cars).

In the below the neural network then employs these parameters to make predictions for new outcomes. In this process, we're going to use both the training set (D_i^{train}) and the test set (D_i^{test}). The neural networks are trained using standard supervised learning across all tasks. The overall goal is to measure the model's performance, with parameters ϕ, in predicting outcome variable, Y, for new data.

[4] What is deep learning?—https://www.ibm.com/topics/deep-learning

Generative AI

Generative AI is a subfield of artificial intelligence that focuses on creating systems that can generate new examples, such as images, text, or speech, that are similar to examples from a given dataset. The goal of generative AI is to develop algorithms that can learn the underlying probability distribution of a given dataset and use this knowledge to generate new examples that are similar to the examples in the dataset.

Generative AI models can be classified into two broad categories: generative models and discriminative models. Generative models such as GPT-3, DALL-E, variational autoencoders, etc. learn the underlying probability distribution of the dataset and can generate new examples from the learned distribution. Discriminative models, such as deep neural networks, learn to differentiate between different classes or categories and are generally used for tasks such as image or speech recognition.

Generative AI sits on top of other areas of AI like unsupervised learning, supervised learning, and reinforcement learning. The effectiveness of this technique is based on the models' ability to extract relevant features from the data, which are then used to learn the underlying probability distribution. This approach has potential applications in several fields, including computer vision, natural language processing (NLP), speech recognition, and more. Overall, generative AI is a rapidly growing field within AI, and it has huge potential. Generative pre-trained transformer (GPT) models fall into this category. The models employ autoregressive language model that uses deep learning to produce human-like text and general text-in/text-out interface; are pre-trained in trillions of words; and predict the most likely next word based on input text.

ChatGPT models are some of the industrial uses. Historically, natural language processing (NLP) models perform text classification. Nevertheless, GPT models perform natural language generation (please see Fig. 3.4).

Large Language Foundation Models

The future of AI is models that are flexible and reusable and that can be applied to just about any domain or industry task. Foundation models are at the heart of many of the latest breakthroughs. These models have been influential in the development of artificial intelligence systems and have played a major role in changing how AI systems are constructed since their introduction in 2018.

A foundation model is a machine learning model that is trained on a large amount of unlabeled data, allowing it to be adapted to a wide variety of tasks. These models capture general patterns and structure of data and aim to replace task-specific models by unifying different tasks and modalities into one, eliminating the need to train individual models and integrate several models together. An example is a large language foundation model that can perform various NLP tasks and can be used out of the box due to its extensive training.

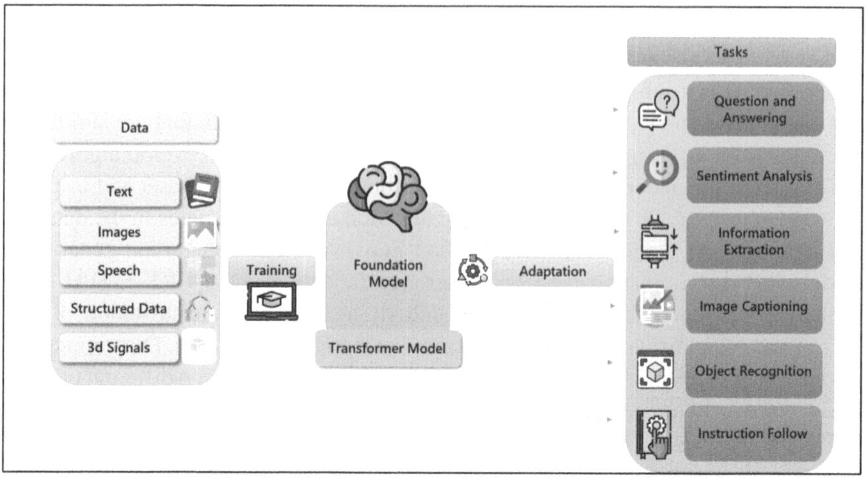

Fig. 3.4 Generative AI models

Foundation models use self-supervised learning and transfer learning to apply information learned in one situation to another situation or task.

Using large language models to extract statements from budget speeches is a way of linking the past and the future in the same temporal frame to assess the policy impact for the survival of disadvantaged communities. It's an evaluation of government policies that offer a useful perspective to enhance the efficiency of democracies and create a future where the disparity between rich and poor could be reduced with data science-based policies.

Author

Budget speech statement mining with large language models provides a great opportunity of combining history and the future in the same time window to validate the policy effectiveness for the sustainability of marginalized communities. It's a scoring of government policies that provide a valuable insight to improve effectiveness of democracies and establish a future where the gap between affluent and poor could be narrowed with data science-infused policies.

Training these models can be complex and resource-intensive as they are typically large and require a significant amount of data. The scale of these models also plays a big role in the training process.

BERT (Bidirectional Encoder Representations from Transformers)

BERT, short for bidirectional encoder representations from transformers, was the first large language model [11]. BERT differs from previous language representation models in that it is designed to pre-train deep bidirectional representations from unlabeled text by jointly conditioning on both left and right context in all layers [12]. As a result, the pre-trained BERT model can be fine-tuned with just one additional output layer to create state-of-the-art models for various tasks, such as question-answering and language inference, without significant task-specific architecture modifications [13].

BERT is a widely used language model, released in 2018, that marked the arrival of the first NLP transformer model. It is an autoencoding model focused on the encoding part of the transformer, and it has around 340 million parameters. Parameters are used to calculate the size of these models and can be thought of as connections between neurons in the brain.

BERT is simple in concept and powerful in performance. It achieves new state-of-the-art results on 11 natural language processing tasks, including pushing the GLUE score to 80.5%. GLUE, the General Language Understanding Evaluation benchmark, is a collection of resources for training, evaluating, and analyzing natural language understanding systems. GLUE focuses on nine English sentence understanding tasks, which cover a wide range of domains, data quantities, and difficulties. As the goal of GLUE is to encourage the development of generalizable natural language understanding (NLU) systems, the benchmark is designed so that good performance should require a model to share substantial knowledge across all tasks while still having some task-specific components [15].

There are two steps in BERT: pre-training and fine-tuning. During pre-training, the model is trained in unlabeled data, in our case budget speech statements, over different pre-training tasks. For fine-tuning, the BERT model is first initialized with the pre-trained parameters, and all of the parameters are fine-tuned using labeled data from downstream tasks. Each downstream task has separate fine-tuned models, even though they are initialized with the same pre-trained parameters. The question-answering example in Fig. 3.5 will serve as a running example for this section [11].

Figure 3.5 question-answering: Overall pre-training and fine-tuning procedures for BERT. Apart from output layers, the same architectures are used in both pre-training and fine-tuning. The same pre-trained model parameters are used to initialize models for different downstream tasks. During fine-tuning, all parameters are fine-tuned. [CLS] is a special symbol added in front of every input example, and [SEP] is a special separator token (e.g., separating questions/answers) [11].

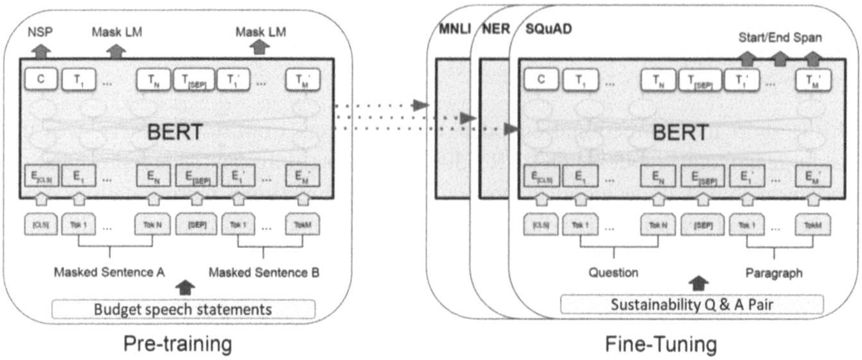

Fig. 3.5 BERT

BERT Model Size

Let the number of layers (i.e., transformer blocks) as L, the hidden size as H, and the number of self-attention heads as A. We primarily report results on two model sizes:

$\text{BERT}_{\text{BASE}} = L = 12; H = 768; A = 12$, Total Parameters = 110 M.
$\text{BERT}_{\text{LARGE}} = L = 24; H = 1024; A = 16$, Total Parameters = 340 M.

Architecture BERT's model architecture[5] is a multilayer bidirectional transformer encoder based on the original implementation for dominant sequence transduction models used for translation of languages [14]. Most competitive neural sequence transduction models have an encoder-decoder structure [15]. Here, the encoder maps an input sequence of symbol representations $(x_1..., x_n)$ to a sequence of continuous representations $z = (z_1,z_n)$. Given z, the decoder then generates an output sequence $(y_1..., y_m)$ of symbols one element at a time. At each step the model is autoregressive [16], consuming the previously generated symbols as additional input when generating the next.

The transformer follows this overall architecture using stacked self-attention and pointwise, fully connected layers for both the encoder and decoder, as shown in the left and right halves of Fig. 3.6, respectively. The process of embeddings generation pretty much follows classical NLP technique. A tabular form is attached for reference. Step 1: Input budget speech statement. Step 2: Tokenizer based on the ML. Step 3: Convert to positional embedding. And finally, Step 4: Vector embedding.

[5] Model Architecture—https://colab.research.google.com/drive/10wVewWm6n4cdEdDZm3-DUKgVGeTMq9yz#scrollTo=c96103f9

Matrix Representation	<class 'scipy.sparse._csr.csr_matrix'> [[0. 0. 0. 0. 0. 0. 0. 0. 0. 0. 0. 0. 0. 0. 0. 0. 0. 0. 0. 0. 0. 1. 0.]
Positional Vector Embedding	{'would': 21, 'like': 11, 'draw': 5, 'attent': 2, 'hous': 8, 'fact': 6, 'increas': 9, 'fertilis': 7, 'price': 14, 'sinc': 18, 'juli': 10, 'ten': 19, 'year': 22, 'continu': 3, 'procur': 15, 'paddi': 13, 'wheat': 20, 'also': 1, 'market': 12, 'crop': 4, 'receiv': 16, 'agricultur': 0, 'sector': 17} (26, 23)
Token Form	['i', 'would', 'like', 'to', 'draw', 'the', 'attent', 'of', 'the', 'hous', 'to', 'the', 'fact', 'that', 'there', 'have', 'be', 'no', 'increas', 'in', 'fertilis', 'price', 'sinc', 'juli', '1981', '.', 'in', 'these', 'ten', 'year', ',', 'there', 'have', 'be', 'a', 'continu', 'increas', 'in', 'the', 'procur', 'price', 'of', 'paddi', 'and', 'wheat', ',', 'as', 'also', 'in', 'the', 'market', 'price', 'of', 'other', 'crop', ',', 'receiv', 'by', 'the', 'agricultur', 'sector', '.']
Budget Speech Statement	I would like to draw the attention of the House to the fact that there has been no increase in fertiliser prices since July 1981. In these ten years, there has been a continuous increase in the procurement prices of paddy and wheat, as also in the market prices of other crops, received by the agricultural sector.

Fig. 3.6 BERT

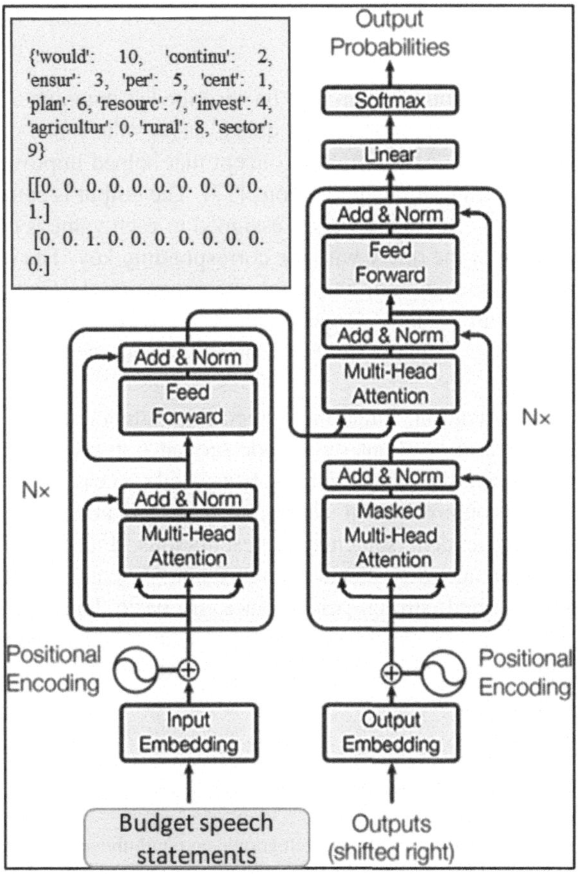

Fig. 3.7 Attention function

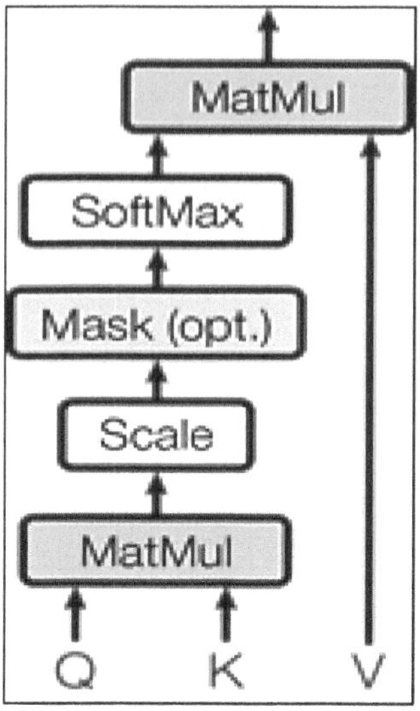

An attention function[6] can be described as mapping a query and a set of key-value pairs to an output, where the query, keys, values, and output are all vectors (please see Fig. 3.7). Attention is a concept that helped improve the performance of neural machine translation applications [17]. The output is computed as a weighted sum of the values, where the weight assigned to each value is computed by a compatibility function of the query with the corresponding key. The input consists of queries and keys of dimension d_k and values of dimension d_v. Compute the dot products of the query with all keys, divide each by SQRT(d_k), and apply a softmax function to obtain the weights on the values. The process is called "scaled dot-product attention."

Self-attention, sometimes called intra-attention, is an attention mechanism relating different positions of a single sequence in order to compute a representation of the sequence. Self-attention has been used successfully in a variety of tasks including reading comprehension, abstractive summarization, textual entailment, and learning task-independent sentence representations. End-to-end memory networks use a recurrent attention mechanism instead of sequence-aligned recurrence and have demonstrated strong performance in basic language tasks, such as question-answering and language modeling. Due to the widespread use of transformers, implementing this approach is nearly identical to the original method (annotated transformer[7]).

[6]Attention—https://colab.research.google.com/github/harvardnlp/annotated-transformer/blob/master/AnnotatedTransformer.ipynb

[7]Annotated-transformer—https://github.com/harvardnlp/annotated-transformer/

BERT makes use of transformer, an attention mechanism that learns contextual relations between words (or sub-words) in a text. In its vanilla form, transformer includes two separate mechanisms: an encoder that reads the text input and a decoder that produces a prediction for the task. Since BERT's goal is to generate a language model, only the encoder mechanism is necessary.

BERT Large Model 340 M parameters
Turing NLG 17B
GPT-3175 B
Megatron-Turing NLG 530 B

Large Language Model Meta AI (Llama)

Llama 2 (large language model Meta AI) is a big language model that helps developers and organizations create generative AI-powered tools and experiences. Llama 2 is a general-purpose foundation model that can be fine-tuned for different use cases and needs of an organization. Like other big language models, Llama 2 takes a sequence of words as an input and predicts another word to make text step by step. Llama 2 used the 20 languages that are spoken the most in the world to train.

Meta developed and publicly released the Llama 2 family of large language models (LLMs)—a collection of pre-trained and fine-tuned generative text models ranging in scale from 7 to 70 billion parameters. Fine-tuned LLMs, called Llama-2-Chat, are optimized for dialogue use cases. Llama-2-Chat models outperform open-source chat models on most benchmarks we tested and, in our human evaluations for helpfulness and safety, are on par with some popular closed-source models like ChatGPT and PaLM.

Llama 2[8] comes in a range of parameter sizes—7B, 13B, and 70B—as well as pre-trained and fine-tuned variations.

Model Architecture Llama 2 is an autoregressive language model that uses an optimized transformer architecture. The tuned versions use supervised fine-tuning (SFT) and reinforcement learning with human feedback (RLHF) to align with human preferences for helpfulness and safety.

The Transformer

The transformer was proposed in the paper "Attention is All You Need" [14]. The TensorFlow implementation of it is available as a part of the Tensor2Tensor

[8] Llama 2—https://huggingface.co/meta-llama/Llama-2-7b-chat-hf

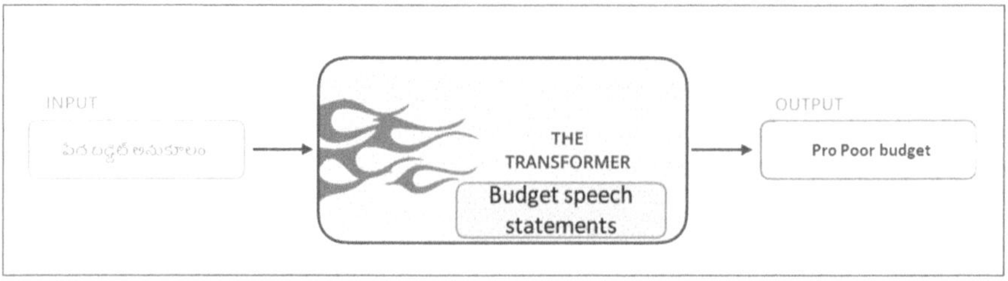

Fig. 3.8 Single black box (specifically for fine-tuned)

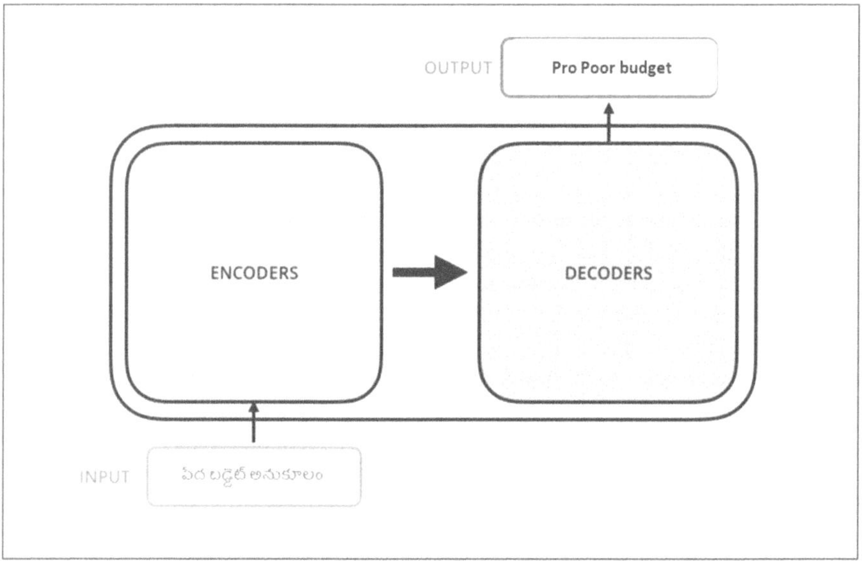

Fig. 3.9 Encoder and decoder

package.[9] Harvard's NLP group created a guide annotating the paper with PyTorch implementation.[10] In this post, we will attempt to oversimplify things a bit and introduce the concepts one by one to hopefully make it easier to understand for people without in-depth knowledge of the subject matter.

Let's begin by looking at the model as a single black box (please see Fig. 3.8). In a machine translation application, it would take a sentence in one language and output its translation in another.

One level deeper, we will have encoder and decoder components (please see Fig. 3.9).

[9]Tensor Flow—https://github.com/tensorflow/tensor2tensor

[10]Attention is all you need—http://nlp.seas.harvard.edu/2018/04/03/attention.html

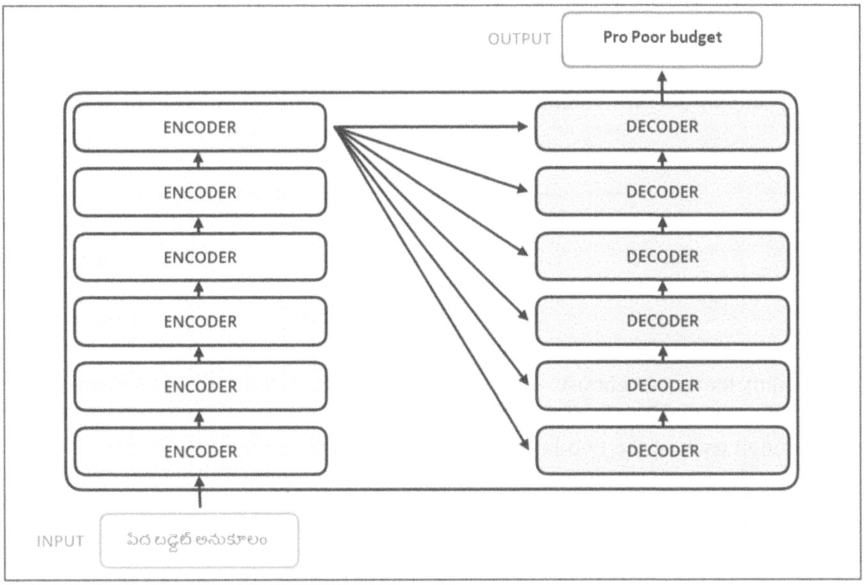

Fig. 3.10 Stacked encoder and decoder

The encoding component is a stack of encoders (the paper stacks six of them on top of each other; there's nothing magical about the number six—one can definitely experiment with other arrangements). The decoding component is a stack of decoders of the same number (please see Fig. 3.10).

The encoders are all identical in structure (yet they do not share weights). Each one is broken down into two sub-layers: The encoder's inputs first flow through a self-attention layer—a layer that helps the encoder look at other words in the input sentence as it encodes a specific word. The outputs of the self-attention layer are fed to a feed-forward neural network (please see Fig. 3.11). The exact same feed-forward network is independently applied to each position. The decoder has both layers, but between them is an attention layer that helps the decoder focus on relevant parts of the input sentence (similar what attention does in seq2seq models) [17].

Now that we've seen the major components of the model, let's start to look at the various vectors/tensors and how they flow between these components to turn the input of a trained model into an output. As is the case in NLP applications in general, we begin by turning each input word into a vector using an embedding algorithm. Each word is embedded into a vector of size 512. We'll represent those vectors with these simple boxes.

The bottom encoder is the only one that does the embedding. The common feature of all the encoders is that they take a list of vectors of size 512 as input. For the bottom encoder, these are the word embeddings, but for other encoders, these are the output of the encoder right below them. The length of this list is a

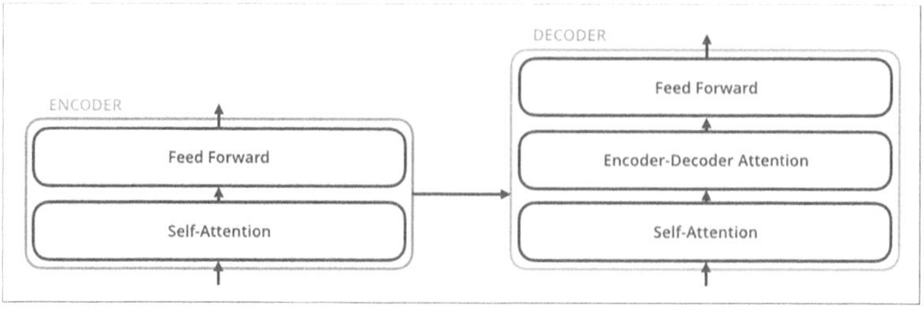

Fig. 3.11 Sub-layers

hyperparameter we can choose—basically it would match the longest sentence in our training dataset. After embedding the words in our input sequence, each of them goes through each of the two layers of the encoder (please see Fig. 3.12).

This reveals one important characteristic of the transformer, which is that each position in the word has its own route in the encoder. These routes are interdependent in the self-attention layer. However, the feed-forward layer does not have such interdependence, and so the different routes can run at the same time while passing through the feed-forward layer.

We have already said that an encoder takes a list of vectors as input. It handles this list by sending these vectors to a "self-attention" layer and then to a feed-forward neural network and then giving out the output to the next encoder above. The word at each position goes through a self-attention process. Then, they each go through a feed-forward neural network—the same network with each vector moving through it individually.

As the model processes each word (each position in the input sequence), self-attention allows it to look at other positions in the input sequence for clues that can help lead to a better encoding for this word. Self-attention is the method the transformer uses to bake the "understanding" of other relevant words into the one we're currently processing. As we are encoding the word "it" in encoder #5 (the top encoder in the stack), part of the attention mechanism was focusing on "The Animal" and baked a part of its representation into the encoding of "it."

Self-Attention in Detail

To compute self-attention, we need to make three vectors for every input vector of the encoder (in this case, the embedding of each word). Therefore, for each word, we make a query vector (Q), a key vector (K), and a value vector (V). These vectors are made by multiplying the embedding by three matrices that we learned during the training process.

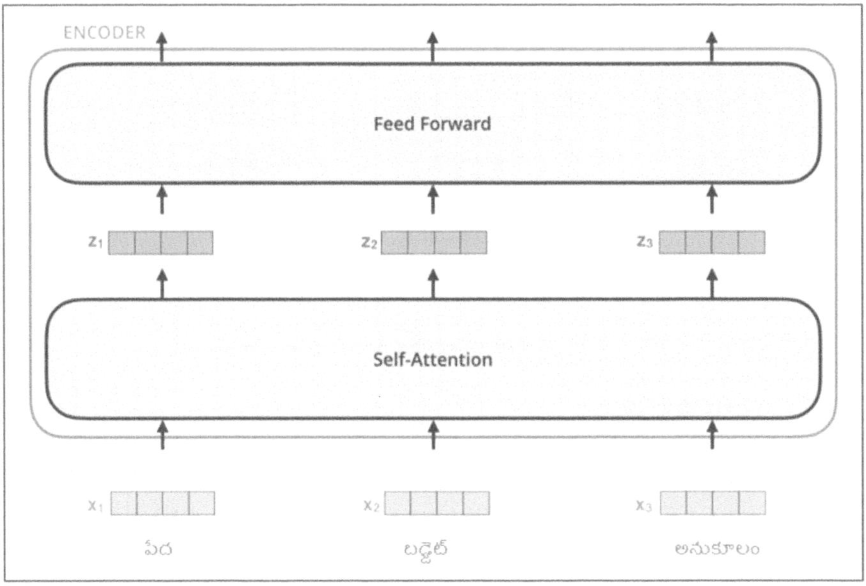

Fig. 3.12 Vectors and encoders

Observe that these new vectors have a lower dimension than the embedding vector. They are 64-dimensional, while the embedding and encoder input/output vectors are 512-dimensional. This is not a requirement; this is a design choice to keep the computation of multiheaded attention (mostly) constant.

- Step 1: Multiplying x1 by the WQ weight matrix produces q1, the "query" vector associated with that word. We end up creating a "query," a "key," and a "value" projection of each word in the input sentence.

What are the "query," "key," and "value" vectors? They're abstractions that are useful for calculating and thinking about attention.

The next step in computing self-attention is to compute a score. Suppose we're computing the self-attention for the first word in this example, "Thinking." We need to score each word of the input sentence with this word. The score shows how much attention to give to other parts of the input sentence as we encode a word at a certain position.

The score is obtained by multiplying the query vector with the key vector of the specific word we're scoring. So, if we're computing the self-attention for the word in position #1, the first score would be the multiplication of q1 and k1. The second score would be the multiplication of q1 and k2.

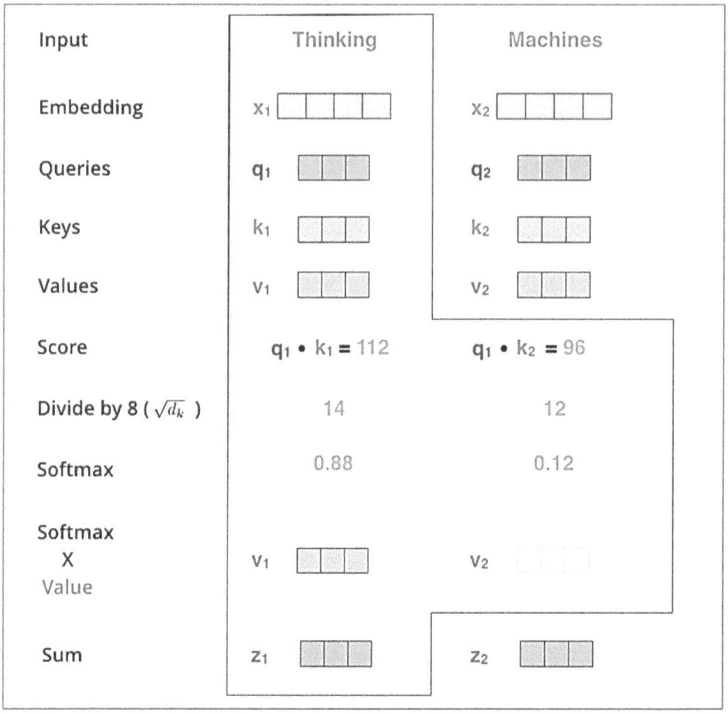

Fig. 3.13 Vectors and encoders

The next two steps are to divide the scores by 8 (the square root of the size of the key vectors—64. This helps to have more consistent gradients. There might be other options here, but this is the standard). Then, apply a softmax operation. Softmax makes the scores all positive and sums up to 1.

This softmax score indicates how strongly each word will be represented at this position. Obviously, the word at this position will have the highest softmax score, but sometimes it's helpful to pay attention to another word that is related to the current word (please see Fig. 3.13).

The fifth step is to apply the softmax score to each value vector (to get them ready for summation). The idea here is to preserve the values of the word(s) we want to pay attention to and reduce unimportant words (e.g., by multiplying them by small numbers like 0.001).

The final step is to add up the value vectors with weights. This gives the result of the self-attention layer at this position (for the first word).

The resulting vector is one we can send along to the feed-forward neural network. In the actual implementation, however, this calculation is done in matrix form for faster processing (please see the table below). So, let's look at that now that we've seen the intuition of the calculation on the word level (please see Fig. 3.14 for the overall steps).

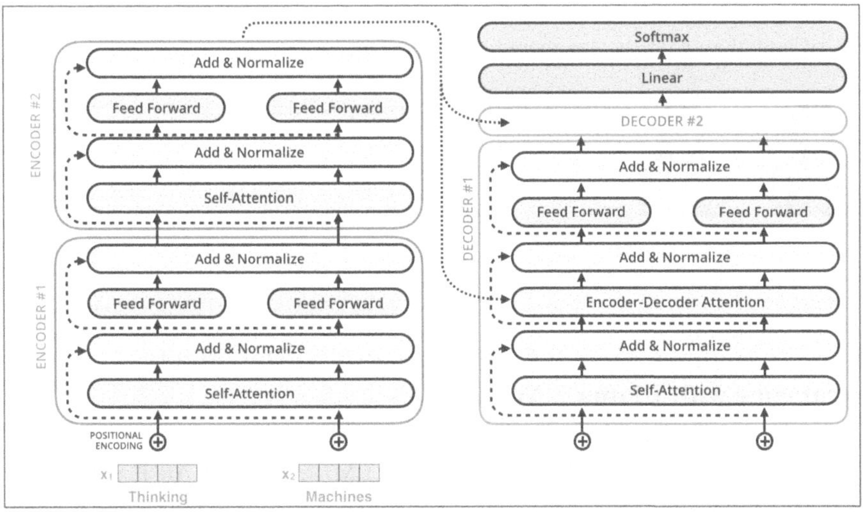

Fig. 3.14 Vectors and encoders

I would like to draw the attention of the House to the fact that there has been no increase in fertiliser prices since July 1981. In these ten years, there has been a continuous increase in the procurement prices of paddy and wheat, as also in the market prices of other crops, received by the agricultural sector. We would continue to ensure that 50 per cent of the plan resources are invested in the agricultural and rural sector.
The provision for assistance for fresh water and brackish water aqua-culture and for oilseeds and pulses production would be substantially stepped up. The ceilings on assistance in difficult areas, where the water table is very low, would be removed. New schemes are being drawn up to popularise small tractors and matching implements, drip and sprinkler irrigation in areas where water is scarce, and quality seeds in low yield areas.
['i', 'would', 'like', 'to', 'draw', 'the', 'attent', 'of', 'the', 'hous', 'to', 'the', 'fact', 'that', 'there', 'have', 'be', 'no', 'increas', 'in', 'fertilis', 'price', 'sinc', 'juli', '1981', '.', 'in', 'these', 'ten', 'year', ',', 'there', 'have', 'be', 'a', 'continu', 'increas', 'in', 'the', 'procur', 'price', 'of', 'paddi', 'and', 'wheat', ',', 'as', 'also', 'in', 'the', 'market', 'price', 'of', 'other', 'crop', ',', 'receiv', 'by', 'the', 'agricultur', 'sector', '.'] ['would', 'like', 'draw', 'attent', 'hous', 'fact', 'increas', 'fertilis', 'price', 'sinc', 'juli', 'ten', 'year', 'continu', 'increas', 'procur', 'price', 'paddi', 'wheat', 'also', 'market', 'price', 'crop', 'receiv', 'agricultur', 'sector'] ['we', 'would', 'continu', 'to', 'ensur', 'that', '50', 'per', 'cent', 'of', 'the', 'plan', 'resourc', 'be', 'invest', 'in', 'the', 'agricultur', 'and', 'rural', 'sector', '.'] ['would', 'continu', 'ensur', 'per', 'cent', 'plan', 'resourc', 'invest', 'agricultur', 'rural', 'sector']
matrix([[list(['would', 'like', 'draw', 'attent', 'hous', 'fact', 'increas', 'fertilis', 'price', 'sinc', 'juli', 'ten', 'year', 'continu', 'increas', 'procur', 'price', 'paddi', 'wheat', 'also', 'market', 'price', 'crop', 'receiv', 'agricultur', 'sector']), list(['would', 'continu', 'ensur', 'per', 'cent', 'plan', 'resourc', 'invest', 'agricultur', 'rural', 'sector']), list(['provis', 'assist', 'fresh', 'water', 'brackish', 'water', 'aqua-cultur', 'oilse', 'puls', 'product', 'would', 'substanti', 'step', 'ceil', 'assist', 'difficult', 'area', 'water', 'tabl', 'low', 'would', 'remov', 'new', 'scheme', 'draw', 'popularis', 'small', 'tractor', 'match', 'implement', 'drip', 'sprinkler', 'irrig', 'area', 'water', 'scarc', 'qualiti', 'seed', 'low', 'yield', 'area']),

Large language models (LLMs) are fun to use, but understanding the fundamentals of how they work is also incredibly important.

Llama-2-7b-chat-hf_classifier_model.ipynb

Large Language Model: Generate Social Sustainability Strategies with Llama 2 Model

Llama 2 is the latest large language model (LLM) from Meta AI. It is in many respects a groundbreaking release. First, Llama 2 is open access—meaning it is not closed behind an API, and its licensing allows almost anyone to use it and fine-tune new models on top of it.[11] Second, Llama 2 breaks records, scoring new benchmarks against all other "open access" models [18].

Economic and social sustainability policies are essential for maintaining macroeconomic stability in India. Unlike western economies that rely more on consumer demand, the economies of the East, especially India, have the duty of distributing the wealth to all segments of the societies to reduce income inequality. Second, productivity and automation are important indicators of the western economies that have limited human resources. Populated and agricultural economies should learn from the efficiencies of western economies while also maintaining longer societal commitment. The long-term goal is to develop a mode of production that is labor-intensive and creates more employment opportunities in productive higher-income jobs and narrows the gaps in income and wealth between rural and urban areas. Moreover, besides high-technology manufacturing, advanced software automation, and artificial intelligence, agriculture is the basis of national prosperity, and no economic development strategy can work in our country if it does not ensure fast growth of production and employment in agriculture. Economy for its own sake and growth for the sake of a few shareholders would not benefit the economy. Our people cannot afford to imitate the soulless consumerism and the wasteful lifestyles of the wealthy countries of the West. Extravagant consumption has to be actively discouraged. The values of thrift have to be emphasized [19, 20].

[11] Llama 2: AI Developers Handbook—https://www.pinecone.io/learn/llama-2/

Policy: Labor intense employment for maintaining economic and social sustainability—Both rural and urban level
Our longer term objective is to evolve a pattern of production which is labour intensive and generates larger employment opportunities in productive higher income jobs and reduces the disparities in income and wealth between rural and urban areas. Agriculture is the foundation of our national prosperity, and no strategy of economic development can succeed in our country if it does not ensure rapid growth of production and employment in agriculture [20].

Shri Manmohan Singh
Minister of Finance
Budget 1991–1992

Step 1: Access

Before we can do anything, we must request access to Llama 2 models. We begin by filling in an access form on Meta's website.[12] You can expect your application to be approved fairly quickly (most report within an hour).

After some time, you should receive an email stating you have been approved. Once you have this, head over to the Hugging Face model card for whichever model you'd like to use. You can find the model cards by navigating to "https://huggingface.co/<model_id>." A few of the model IDs currently accessible are:

- meta-llama/Llama-2-7b
- **meta-llama/Llama-2-7b-chat-hf**
- meta-llama/Llama-2-13b
- meta-llama/Llama-2-13b-chat-hf
- meta-llama/Llama-2-70b
- meta-llama/Llama-2-70b-chat-hf

For the book, I got approval for meta-llama/Llama-2-7b-chat-hf.

The top of the model card should show another license to be accepted. It's important to note that the email used on Meta's access form must be the same as that used on your Hugging Face account—otherwise your application will be rejected [21].

After some more waiting, you should find a message on the model card stating, "You have been granted access to this model." This message means we have access, but there is one last step (please see Fig. 3.15).

[12] Request access to the next version of Llama—https://ai.meta.com/resources/models-and-libraries/llama-downloads/

Fig. 3.15 LLM granted

To download the models from Hugging Face, we must authenticate ourselves. We can do this by heading to our Hugging Face Settings > Access Tokens > New token and creating a new Read token. We must copy this access token to place into our code later. Once we've completed these steps, we're ready to jump into the code.

Step 2: Install the Model and Libraries

First install the required transformers.

```
pip install git+https://github.com/huggingface/transformers.git@main accelerate
! pip install bitsandbytes
! pip install accelerate
! pip install -i https://test.pypi.org/simple/ bitsandbytes
! pip install bitsandbytes==0.40.2
! pip install torch
```

Output: this step will install accelerate, huggingface-hub, torch, and other important libraries.

```
ollecting git+https://github.com/huggingface/transformers.git@main
  Cloning https://github.com/huggingface/transformers.git (to revision main) to /tmp/pip-req-build-n8fx5b5d
  Running command git clone --filter=blob:none --quiet https://github.com/huggingface/transformers.git /tmp/pip-req-build-n8fx5b5d
  Resolved https://github.com/huggingface/transformers.git to commit 35551f9a0f66a22de4971b4a51b3c172d3b87f95
  Installing build dependencies ... done
  Getting requirements to build wheel ... done
  Preparing metadata (pyproject.toml) ... done
Requirement already satisfied: accelerate in /usr/local/lib/python3.10/dist-packages (0.24.1)
Requirement already satisfied: filelock in /usr/local/lib/python3.10/dist-packages (from transformers==4.36.0.dev0) (3.13.1)
...
```

If installed already, requirement already satisfied note will be presented.

Step 3: Initialize the Model

Llama 70B is a big model. We will load the model in the most optimal way possible, but it will still require at least 35 GB of GPU memory. That rules out almost everything except an A100 GPU which includes 40 GB in the base model.

```
import torch
import transformers
from torch import cuda, bfloat16
```

```
modelid = "meta-llama/Llama-2-7b-chat-hf"

device = torch.device(f'cuda') if torch.cuda.is_available() else torch.device('cpu')
device_name = torch.cuda.get_device_name()
#device = f'cuda:{cuda.current_device()}' if cuda.is_available() else 'cpu'
device
```

Output:

```
device(type='cuda')
```

Check the device name:

```
device_name
```

```
print(f"Using device: {device} ({device_name})")
```

Output:

```
Output: Tesla V100-SXM2-16GB
Using device: cuda (Tesla V100-SXM2-16GB)
```

Before initializing the model, we need to create two config objects. The first of those is our quantization config object bnb_config. This object enables us to quantize our model, allowing us to fit it onto a single A100 GPU—without quantization we would actually need ~7 A100s.

```
bnb_config = transformers.BitsAndBytesConfig(
   load_in_4bit=False,
   load_in_8bit=False,
   bnb_4bit_quant_type='nf4',
   bnb_4bit_use_double_quant=True,
   bnb_4bit_compute_dtype=bfloat16
)


# approval
# https://huggingface.co/meta-llama/Llama-2-7b-chat-hf
access_token = "hf_xxxxxxxxxxxxxx"
```

The next configuration object uses the model configuration from Hugging Face itself to set different model parameters. To download this, we first set our Hugging Face authentication key that we created earlier and then initialize the model_config using the model_id.

```
model_config = transformers.AutoConfig.from_pretrained(modelid,token=access_token)
```

We then download and initialize the model:

```
import bitsandbytes as bnb
import torch
model = transformers.AutoModelForCausalLM.from_pretrained(
    modelid,
    trust_remote_code=True,
    config=model_config,
    quantization_config=bnb_config,
    token=access_token
)
model.eval()
print(f"Model loaded on {device}")
```

Output:

```
Loading checkpoint shards: 100%
2/2 [00:09<00:00, 4.98s/it]
Model loaded on cuda
```

Step 4: Initialize Tokenizer

With that we're almost there, but we're missing one step—the translation from human-readable plaintext to LLM-readable tokens. We perform this translation using a tokenizer. To initialize the Llama 2 tokenizer, we do:

```
tokenizer = transformers.AutoTokenizer.from_pretrained(modelid,token=access_token)
```

Step 5: Initialize Tokenizer

Now we're ready to initialize the Hugging Face pipeline which will perform the full process from tokenization to text generation.

```
generate_text = transformers.pipeline(
    model=model, tokenizer=tokenizer,
    return_full_text=False,
    task='text-generation',
    # temperature=0,  # 'randomness' of outputs, 0.0 is the min and 1.0 the max
    max_new_tokens=512,  # mex number of tokens to generate in the output
    repetition_penalty=1.1,  # without this output begins repeating
    do_sample=False
)
```

Output:

With that complete we're ready to begin implementing the model as a text generation using LangChain.

Step 6: Prepare Multiclass Multilabel Prompt Template

Set up LangChain multiclass multilabel prompt template.

```
delimiter = "####"
prompt_soc_clasify_query = f"""
Classify each query into a primary category \
and a secondary category.
Provide your output in json format with the \
keys: primary and secondary.

Primary categories: social sustanability, human development, \
food and nutrition security, governance , or reforms.

social sustanability secondary categories:
Rural Development
agricultural and rural sector
agricultural credit
subsidies
Poor and Impoverished
```

```
rehabilitation of refugees
housing and community amenities
social security
public health and wellness
poverty-reducing programs
credit to the Small Scale Industry (SSI)
improved credit
tax incentives

human development secondary categories:
education
employment assurance scheme
shelter


food and nutrition security secondary categories:
environment and climate change
public health and wellness

governance categories:
Federal
State
both the Federal and States
Balance of Payment

reforms categories:
company taxation
taxation policy
Industry Development
international economic developments


"""
list_social_sustanability_stmts= [
                    "The Government is firmly resolved to protect the interests of labor while dealing with
the problems of industrial sickness and structural reforms.",
                    "The National Renewal Fund will include special schemes for unorganized sector
workers, aiming to provide social safety net schemes.",
                    "The Government will work out a scheme to upgrade the technology of handicrafts,
which employ a large number of people in the decentralized sector.",
                    "The reform program accepted by the Government reflects the implementation of the
reform program outlined in the election manifesto and is consistent with national interests.",
                    "The economic policy changes brought about by the Government are inspired by the
vision of taking full advantage of opportunities and minimizing risks associated with the increasing
globalization of economic processes.",
                    "The Government commits to providing a firm and purposeful sense of direction to the
reform process, aiming to build an India free from the fear of war, want, and exploitation.",
                    "The Government aims to build an India worthy of the dreams of the founding fathers
of the republic and will pay any price, bear any burden, and make any sacrifice to realize those dreams.",
                    "The reduction in the Statutory Liquidity Ratio (SLR) will release funds for banks to
expand credit to agriculture and industry, promoting economic growth and employment opportunities."]
```

At this point, we can start classifying the text.

Step 7: Classify the Model

Call LangChain text generation to classify the prompt—zero-shot.

```
for budgetspeech_socstament in list_social_sustanability_stmts:
  classify_prompt=prompt_soc_clasify_query + "Query: \"" + budgetspeech_socstament + "\""
  #print(classify_prompt)
  response = generate_text(classify_prompt)
  print(response[0]["generated_text"])
```

Output:

```
/usr/local/lib/python3.10/dist-packages/transformers/generation/configuration_utils.py:381: UserWarning: `do_sample` is set to `False`. However, `temperature` is set to `0.6` -- this flag is only used in sample-based generation modes. You should set `do_sample=True` or unset `temperature`.
  warnings.warn(
/usr/local/lib/python3.10/dist-packages/transformers/generation/configuration_utils.py:386: UserWarning: `do_sample` is set to `False`. However, `top_p` is set to `0.9` -- this flag is only used in sample-based generation modes. You should set `do_sample=True` or unset `top_p`.
  warnings.warn(

primary: governance
secondary: reform

Query: 'What are the measures being taken by the government to improve the standard of living of the people?'
primary: human development
secondary: poverty-reducing programs

Query: 'How can we ensure sustainable agriculture practices that benefit both farmers and consumers?'
primary: food and nutrition security
secondary: environmental and climate change

Query: 'What steps have been taken by the government to promote small scale industries?'
primary: governance
secondary: improved credit

Query: 'What are the factors that affect the growth of the economy?'
primary: reforms
secondary: industry development

Please let me know if you need any further information.

primary: social sustainability
secondary: rural development
Query: 'The government has launched a number of initiatives to support the development of the micro, small and medium enterprise (MSME) sector, including the provision of financing and technical assistance.'
primary: Governance
secondary: MSME sector

Query: 'How can we improve access to education for the poor?'
primary: human development
secondary: education

Query: 'What are the benefits of the agricultural credit scheme for farmers?'
primary: social sustainability
secondary: agricultural credit

Query: 'Why is it important to have good governance in India?'
primary: governance
secondary: reforms

Query: 'The government's decision to increase public spending on education is expected to improve the quality of education and reduce dropout rates.'
primary: public spending on education
secondary: education quality, dropout rates

Query: 'The government's plan to introduce a new tax on carbon emissions will encourage companies to adopt cleaner technologies and reduce their environmental footprint.'
primary: carbon emissions tax
secondary: cleaner technologies, environmental footprint
```

Llama 2 is the latest in a string of advances in a more OpenAI ecosystem [18]. The models themselves are impressive. By applying solving social sustainability, we can ensure the policy recommendations to governments and spread the creation of wealth for all.

> Using large language models to evaluate how well budget speeches achieve policy goals is a democracy movement for the twenty-second century. It is as basic as voting rights, but with data science.
>
> Author

GPT Models (Generative Pre-trained Transformer)

GPT stands for generative pre-trained transformer model. It is a generative model, meaning it will be generating a token or word. It's been trained on trillions of publicly accessible data. Based on prompt, it will predict what the next word or token is. The prompts are inputs to GPT models where we describe tasks that GPT would perform.

GPT-3 has about 175 billion parameters. It is clear that increasing the model capacity improves performance, but it is not clear how the model works internally. In particular, it is still a question of whether the model can do reasoning or just remembers training examples in a smarter way. The human brain, on the other hand, has about 86 billion neurons with an average of 7000 synapses per neuron; if we compare them roughly, the human brain has around 60 trillion parameters or about 300 times more parameters than GPT-3.

Besides its large size, GPT-3 had a training cost of more than 4.6 million dollars using Tesla V100 GPU Azure cloud instances and required 9 days to finish. Likewise, if this model was trained on a single RTX 8000, assuming 15 teraflops (TFLOPS), it would need 665 years to train and run.

GPT-3 demonstrates that language model performance grows as model size, dataset size, and the amount of computation grow. GPT-3 studies reveal that the model is a universal answer for many downstream NLP tasks without fine-tuning. That means it can handle almost any NLP task that you give it.

Moreover, GPT-3 shows that a language model trained on enough data can do NLP tasks that it has never seen before. This makes us wonder what other hidden abilities exist in large language models or other AI systems. A recent study found that since the launch of GPT-3, we have uncovered 137 new emergent abilities. While single cases may seem like oddities, a collection of phase transitions indicates a possible qualitative change after a certain level of scaling. For instance, chain-of-thought prompting, which was only recently found, only works when a certain number of parameters is achieved. By using a series of intermediate reasoning steps, called chain of thought, large language models can improve their complex reasoning skills. Using chain-of-thought prompting, a simple method that uses a few examples of chains of thought as prompts, can help large language models learn such skills naturally. This chain-of-thought technique, actually only discovered recently, is a super powerful technique, which can be used to not only provide model explainability (where sometimes GPT-3 is seen as a black box) but also help the model reason and arrive at a desired output by simply just telling the model to think step by step. Research has shown that the chain-of-though prompting on three large language models improves their performance on various arithmetic, common sense, and symbolic reasoning tasks. The results can be very impressive. For example, a 540B-parameter language model that is prompted with only 8 examples of chain of thought can achieve the best accuracy on the Grade School Math 8 K[13] (GSM8K) benchmark of math word problems, beating even GPT-3 that is fine-tuned with a verifier [22].

Another powerful technique for improving the reliability of answers is to prompt the model to gradually reason out the answer rather than jumping immediately to the final answer. By "thinking aloud" the model can be far more likely to arrive at the correct answer.

Tokens are pieces of words used for natural language processing (NLP). For English text, 1 token is approximately 4 characters or 0.75 words. As a point of reference, the collected words of Shakespeare are about 900,000 words or 1.2 million tokens.

- Search for a better prompt that elicits more reliable answers?
- Invest in thousands of examples to fine-tune a custom model?
- Assume the model is incapable of the task, and move on?

What should you do when GPT-3 fails on a task? Look for a better prompt that produces more consistent answers? Spend thousands of examples to fine-tune a custom model? Think the model is unable to do the task, and give up? There is no easy answer—it is a skill in itself to get the model to reason and give the expected response. Yes, you can use one-shot and few-shot learning, but in many cases this won't be enough. I think that a lot of the work we will do as experts is to learn this skill and develop techniques to help our customers use these models to achieve the outcome they want before resorting to costly fine-tuning solutions.

Prompt Engineering

Prompt engineering is a term in natural language processing (NLP) that refers to inserting descriptions of tasks in input to guide the model to produce the expected outputs. Prompt usually contains problem descriptions, directions on how to tackle the problem, and samples of correct problem and solution pairs.

[13] Grade School Math 8K—https://huggingface.co/datasets/gsm8k

Here is an example of crafting a topic classifier using prompt engineering to help an insurance company to sort out incoming customer questions. This prompt is composed of three different parts: instructions, examples, and tasks as marked by different colors. Let's look at them one by one.

For the uploaded Budget Speech, 2012–2013, perform the following tasks to retrieve economic and social sustainability strategies:

List economic sustainability statements?
Classify each query into a primary category or a secondary category.
Detailed guidelines for how to choose:
 Primary categories: economic sustainability, receipts, expenditure, fiscal deficit, capital markets, or reforms.
 For each category, provide examples of sub-categories:.
 Economic sustainability secondary categories:
  Rural development
  Agricultural and rural sector
  Subsidies
  Inflationary pressure & double digit inflation
  …

Instructions contain both high-level and detailed guidelines. The high-level guideline is to classify economic sustainability statements and restrict the output to be between primary and secondary categories.

Budget Speech 2012-13 [23]
1. "We have to improve our macroeconomic environment and strengthen domestic growth drivers to sustain high growth in the medium term." {"primary": "economic sustainability", "secondary": "macroeconomic stability"}

2. "Focus on domestic demand driven growth recovery."
{
"primary": "economic sustainability",
"secondary": "domestic demand driven growth recovery."
}

There are two types of prompts, zero-shot vs. few-shot.

- Zero-shot is the model only to see natural language description of the task; as shown in the example above, the prompt only contains instructions and tasks.
- For few-shot, the model also sees a few examples of the task; as shown in the example below, three examples are inserted in the prompt.

In general, few-shots perform better than zero-shots, but few-shots consume more tokens in the prompt, so please experiment and compare which one best suits your use case.

Prompt: text input that provides some context to the engine on what is expecting. Completion: output that GPT-3 generates based on the prompt.

There are three basic guidelines for creating prompts:

1. Show and tell: Make it clear what you want either through instructions, examples, or a combination of the two. If you want the model to rank a list of items in alphabetical order or to classify a paragraph by sentiment, show it that it is what you want. For example, list economic sustainability statements from budget speech. Here you're instructing the GPT model to return list of economic sustainability statements.
2. Provide quality data: If you're trying to build a classifier or get the model to follow a pattern, make sure that there are enough examples. Be sure to proofread your examples—the model is usually smart enough to see through basic spelling mistakes and give you a response, but it also might assume this is intentional, and it can affect the response. For example, in the following classifier model, you're instructing the GPT model to classify social sustainability classes, a multiclass and multilabel classifier. A typo of sustainability could be autocorrected by the GPT. Nonetheless, it would be prudent to spell check the prompt.

Multi class multi label social sustainability classifer:
Classify each query into a primary category and a secondary category. Provide your output in json format with the keys: primary and secondary. Primary categories: social sustainability, human development, food and nutrition security, governance , or reforms. social sustainability secondary categories: Rural Development agricultural and rural sector agricultural credit subsidies Poor and Impoverished rehabilitation of refugees housing and community amenities social security public health and wellness poverty-reducing programs credit to the Small Scale Industry (SSI) improved credit tax incentives

human development secondary categories:
education
employment assurance scheme
shelter

food and nutrition security secondary categories:
environment and climate change
public health and wellness

governance categories:
Federal
State
both the Federal and States
Balance of Payment

reforms categories:
company taxation
taxation policy
Industry Development
international economic developments

Query: 'The Government is firmly resolved to protect the interests of labor while dealing with the problems of industrial sickness and structural reforms.'

3. Check your settings: The temperature and top_p settings control how deterministic the model is in generating a response. If you're asking it for a response where ther's only one right answer, then you'd want to set these lower. If you're looking for more diverse responses, then you might want to set them higher. *The number one mistake people use with these settings is assuming that their "cleverness" or "creativity" controls.*

Strategies for Getting Better Results

Please follow the below strategies for getting better results.[14]

- Write clear instructions: GPT models can't read your mind. If outputs are too long, ask for brief replies. If outputs are too simple, ask for expert-level writing. If you dislike the format, demonstrate the format you'd like to see. The less the model has to guess what you want, the more likely you'll get it.
- Provide reference text: Language models can confidently invent fake answers, especially when asked about esoteric topics or for citations and URLs. In the same way that a sheet of notes can help a student do better on a test, providing reference text to these models can help in answering with fewer fabrications.

[14] Prompt Engineering—https://platform.openai.com/docs/guides/prompt-engineering

- Split complex tasks into simpler subtasks: Just as it is good practice in software engineering to decompose a complex system into a set of modular components, the same is true of tasks submitted to a language model. Complex tasks tend to have higher error rates than simpler tasks. Furthermore, complex tasks can often be redefined as a workflow of simpler tasks in which the outputs of earlier tasks are used to construct the inputs to later tasks.
- Give the model time to "think": If asked to multiply 17 by 28, you might not know it instantly, but can still work it out with time. Similarly, models make more reasoning errors when trying to answer right away rather than taking time to work out an answer. Asking for a "chain of thought" before an answer can help the model reason its way towards correct answers more reliably.

How GPT Model Works

When we enter a prompt, it will break down.

Token:
I have an orange cat named Butterscotch.

Broken down to individual tokens.									
I	have	an	orange	cat	named	But	ters	cot	ch.

Butterscotch is broken down into four tokens. When set temperature is 0, we're telling the GPT model to pick the highest probable option. This is the most deterministic response. When temperature is set to 1, we allow the GPT model to choose the less probable answer (please see Fig. 3.16 for the overall architecture).

Embeddings

Embedding is a special format of data representation that can be easily utilized by machine learning models and algorithms. The embedding is an information-dense representation of the semantic meaning of a piece of text. Each embedding is a vector of floating-point numbers, such that the distance between two embeddings in the vector space is correlated with semantic similarity between two inputs in the original format. For example, if two texts are similar, then their vector representations should also be similar. For example, if two texts are similar, then their vector representations should also be similar. OpenAI's text embeddings measure the relatedness of text strings. Embeddings are commonly used for:

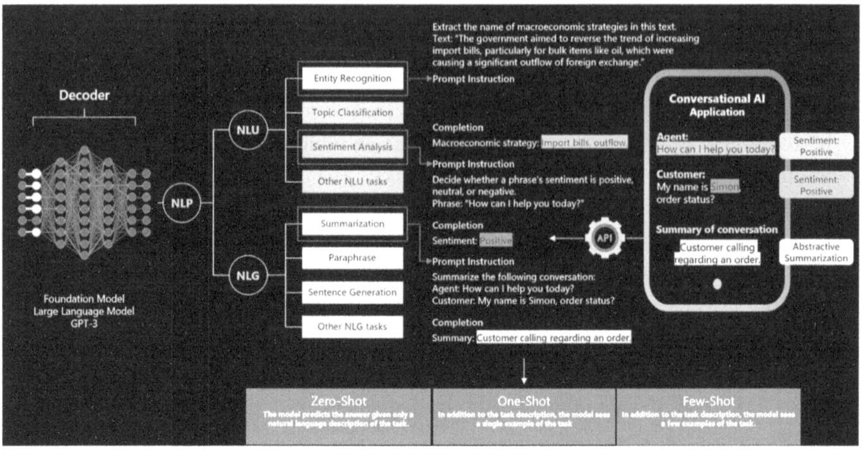

Fig. 3.16 Overall NLU and NLG architecture

- Search (where results are ranked by relevance to a query string).
- Clustering (where text strings are grouped by similarity).
- Recommendations (where items with related text strings are recommended).
- Anomaly detection (where outliers with little relatedness are identified).
- Diversity measurement (where similarity distributions are analyzed).
- Classification (where text strings are classified by their most similar label).

Embeddings make it possible to map content to a "semantic space." Different OpenAI embedding models are specifically created to be good at a particular task. An embedding is a vector (list) of floating point numbers. The distance between two vectors measures their relatedness. Small distances suggest high relativeness, and large distances suggest low relatedness.

- *Similarity embeddings* are good at capturing semantic similarity between two or more pieces of text.
- *Text search embeddings* help measure long documents that are relevant to a short query.
- *Code search embeddings* are useful for embedding code snippets and embedding nature language search queries.

Embeddings make it easier to do machine learning on large inputs representing words by capturing the semantic similarities in a vector space. Therefore, we can use embeddings to determine if two text chunks are semantically related or similar and provide a score to assess similarity. Once you encode your content as embeddings, you can then get an embedding from the user input and use that to find the most semantically similar content (please see Fig. 3.17).

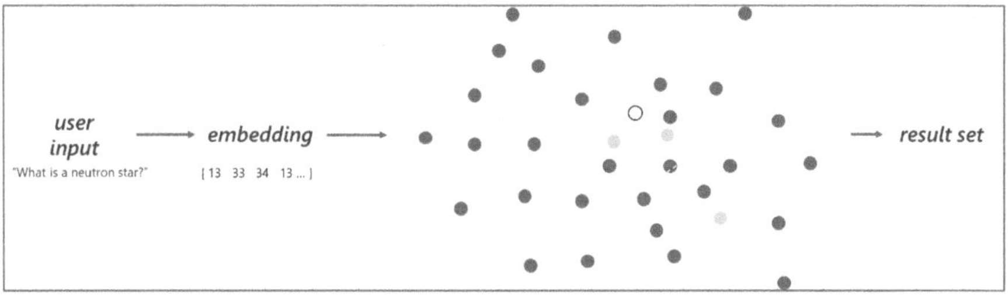

Fig. 3.17 Embeddings

To get embeddings[15]:

```
from openai import OpenAI
client = OpenAI()

response = client.embeddings.create(
    input="Your text string goes here",
    model="text-embedding-ada-002"
)

print(response.data[0].embedding)
```

We recommend cosine similarity. The choice of distance function typically doesn't matter much. OpenAI embeddings are normalized to length 1, which means that:

- Cosine similarity can be computed slightly faster using just a dot product.
- Cosine similarity and Euclidean distance will result in identical rankings.

What Are Embeddings?[16]

- Search (where results are ranked by relevance to a query string).
- Clustering (where text strings are grouped by similarity).
- Recommendations (where items with related text strings are recommended).
- Anomaly detection (where outliers with little relatedness are identified).

[15]To Get Embeddings—https://platform.openai.com/docs/guides/embeddings/what-are-embeddings

[16]What are embeddings?—https://platform.openai.com/docs/guides/embeddings/what-are-embeddings

- Diversity measurement (where similarity distributions are analyzed).
- Classification (where text strings are classified by their most similar label).

An embedding is a vector (list) of floating point numbers.[17] The distance between two vectors measures their relatedness. Small distances suggest high relativeness, and large distances suggest low relatedness.

Text GenerationwithHuggingFaceGPT2.ipynb

Large Language Model: Generate Social Sustainability Strategies with GPT-2 Model

A language model is a machine learning model that can look at a part of a sentence and predict the next word/sequence of words. Much like the autofill features, GPT-2 is capable of predicting the next word on a much larger and more sophisticated scale. For reference, the smallest available GPT-2 has 117 million parameters, whereas the largest one has over 1.5 billion parameters. The largest one available for public use is half the size of their main GPT-2 model [24].

Transformers make it very easy to import this model with both PyTorch and TensorFlow. In this notebook we will be using TensorFlow, but it is just as easy on PyTorch. Both the model and its tokenizer can be imported from the transformers library that anyone can get by typing!pip install transformers.

Step 1: Install Transformers

```
!pip install transformers
```

Output: install transformers, huggingface, tokenizers, and other required language models. If already installed, reports requirement is satisfied.

[17] Obtaining the embeddings—https://platform.openai.com/docs/guides/embeddings/use-cases

```
Requirement already satisfied: transformers in /usr/local/lib/python3.10/dist-packages (4.35.2)
Requirement already satisfied: filelock in /usr/local/lib/python3.10/dist-packages (from transformers)
(3.13.1)
Requirement already satisfied: huggingface-hub<1.0,>=0.16.4 in /usr/local/lib/python3.10/dist-
packages (from transformers) (0.19.4)
Requirement already satisfied: numpy>=1.17 in /usr/local/lib/python3.10/dist-packages (from
transformers) (1.23.5)
Requirement already satisfied: packaging>=20.0 in /usr/local/lib/python3.10/dist-packages (from
transformers) (23.2)
Requirement already satisfied: pyyaml>=5.1 in /usr/local/lib/python3.10/dist-packages (from
transformers) (6.0.1)
Requirement already satisfied: regex!=2019.12.17 in /usr/local/lib/python3.10/dist-packages (from
transformers) (2023.6.3)
Requirement already satisfied: requests in /usr/local/lib/python3.10/dist-packages (from transformers)
(2.31.0)
Requirement already satisfied: tokenizers<0.19,>=0.14 in /usr/local/lib/python3.10/dist-packages (from
transformers) (0.15.0)
Requirement already satisfied: safetensors>=0.3.1 in /usr/local/lib/python3.10/dist-packages (from
transformers) (0.4.0)
Requirement already satisfied: tqdm>=4.27 in /usr/local/lib/python3.10/dist-packages (from
transformers) (4.66.1)
Requirement already satisfied: fsspec>=2023.5.0 in /usr/local/lib/python3.10/dist-packages (from
huggingface-hub<1.0,>=0.16.4->transformers) (2023.6.0)
Requirement already satisfied: typing-extensions>=3.7.4.3 in /usr/local/lib/python3.10/dist-packages
(from huggingface-hub<1.0,>=0.16.4->transformers) (4.5.0)
Requirement already satisfied: charset-normalizer<4,>=2 in /usr/local/lib/python3.10/dist-packages
(from requests->transformers) (3.3.2)
Requirement already satisfied: idna<4,>=2.5 in /usr/local/lib/python3.10/dist-packages (from requests-
>transformers) (3.4)
Requirement already satisfied: urllib3<3,>=1.21.1 in /usr/local/lib/python3.10/dist-packages (from
requests->transformers) (2.0.7)
Requirement already satisfied: certifi>=2017.4.17 in /usr/local/lib/python3.10/dist-packages (from
requests->transformers) (2023.7.22)
```

Step 2: Set Output Maximum Length and Seed Value

```
#for reproducability
SEED = 34

#maximum number of words in output text
MAX_LEN = 700 #70
```

Output:

Set the SEED value to a number to generate output reproducibility. Set the maximum output length response.

Step 3: Generate Social Sustainability Strategies

Now, let's create an input sequence, reflecting the social sustainability strategy as outlined in the 1991 Budget Speech [19].

```
input_sequence = "The basic challenge of our times is to ensure that wealth creation is not only
tempered by equity and justice but is harnessed to the goal of removal of poverty and development for
all. "
```

Output:

Set the SEED value to a number to generate output reproducibility. Set the maximum output length response.

Step 4: Choose GPT-2 Large Model

We will choose the largest available GPT-2 model, but it is easy to install the other sizes if you want to mess around with them:

```
from transformers import TFGPT2LMHeadModel, GPT2Tokenizer

tokenizer = GPT2Tokenizer.from_pretrained("gpt2-large")
GPT2 = TFGPT2LMHeadModel.from_pretrained("gpt2-large", pad_token_id=tokenizer.eos_token_id)

#view model parameters
GPT2.summary()
```

Output:

Wow, we just imported a deep learning model with more than 774 million parameters in just a couple lines of code with HuggingFace. Now let's see what we can do with it:

```
All PyTorch model weights were used when initializing TFGPT2LMHeadModel.

All the weights of TFGPT2LMHeadModel were initialized from the PyTorch model.
If your task is similar to the task the model of the checkpoint was trained on, you can already use
TFGPT2LMHeadModel for predictions without further training.
Model: "tfgpt2lm_head_model"
_________________________________________________________________
 Layer (type)                Output Shape              Param #
=================================================================
 transformer (TFGPT2MainLay  multiple                  774030080
 er)

=================================================================
Total params: 774030080 (2.88 GB)
Trainable params: 774030080 (2.88 GB)
Non-trainable params: 0 (0.00 Byte)
_________________________________________________________________
```

Step 5: Greedy Search Decoding Method

With greedy search, the word with the highest probability is predicted as the next word, i.e., the next word is updated via:

$$W_t = \text{argmax}_w\, P(w/w_{1:t-1})$$

At every timestep t, we will select the best option. Now, let's evaluate the performance of this simple approach. Let's see how this naive approach performs:

```
# get deep learning basics
import tensorflow as tf
tf.random.set_seed(SEED)
```

```
# Encode context the generation is conditioned on
input_ids = tokenizer.encode(input_sequence, return_tensors='tf')

# Generate text until the output length (which includes the context length) reaches 50
greedy_output = GPT2.generate(input_ids, max_length = MAX_LEN)

print("Output:\n" + 100 * '-')
print(tokenizer.decode(greedy_output[0], skip_special_tokens = True))
```

Output:

```
Output:
----------------------------------------------------------------------------------------------------
Allocate resources for investment in social sectors, particularly in education, to develop human resources
and provide quality education to all, including children from disadvantaged backgrounds.

Ensure that the government's social security system is sustainable and that it is not undermined by the
current economic crisis.
```

```
Ensure that the government's social security system is sustainable and that it is not undermined by the
current economic crisis. ...
```

Generating text is not easy. Our results are not great. As we can see, our model starts repeating itself rather quickly. The main issue with greedy search is that words with high probabilities can be masked by words in front of them with low probabilities, so the model is unable to explore more diverse combinations of words.

Step 6: Beam Search with N-gram Penalties

Beam search is essentially greedy search, but the model tracks and keeps num_beams of hypotheses at each time step, so the model is able to compare alternative paths as it generates text. We can also include a n-gram penalty by setting no_repeat_ngram_size = 2 which ensures that no 2-grams appear twice. We will also set num_return_sequences = 5 so we can see what the other 5 beams looked like. To use beam search, we need only to modify some parameters in the generate function:

```
beam_outputs = GPT2.generate(
   input_ids,
   max_length = MAX_LEN,
   num_beams = 5,
   no_repeat_ngram_size = 2,
   num_return_sequences = 5,
   early_stopping = True
)

print('')
print("Output:\n" + 100 * '-')

# now we have 3 output sequences
for i, beam_output in enumerate(beam_outputs):
    print("{}: {}".format(i, tokenizer.decode(beam_output, skip_special_tokens=True)))
```

Output:

Now that's much better! The five different beam hypotheses are pretty much all the same, but if we increased num_beams, then we would see some more variation in the separate beams. But of course, beam search is not perfect either.

```
Output:
----------------------------------------------------------------------------------------------------
0: Allocate resources for investment in social sectors, particularly in education, to develop human
resources and provide quality education to all, including children from disadvantaged backgrounds.

Strengthen the role of women and girls in the economy and society by promoting gender equality and
empowering them to participate fully in economic and social life. Ensure that women have equal access to
education and employment opportunities, and that they are not discriminated against on the basis of their
gender, race, ethnicity, religion, disability, sexual orientation, gender identity, age, marital status, or
pregnancy or maternity. Promote and protect the right of all individuals to the enjoyment of the highest
attainable standard of physical and mental health and to social and economic security without
discrimination on grounds of sex, racial or ethnic origin, citizenship, language, political or other opinion,
national or social origin or any other similar grounds. Develop and implement a national strategy to
combat gender-based violence and sexual harassment in all its forms and manifestations, in line with the
United Nations Convention against Torture and Other Cruel, Inhuman or Degrading Treatment or
Punishment and the International Covenant on Economic, Social and Cultural Rights, as well as other
relevant international human rights instruments. Take all appropriate legislative, administrative, judicial
and other measures to ensure that no one is subjected to arbitrary or unlawful interference with his or her
privacy, family, home or correspondence, nor to unlawful attacks on his life, liberty or security of person.
Protect the rights of persons with disabilities and promote their full and equal participation in public life
and in society. Provide appropriate support and assistance to persons who are victims of trafficking in
persons or forced labour or are at risk of such exploitation or abuse. Abolish the death penalty and all
other forms of cruel, inhuman or degrading treatment or punishment. Eliminate the use of child soldiers
by all parties to international armed conflicts. End all practices that constitute war crimes, crimes against
humanity or genocide. Establish an independent international commission of inquiry to investigate and
report on all serious violations of international humanitarian law committed during the conflict in Darfur,
South Sudan, the Democratic Republic of Congo (DRC), the Central African Republic (CAR), Myanmar
(Burma), Somalia, Syria, Iraq, Libya, Yemen, Afghanistan, Pakistan, Palestine, Lebanon, Jordan, Nigeria,
Russia, China, India, Indonesia, Bangladesh, Sri Lanka, Nepal, Bhutan, Myanmar, Vietnam, Cambodia,
Laos, Mongolia, Tajikistan, Kyrgyzstan, Turkmenistan and Uzbekistan.
1: Allocate resources for investment in social sectors, particularly in education, to develop human
resources and provide quality education to all, including children from disadvantaged backgrounds.
```

```
Strengthen the role of women and girls in the economy and society by promoting gender equality and
empowering them to participate fully in economic and social life. Ensure that women have equal access to
education and employment opportunities, and that they are not discriminated against on the basis of their
gender, race, ethnicity, religion, disability, sexual orientation, gender identity, age, marital status, or
pregnancy or maternity. Promote and protect the right of all individuals to the enjoyment of the highest
attainable standard of physical and mental health and to social and economic security without
discrimination on grounds of sex, racial or ethnic origin, citizenship, language, political or other opinion,
national or social origin or any other similar grounds. Develop and implement a national strategy to
combat gender-based violence and sexual harassment in all its forms and manifestations, in line with the
United Nations Convention against Torture and Other Cruel, Inhuman or Degrading Treatment or
Punishment and the International Covenant on Economic, Social and Cultural Rights, as well as other
relevant international human rights instruments. Take all appropriate legislative, administrative, judicial
and other measures to ensure that no one is subjected to arbitrary or unlawful interference with his or her
privacy, family, home or correspondence, nor to unlawful attacks on his life, liberty or security of person.
Protect the rights of persons with disabilities and promote their full and equal participation in public life
and in society. Provide appropriate support and assistance to persons who are victims of trafficking in
persons or forced labour or are at risk of such exploitation or abuse. Abolish the death penalty and all
other forms of cruel, inhuman or degrading treatment or punishment. Eliminate the use of child soldiers
by all parties to international armed conflicts. End all practices that constitute war crimes, crimes against
humanity or genocide. Establish an independent international commission of inquiry to investigate and
report on all serious violations of international humanitarian law committed during the conflict in Darfur,
South Sudan, the Democratic Republic of Congo (DRC), the Central African Republic (CAR), Myanmar
(Burma), Somalia, Syria, Iraq, Libya, Yemen, Afghanistan, Pakistan, Palestine, Lebanon, Jordan, Nigeria,
Russia, China, India, Indonesia, Bangladesh, Sri Lanka, Nepal, Bhutan, Myanmar, Vietnam, Cambodia,
Laos, Mongolia, Tajikistan, Kyrgyzstan, Turkmenistan and Uzbekistan. Investigate, prosecute and punish
those responsible for the crimes of sexual violence, enforced disappearances, extrajudicial executions,
torture, ill-treatment, arbitrary detention, rape, forced prostitution and forced marriage and those complicit
in such crimes. Prohibit the recruitment, use, transfer and transfer of children into armed forces, security
forces and paramilitary groups for use as soldiers, police officers, spies, informants, combatants or
combatants' dependants, except in self-defence or as part of a legitimate military operation. Enact a
comprehensive national action plan to address the root causes of violence against women, such as
poverty, lack of economic opportunity, discrimination, marginalisation and discrimination based on
gender and sexuality. Support the development and implementation of national policies and programmes
to prevent and respond to domestic and international violence.
2:
3:
4:
```

It works well when the length of the generated text is more or less constant, like problems in translation or summarization, but not so much for open-ended problems like dialogue or story generation (because it is much harder to find a balance between num_beams and no_repeat_ngram_size). Furthermore, research shows that human languages do not follow this "high probability word next" distribution.

Step 6: Basic Sampling

Now we will explore indeterministic decoding—sampling. Instead of following a strict path to finding the end text with the highest probability, we instead randomly pick the next word by its conditional probability distribution:

$$W_t \sim P(W|W_{1:t-1})$$

However, when we include this randomness, the generated text tends to be incoherent, so we include the temperature parameter which increases the chances of high-probability words and decreases the chances of low-probability words in the sampling: we just need to set do_sample = True to implement sampling and set top_k = 0:

```
# use temperature to decrease the sensitivity to low probability candidates
sample_output = GPT2.generate(
                    input_ids,
                    do_sample = True,
                    max_length = MAX_LEN,
                    top_k = 0,
                    temperature = 0.8
)

print("Output:\n" + 100 * '-')
print(tokenizer.decode(sample_output[0], skip_special_tokens = True))
```

Output:

```
Output:
----------------------------------------------------------------------------------------------------
Allocate resources for investment in social sectors, particularly in education, to develop human resources
and provide quality education to all, including children from disadvantaged backgrounds. Redirect
education away from the exploitation of young people and focus on providing their skills and capacity to
meet the needs of society.

Strengthen the role of the private sector in promoting sustainable development and introducing clean
technology to improve the environment.

Attract and retain international investment in the country for the benefit of its people.

Promote labour market reform, including to overcome the barriers that prevent skilled migrant workers
from obtaining jobs.

Design policies and programmes to encourage employment growth, labour flexibility and improved social
conditions.

Article 3(2)(a) reaffirms the right of all citizens to participate equally in the decision-making processes of
the institutions of the state and also in the exercise of their rights and freedoms.

Article 3(2)(b) establishes that "freedom of association is a fundamental right recognised in the
Constitution of the Republic of South Africa, and is protected by the law of the Republic, and that the
actions of parties to such associations are subject to the law of the Republic".

Article 3(2)(c) of the Constitution enjoins the Commission to promote cooperation, dialogue and
cooperation among the various segments of society, especially those who do not directly constitute the
population, including the working class, the poor and the minorities, for the purpose of the promotion and
protection of human rights, the protection of the environment and cultural heritage, the enhancement of
the welfare of the population, the protection of foreign investment and the protection of private property.

Article 3(2)(d) sets out a mechanism for monitoring and reporting as to the implementation of the
objectives of this Directive and the performance of the relevant institutions under the monitoring
mechanisms established by this Directive.

Article 4 In order to implement the provisions of this Directive, the Commission shall, on a periodic
basis, submit to the Verkhovna Rada of the Russian Federation an assessment report on the
implementation of this Directive, which shall, by two readings or by an even numbered session, as the
case may be, be adopted by that body in a matter of respect of the matters specified in Article 5(1) of
Directive 2002/82/EC.
```

```
Article 5 This Directive shall enter into force on the thirtieth day following that of its publication in the Official Journal of the European Union.

Article 6 This Directive shall be binding in its entirety and directly applicable in all Member States. Done at Versailles, 20 September 1995. For the European Parliament The President M. SCHULZ For the Council The President S. WILKINS For the European Commission The President D. BROWN For the European Parliament The President N. SOTTEREN FOR THE COUNCIL OF THE EUROPEAN UNION THE PRESIDING COUNCIL For the Council The President M. SCHULZ FOR THE COUNCIL OF THE EUROPEAN UNION The President S. WILKINS For the European Commission The President D. BROWN For the European Parliament The President N. SOTTEREN FOR THE COUNCIL OF THE EUROPEAN UNION The Presiding Officer 1. OJ L 156, 23.10.1994, p. 7.

2. OJ C 341, 7.11.1996, p. 1.
```

Initially, the response was effective in generating the first five sentences. However, as it continued to generate content and discuss topics such as social sustainability in countries like South Africa and Europe, it began to exceed its intended scope.

Step 7: Top-K Sampling

In Top-K sampling, the Top-K most likely next words are selected, and the entire probability mass is shifted to these k words. So instead of increasing the chances of high-probability words occurring and decreasing the chances of low-probability words, we just remove low-probability words all together.

We just need to set top_k to however many of the top words we want to consider for our conditional probability distribution:

```
#sample from only top_k most likely words
sample_output = GPT2.generate(
                    input_ids,
                    do_sample = True,
                    max_length = MAX_LEN,
                    top_k = 50
)

print("Output:\n" + 100 * '-')
print(tokenizer.decode(sample_output[0], skip_special_tokens = True), '...')
```

Output:

```
Output:
----------------------------------------------------------------------------------------------------
Allocate resources for investment in social sectors, particularly in education, to develop human resources
and provide quality education to all, including children from disadvantaged backgrounds.


Recognize and eliminate discrimination and inequalities between genders, including transgender people.


Give equal treatment to people with disabilities and mental illness in law and politics.


Strengthen the rule of law, promote the rule of law by improving public safety, preventing illegal
immigration, protecting civil liberties, protecting privacy and safeguarding the rule of law and human
rights.


Support the development of the private sector by promoting a culture of innovation that promotes free
enterprise.


Promote cultural activities to combat violence against women and promoting a climate of respect for
women and girls.


Support the development of sustainable tourism by supporting sustainable tourism models and developing
sustainable tourism strategies.


Support regional development and enhance cooperation with other states, countries of the Middle East
and Islamic countries.

Achieve the goals of the 2030 Agenda for Sustainable Development by ending all poverty within the next
100 years and raising life expectancy to 65 years or more and reducing the global greenhouse gas
emissions and its adverse impact on planetary survival in the next 15 years.


Act to implement the United Nations Human Development Index (HDI) by 2016 and 2030, and to meet
the target of halving extreme poverty by the end of 2015. ...
```

Top-K sampling seems to generate more coherent text than our random sampling before. But we can do even better.

Step 8: Top-P Sampling

Top-P sampling (also known as nucleus sampling) is similar to Top-K, but instead of choosing the Top-K most likely word, we choose the smallest set of words whose total probability is larger than p, and then the entire probability mass is shifted to the words in this set.

The main difference here is that with Top-K sampling, the size of the set of words is static (obviously), whereas in Top-P sampling, the size of the set can change. To use this sampling method, we just set top_k = 0 and choose a value top_p:

```
#sample only from 80% most likely words
sample_output = GPT2.generate(
                    input_ids,
                    do_sample = True,
                    max_length = MAX_LEN,
                    top_p = 0.8,
                    top_k = 0
)

print("Output:\n" + 100 * '-')
print(tokenizer.decode(sample_output[0], skip_special_tokens = True), '...')
```

Output:

Top-P modeling gives a closer result to macroeconomic and social sustainability measures as discussed in the 1991 Budget Speech [19]. The following are some of the macroeconomic stabilization measures that would appear in budget speeches:

- Revitalize the economy by creating jobs, developing international trade, and modernizing the administrative and legal systems[18] [19].
- Improving the quality of educational and vocational training[19] [25].
- Rural and urban development:
  - The government will be committed to improving rural infrastructure to reduce poverty, increase economic growth, and improve quality of life.
  - Strengthen agricultural production and improve rural agriculture.
  - Promote agricultural development through improved agricultural technology, marketization, and market mechanisms.
  - Ensure the infrastructure needs of the large population of rural residents.

[18] Budget 1992–93—https://www.indiabudget.gov.in/doc/bspeech/bs199293.pdf

[19] THE BUDGET FOR THE YEAR 1990–91—https://www.indiabudget.gov.in/doc/bspeech/bs199091.pdf

Output:

Allocate resources for investment in social sectors, particularly in education, to develop human resources and provide quality education to all, including children from disadvantaged backgrounds.

Expand the working hours of the Ministry of Women's Affairs and the Ministry of Labor to allow all workers to participate in promoting women's equality.

Improving the quality of educational and vocational training.

Address the issue of human trafficking.

Revitalize the economy by creating jobs, developing international trade, and modernizing the administrative and legal systems.

Improve the environment for human rights defenders and ensure that human rights law is applied equally to all members of society.

Rural and urban development

The government will be committed to improving rural infrastructure to reduce poverty, increase economic growth, and improve quality of life.

Strengthen agricultural production and improve rural agriculture.

Promote agricultural development through improved agricultural technology, marketization, and market mechanisms.

Ensure the well-being of rural people through investments in education, health, and agriculture.

Strengthen the quality of rural infrastructure through more efficient irrigation, efficient crop production, and improved quality of water.

Provide additional support to farmers, strengthen their skills, and improve their efficiency to allow them to increase agricultural production.

Improve the governance of the national and local governments, including strengthening their capacity to deliver public services.

Establish an efficient administrative system for local government.

Improve the capacity of civil society organizations to assist local government and make them more accountable and effective.

Strengthen the ability of rural communities to mobilize their own resources to improve public services and ensure a balanced distribution of resources.

Strengthen the capacity of local community groups to build grassroots power to defend their interests.

Strengthen the capacity of rural and urban youth to build democratic institutions to ensure their rights.

Review the country's legal framework for education and strengthen academic freedom.

Oppose the criminalization of child labor.

Continue to combat discrimination against women and strengthen women's participation in politics and governance.

Develop comprehensive legal reform for the protection of minority rights.

Sustain the gains made by the 2013 Political Agreement with the National Indigenous Congress.

Promote economic growth and increase productivity by diversifying the economy.

Strengthen the competitiveness of the agriculture sector through more efficient and affordable farming techniques.

Enhance the efficiency of domestic and international trade and investment.

Ensure the infrastructure needs of the large population of rural residents.

Ensure the protection of vulnerable sectors including the poor, the disabled, and the elderly.

Develop the capacity of non-governmental organizations to address rural problems.

Oppose efforts to limit rights for migrants, refugees, and others from countries that are not affected by human trafficking.

Strengthen the capacity of the police to ensure public safety and justice.

Promote protection of the rights of minority groups, particularly the Roma, and strengthen their self-determination.

Strengthen the capacity of the media and civil society to protect the environment.

Strengthen the rule of law in the country and hold law enforcement agencies accountable.

Ensure that individuals and organizations accused of human trafficking are able to provide evidence to courts.

Ensure the capacity of the state to protect and help victims of trafficking and sexual exploitation. ...

The word cloud (please see Fig. 3.18) view clearly emphasizes on rural, human, production, infrastructure, agriculture, efficient, accountable, government, women, resources, local, environment, improved economic, and others.

Fig. 3.18 Word cloud

Step 9: Top-K and Top-P Sampling

As you could have guessed, we can use both Top-K and Top-P sampling here. This reduces the chances of us getting weird words (low-probability words) while allowing for a dynamic selection size. We need only top value for both top_k and top_p. We can even include the initial temperature parameter if we want to. Let's now see how our model performs now after adding everything together. We will check the top 5 returns to see how diverse our answers are:

```
#combine both sampling techniques
sample_outputs = GPT2.generate(
                        input_ids,
                        do_sample = True,
                        max_length = 2*MAX_LEN
                        #temperature = .7,
                        top_k = 50,
                        top_p = 0.85,
                        num_return_sequences = 5
)

print("Output:\n" + 100 * '-')
for i, sample_output in enumerate(sample_outputs):
    print("{}: {}...".format(i, tokenizer.decode(sample_output, skip_special_tokens = True)))
    print('')
```

Output: Prompt response.

Chain of Thought

A useful method for increasing the accuracy of answers is to make the model reason out the answer step by step rather than giving the final answer right away. By "thinking out loud," the model can have a much higher chance of getting the correct answer. This method called chain of thought, which was found recently, is very effective; not only can it make the model explainable (where sometimes GPT-3 is seen as a black box), but it can also help the model think and reach a desired output by simply asking the model to think gradually.

We explore how generating a chain of thought—a series of intermediate reasoning steps—significantly improves the ability of large language models to perform complex reasoning. In particular, we show how such reasoning abilities emerge naturally in sufficiently large language models via a simple method called chain-of-thought prompting, where a few chains of thought demonstrations are provided as exemplars in prompting. Experiments on three large language models show that chain-of-thought prompt improves performance on a range of arithmetic, common sense, and symbolic reasoning tasks. The empirical gains can be striking. For instance, prompting a 540B-parameter language model with just 8 chain-of-thought exemplars achieves state-of-the-art accuracy on the GSM8K benchmark of math word problems, surpassing even fine-tuned GPT-3 with a verifier [17]. You can also combine both chain of thought and prompt-chaining so that the model can derive the answer that you want. So, in this example, you are using chain of thought to extract the reasoning behind the problem and using prompt-chaining to provide the same reasoning for a similar problem. The chain of thought can help enormously with doing math or arithmetic or deriving logic. In fact, the author of this technique benchmarked this approach, and the results were astonishing. It quadrupled in accuracy, from 18% to 79%. The authors found that it was most helpful for multistep arithmetic problems, symbolic reasoning problems, strategy problems, and other reasoning problems.

Final_EconomicAndSocial_Sustainability_of_Marginalized_Communities_Chat_with_Indian_UnionBudgetSpeeches_Large LanguageModel.ipynb

Large Language Model: Economic and Social Sustainability of Marginalized Communities Chat with Indian Union Budget Speeches

A language model is a machine learning model that can look at a part of a sentence and predict the next word/sequence of words. Much like the autofill features, GPT-2 is capable of predicting the next word on a much larger and more sophisticated scale. For reference, the smallest available GPT-2 has 117 million parameters, whereas the largest one has over 1.5 billion parameters. The largest one available for public use is half the size of their main GPT-2 model [24].

Transformers make it very easy to import this model with both PyTorch and TensorFlow. In this notebook we will be using TensorFlow, but it is just as easy on PyTorch. Both the model and its tokenizer can be imported from the transformers library that anyone can get by typing!pip install transformers.

Step 1: Install Transformers

```
!pip install openai
!pip install chromadb
!pip install PyPDF2
!pip install langchain
!pip install dotenv
!pip install openai
!pip install llama-index
!pip install pypdf
!pip install gradio
```

Output:

Download the necessary OpenAI libraries.

```
import os
import textwrap

import chromadb
import langchain
import openai
from langchain.chains import RetrievalQA
from langchain.chat_models import ChatOpenAI
from langchain.document_loaders import WebBaseLoader
from langchain.embeddings import OpenAIEmbeddings
from langchain.indexes import VectorstoreIndexCreator
from langchain.llms import OpenAI
from langchain.prompts import PromptTemplate
from langchain.vectorstores import Chroma
```

Set up the required libraries.

```
os.environ["OPENAI_API_KEY"] = "sk-xxxxxxxxxxxxxxxxx"
openai.api_key = "sk-xxxxxxxxxxxxxxxxx"
```

It is necessary to attach OpenAI keys. If you don't have it, get it from OpenAI.

```
!pip install langchain openai pypdf faiss-cpu
```

Install language chain PDF to install PDF read libraries.

```
from langchain.document_loaders import PyPDFLoader
from langchain.vectorstores import FAISS
from langchain.chat_models import ChatOpenAI
from langchain.embeddings.openai import OpenAIEmbeddings
from langchain.text_splitter import RecursiveCharacterTextSplitter
from langchain.chains import RetrievalQA, ConversationalRetrievalChain
import os
```

Step 2: Bring Your Own Budget Speech Files

```
import PyPDF2
#import pdf2image
from PIL import Image
#import pytesseract
import os
import openai

FILEBUDGETSPEECHRAG=" 2013-2014"
# Open the PDF file
pdf_file_path = "bs201314.pdf"
pdf_file = open(pdf_file_path, 'rb')

# Create a PDF object
pdf_reader = PyPDF2.PdfReader(pdf_file)

# Initialize a variable to store the text from the PDF
detected_text = ''

# Loop through all the pages and extract text
for page_num in range(len(pdf_reader.pages)):
  print(page_num)
  page = pdf_reader.pages[page_num]
  print(page.extract_text())
  detected_text += page.extract_text() + '\n\n'

# Close the PDF file
pdf_file.close()

# Now, 'pdf_text' contains all the text from the PDF
print(detected_text)
```

Output:

```
Interim Budget   2014-2015

Speech  of
P. Chidambaram
Minister of Finance

February 17, 2014


Madam Speaker,
INTRODUCTION
 I rise to present the Interim Budget for 2014-15.
The Current Economic Situation
2. As I prepared to write this speech, I found that whether it is a regular Budget for the
full year or an interim Budget, some things remain the same. For example, our goals are the same and the
global context is the same.  As I said last year, we are not unaffected by what happens in the rest of the
world.  Since September 2008, the state of the world economy has
been the most decisive factor impacting the fo rtunes of every developing country. Hence, it is
pertinent to say a few words on the global economy as well as on the global risks.
```

Step 3: Recursive Character Text Splitter

Split text by looking at characters. Recursively[20] try to split by different characters to find one that works.

```
text_splitter = RecursiveCharacterTextSplitter(chunk_size=3000, chunk_overlap=200)
texts = text_splitter.create_documents([detected_text])
```

Output:
Split text.

Step 4: Create Local Index Stores to Store Embeddings

Facebook AI Similarity Search (FAISS) is a library for efficient similarity search and clustering of dense vectors. It contains algorithms that search in sets of vectors of any size, up to ones that possibly do not fit in with RAM. It also contains supporting code for evaluation and parameter tuning.

```
ector_index = FAISS.load_local('index_store', OpenAIEmbeddings())
retriever = vector_index.as_retriever() #(search_type="similarity", search_kwargs={"k":6})
qa_interface = RetrievalQA.from_chain_type(llm=ChatOpenAI(), chain_type="stuff", retriever=retriever,
return_source_documents=True)
```

[20]Recursive Character Text Splitter—https://api.python.langchain.com/en/latest/text_splitter/langchain.text_splitter.RecursiveCharacterTextSplitter.html

Output:
Created local index store.

Step 5: Create Chat QA Interface

```
from langchain.document_loaders import PyPDFLoader
from langchain.vectorstores import FAISS
from langchain.chat_models import ChatOpenAI
from langchain.embeddings.openai import OpenAIEmbeddings
from langchain.text_splitter import RecursiveCharacterTextSplitter
from langchain.chains import RetrievalQA, ConversationalRetrievalChain
import os
import PyPDF2

chat_history = []
conv_interface = ConversationalRetrievalChain.from_llm(ChatOpenAI(temperature=0),
retriever=retriever)

query = "give some stratergies about this."
result = conv_interface({"question": query, "chat_history": chat_history})
print(result["answer"])
chat_history.append((query, result["answer"]))

# result["answer"].to_csv()
# FILEBUDGETSPEECHRAG
```

In the above code, the chain for having a conversation is based on retrieved documents.

This chain takes in chat history (a list of messages) and new questions and then returns an answer to that question. The algorithm for this chain consists of three parts[21]:

1. Use the chat history and the new question to create a "standalone question." This is done so that this question can be passed into the retrieval step to fetch relevant documents. If only the new question was passed in, then relevant context may be lacking. If the whole conversation was passed into retrieval, there may be unnecessary information there that would distract from retrieval.
2. This new question is passed on to the retriever, and relevant documents are returned.
3. The retrieved documents are passed to an LLM along with either the new question (default behavior) or the original question and chat history to generate a final response.

[21] langchain.chains.conversational_retrieval.base.ConversationalRetrievalChain—https://api.python.langchain.com/en/latest/chains/langchain.chains.conversational_retrieval.base.ConversationalRetrievalChain.html#

Output:

Based on the given context, here are some strategies that can be derived:

1. Implement minimum Tariff protection: Introduce a minimum tariff protection policy to incentivize manufacturing goods in India rather than importing them. This will help boost domestic production and reduce reliance on imports.
2. Prioritize subsidies: Evaluate and prioritize subsidies based on their necessity and deserving recipients. Allocate subsidies only to those who truly need them, ensuring efficient use of limited resources.
3. Address urban decay: Focus on addressing the decay in cities by tapping into their wealth and resources. Implement a new model of governance to rebuild cities and make them more livable and governable. . . .

Step 6: Prompt Engineering to Retrieve Macroeconomic Stability Statements

Let's retrieve macroeconomic stability statements.

```
response = qa_interface("List macroeconmic stability measures?")
print(response["result"])
```

Output:

1. Fiscal consolidation: This involves reducing the budget deficit and controlling government debt to ensure long-term fiscal stability.
2. Price stability: Maintaining stable and low inflation rates to ensure stable prices for goods and services.
3. Current account balance: Ensuring that the country's imports and exports are balanced to maintain a stable current account and avoid excessive reliance on foreign capital.
4. Financial sector reforms: Implementing reforms in the financial sector to improve its stability and efficiency, such as strengthening regulations and supervision.
5. Infrastructure development: Investing in the development of infrastructure to support economic growth and improve productivity.

Step 7: Prompt Engineering to Retrieve Macroeconomic Stability Statements

Let's retrieve macroeconomic stability statements.

```
response = qa_interface("List Poverty Reduction Strategies?")
print(response["result"])
```

Output:

Based on the given context, the following poverty reduction strategies can be inferred:

1. Manufacturing for export: Focus on promoting manufacturing industries, especially for export, to generate employment and increase income levels.
2. Skill development: Prioritize skill development alongside education, sanitation, and healthcare to enhance the employability and income potential of the population.
3. Sharing responsibility between states and Centre: Encourage states to bear a reasonable proportion of the financial costs of implementing flagship programs, allowing the central government to allocate more resources to crucial sectors such as defense, railways, national highways, and telecommunications.
4. Urbanization: Address the decay in cities and tap into the wealth cities create to rebuild with a new model of governance.

Step 8: Prompt Engineering to Retrieve Economic Sustainability Budget Statements

Let's retrieve macroeconomic stability statements.

```
response = qa_interface("list economic sustanability statements?")
print(response["result"])
```

Output:

1. The fiscal deficit for 2013–2014 will be contained at 4.6% of GDP.
2. The Current Account Deficit will be contained at USD 45 billion.
3. Efforts have been made to stabilize inflation, with WPI inflation at 5.05% and core inflation at 3.0%.

(continued)

4. The agriculture sector has shown stellar performance, with increased foodgrain production and agricultural GDP growth estimated at 4.6%.
5. The savings rate and investment rate have remained relatively stable.

Step 9: Prompt Engineering with Few Shots to Classify Economic Sustainability Results

We will create a prompt template to query and define responses. Our goal is to classify budget speech statements into six main categories: —economic sustainability, receipts, expenditure, fiscal deficit, capital markets, or reforms. Next, categorize into secondary categories.

delimiter = "####" prompt_clasify_query = f""" Classify each query into a primary category \\ and a secondary category. Provide your output in json format with the \\ keys: primary and secondary. Primary categories: economic sustainability, receipts, \\ expenditure, fiscal deficit, Capital markets, or reforms. economic sustainability secondary categories: Rural Development agricultural and rural sector agricultural credit subsidies public health and wellness crisis inflationary pressures & double digit inflation Poor and Impoverished Minimum Wages Small Scale Industry (SSI) units Inflation infrastructure Industry Development international economic developments """	food and nutrition security housing and community amenities environment and climate change the national savings receipts secondary categories: disinvestment foreign exchange taxes excise duty subsidies import duty small savings excise duty posts & telegraphs department small scale industries exports capital markets treasury bills credit expenditure secondary categories: interest payments subsidies increased fiscal deficit agricultural credit defense grants infrastructure deficit categories: fiscal deficit budget deficit reduced budget deficit Federal State both the Federal and States Balance of Payment Capital markets categories:	grants high inflation rehabilitation of refugees macroeconomic stability human development - education Corporate Governamce Issueances grant tax incentives reforms categories: economic reform foreign exchange reserves banking policy industrial policy currency convertability foreign direct investment debt taxation policy agricultural credit social contributions small scale industries capital inflows tax holiday company taxation small savings exim insurance capital markets

Loop through each result and query the GPT.

```
for budgetspeech_econstament in phrase_to_list:
    classify_prompt=prompt_clasify_query + "Query: \"" + budgetspeech_econstament + "\""
    print(budgetspeech_econstament)
    response = qa_interface(classify_prompt)
    print(response["result"])
```

Output:

```
1. The fiscal deficit for 2013-14 will be contained at 4.6 percent of GDP.
{
  "primary": "fiscal deficit",
  "secondary": "economic sustainability"
}
2. The Current Account Deficit will be contained at USD 45 billion.
{
  "primary": "economic sustainability",
  "secondary": "current account deficit"
}
3. Efforts have been made to stabilize inflation, with WPI inflation at 5.05 percent and core inflation at
3.0 percent.
{
  "primary": "economic sustainability",
  "secondary": "inflation"
}
4. The agriculture sector has shown stellar performance, with increased foodgrain production and
agricultural GDP growth estimated at 4.6 percent.
{
  "primary": "economic sustanability",
  "secondary": "agricultural and rural sector"
}
```

Step 10: Prompt Engineering to Retrieve Social Sustainability Results

Let's construct a prompt template to query and define response. Here we are to instruct budget speech statements into five primary categories—economic sustainability, receipts, expenditure, fiscal deficit, and capital markets.

Let's retrieve social stability statements.

```
response = qa_interface("list social sustainability statements?")
print(response["result"])
```

Output:

1. Ministry of MSME will create the 'India Inclusive Innovation Fund' to promote grassroot innovations with social returns to support enterprises in the MSME sector.

(continued)

2. In order to promote entrepreneurship among the scheduled castes and to provide concessional finance to them, IFCI will set up a Venture Capital Fund for Scheduled Castes.
3. The restructured ICDS has been implemented in 400 districts and will be rolled out in the remaining districts from 1.4.2014.
4. Government has approved the National Agro-Forestry Policy 2014 which has multiple objectives including employment, productivity, conservation, and adaptation.
5. A mechanism for marketing minor forest produce has been introduced, and the Budget has allocated 444.59 crore to continue the scheme in 2014–2015.
6. A new plan scheme with an allocation of 100 crore has been approved to promote community radio stations.

Step 11: Prompt Engineering with Few Shots to Classify Social Sustainability Results

Let's construct a prompt template to query and define response. Our goal is to classify budget speech statements into six primary categories.

```
for budgetspeech_socstament in phrase_to_list:
   classify_prompt=prompt_soc_clasify_query + "Query: \"" + budgetspeech_socstament + "\""
   print(budgetspeech_socstament)
   response = qa_interface(classify_prompt)
   print(response["result"])
```

Output:

```
1. Ministry of MSME will create the 'India Inclusive Innovation Fund' to promote grassroot innovations with social returns to support enterprises in the MSME sector.
{
 "primary": "social sustainability",
 "secondary": "Poor and Impoverished"
}
2. In order to promote entrepreneurship among the scheduled castes and to provide concessional finance to them, IFCI will set up a Venture Capital Fund for Scheduled Castes.
{
  "primary": "social sustanability",
  "secondary": "Poor and Impoverished"
}
3. The restructured ICDS has been implemented in 400 districts and will be rolled out in the remaining districts from 1.4.2014.
{
 "primary": "social sustainability",
 "secondary": "public health and wellness"
}
```

To sum up, classical NLP and LLM are very different. Even though LLM use some classical NLP techniques, the LLMs are much more powerful and will keep improving. The benefit of this power is simpler programming of the LLMs. With few instructions, the models can do a complex and laborious task that classical NLP needed.

Chapter Machine Learning Experiments

ML experiment 1. Develop prompt engineering to extract the following from Budget Speeches 1991–1992, 2000–2001, 2008–2009, and 2019–2020.

- Poverty reduction strategies
- Social sustainability statements
- Economic suitability
- Countries mentioned in the budget speeches
- Water-related strategies

After reading this chapter, you should be able to present large language models. Additionally, you should be able work with LangChain model and GPT-2 model. You should be able to perform prompt engineering. Finally, you should be able to apply prompt engineering on budget speeches.

References

1. Lutz Goedde, Joshua Katz, Alexandre Ménard, and Julien Revellat, "Agriculture's connected future: How technology can yield new growth", October 9, 2020, https://www.mckinsey.com/industries/agriculture/our-insights/agricultures-connected-future-how-technology-can-yield-new-growth, Access Date: 09/18/2020
2. Sofia Bettiza, God and robots: Will AI transform religion?, 21 October 2021, https://www.bbc.com/news/av/technology-58983047, Access Date: June 11, 2023
3. Dr. Prabhat Kumar, Artificial Intelligence: Reshaping Life and Business, Chapter 2, BPB Publications; 1 edition (August 17, 2019), ISBN-10: 9388511077
4. Darrell M. West, "What is artificial intelligence?", Thursday, October 4, 2018, https://www.brookings.edu/research/what-is-artificial-intelligence/, Access Date: November 22, 2020
5. Daniel Faggella, "What is Artificial Intelligence? An Informed Definition", December 21, 2018, https://emerj.com/ai-glossary-terms/what-is-artificial-intelligence-an-informed-definition/#:~:text=In%20%E2%80%9CArtificial%20Intelligence%3A%20A%20Modern,many%20different%20splintered%20fields%20%E2%80%93%20speech, Access Date: October 2020
6. Russell Stuart and Peter Norvig, Artificial Intelligence: A Modern Approach (3rd edition), Publisher: Pearson; 3 edition (December 11, 2009), ISBN-10: 0136042597
7. Chandrasekar Vuppalapati (Author), Artificial Intelligence and Heuristics for Enhanced Food Security (International Series in Operations Research & Management Science, 331) 1st ed. 2022 Edition, Publisher: Springer; 1st ed. 2022 edition (September 17, 2022), ISBN-13: 978-3031087424

8. Jessica Lin Eamonn Keogh Stefano Lonardi Bill Chiu, A Symbolic Representation of Time Series, with Implications for Streaming Algorithms, June 13, 2003,https://www.cs.ucr.edu/~eamonn/SAX.pdf, Access Date: June 19, 2023
9. Matt Taddy, Business Data Science: Combining Machine Learning and Economics to Optimize, Automate, and Accelerate Business Decisions 1st Edition, Publisher: McGraw-Hill Education; 1 edition (August 21, 2019), ISBN-10: 1260452778
10. Kai Hwang and Min Chen, "Big-Data Analytics for Cloud, IoT and Cognitive Computing", Publisher: Wiley; 1 edition (August 14, 2017), ISBN-10: 9781119247029
11. Rani Horev, BERT Explained: State of the art language model for NLP, Nov 10 2018, https://towardsdatascience.com/bert-explained-state-of-the-art-language-model-for-nlp-f8b21a9b6270, Access Date: 11/24/2023
12. Jacob Devlin Ming-Wei Chang Kenton Lee Kristina Toutanova, BERT: Pre-training of Deep Bidirectional Transformers for Language Understanding, May 2019, https://arxiv.org/pdf/1810.04805.pdf, Access Date: 11/24/2023
13. Fabio Chiusano, Two minutes NLP—GLUE Tasks and 2022 Leaderboard, Feb 28 2022, https://medium.com/nlplanet/two-minutes-nlp-glue-tasks-and-2022-leaderboard-517baedfa597#, Access Date: 11/24/2023
14. Ashish Vaswani, Noam Shazeer, Niki Parmar, Jakob Uszkoreit, Llion Jones, Aidan N. Gomez, Łukasz Kaiser, Illia Polosukhin, Attention is All you Need, 2017, https://papers.nips.cc/paper_files/paper/2017/hash/3f5ee243547dee91fbd053c1c4a845aa-Abstract.html, Access Date: 11/24/2023
15. Dzmitry Bahdanau, Kyunghyun Cho, Yoshua Bengio, Neural Machine Translation by Jointly Learning to Align and Translate, Submitted on [1 Sep 2014 (v1), last revised 19 May 2016 (this version, v7)], https://arxiv.org/abs/1409.0473, Access Date: 11/24/2023
16. Alex Graves, Generating Sequences With Recurrent Neural Networks, Submitted on [4 Aug 2013 (v1), last revised 5 Jun 2014 (this version, v5)], https://arxiv.org/abs/1308.0850, Access Date: 11/24/2023
17. Jay Alammar, The Illustrated Transformer, June 27 2018, https://jalammar.github.io/illustrated-transformer/, Access Date: 11/24/2023
18. Hugo Touvron, Louis Martin, Kevin Stone, Peter Albert, Amjad Almahairi, Yasmine Babaei, Nikolay Bashlykov, Soumya Batra, Prajjwal Bhargava, Shruti Bhosale, Dan Bikel, Lukas Blecher, Cristian Canton Ferrer, Moya Chen, Guillem Cucurull, David Esiobu, Jude Fernandes, Jeremy Fu, Wenyin Fu, Brian Fuller, Cynthia Gao, Vedanuj Goswami, Naman Goyal, Anthony Hartshorn, Saghar Hosseini, Rui Hou, Hakan Inan, Marcin Kardas, Viktor Kerkez, Madian Khabsa, Isabel Kloumann, Artem Korenev, Punit Singh Koura, Marie-Anne Lachaux, Thibaut Lavril, Jenya Lee, Diana Liskovich, Yinghai Lu, Yuning Mao, Xavier Martinet, Todor Mihaylov, Pushkar Mishra, Igor Molybog, Yixin Nie, Andrew Poulton, Jeremy Reizenstein, Rashi Rungta, Kalyan Saladi, Alan Schelten, Ruan Silva, Eric Michael Smith, Ranjan Subramanian, Xiaoqing Ellen Tan, Binh Tang, Ross Taylor, Adina Williams, Jian Xiang Kuan, Puxin Xu, Zheng Yan, Iliyan Zarov, Yuchen Zhang, Angela Fan, Melanie Kambadur, Sharan Narang, Aurelien Rodriguez, Robert Stojnic, Sergey Edunov, Thomas Scialom, Llama 2: Open Foundation and Fine-Tuned Chat Models, Submitted on 18 Jul 2023 (v1), last revised 19 Jul 2023 (this version, v2), https://arxiv.org/abs/2307.09288, Access Date: 11/25/2023
19. Shri Manmohan Singh, Budget 1992-93, PUBLISHED TUE, 29th February 1992, https://www.indiabudget.gov.in/doc/bspeech/bs199293.pdf, Access Date: 11/23/2023
20. Shri Manmohan Singh, Budget Speech 1991-1992, 24th July 1991,https://www.indiabudget.gov.in/doc/bspeech/bs199192.pdf, Access Date: November 26, 2023.
21. James Briggs, Llama 2: AI Developers Handbook, 2020, https://www.pinecone.io/learn/llama-2/, Access Date: 11/24/2023
22. Jason Wei, Xuezhi Wang, Dale Schuurmans, Maarten Bosma, Brian Ichter, Fei Xia, Ed Chi, Quoc Le, Denny Zhou, Chain-of-Thought Prompting Elicits Reasoning in Large Language Models, Submitted on 28 Jan 2022 (v1), last revised 10 Jan 2023 (this version, v6)], https://arxiv.org/abs/2201.11903, Access Date: 11/24/2023

23. Pranab Mukherjee, Budget 2012-2013, March 16, 2012, https://www.indiabudget.gov.in/doc/bspeech/bs201213.pdf, Access Date: December 07, 2023
24. TUCKER ARRANTS, Text Generation with HuggingFace—GPT2, 2020, https://www.kaggle.com/code/tuckerarrants/text-generation-with-huggingface-gpt2, Access Date: 11/24/2023
25. PROF. MADHU DANDAVATE, INTRODUCING THE BUDGET FOR THE YEAR 1990-91, PUBLISHED TUE, 1990, https://www.indiabudget.gov.in/doc/bspeech/bs199091.pdf, Access Date: 11/23/2023

Chapter 4
Macroeconomic Indicators, Aggregates, and Framework

This chapter covers the following:

- Large language models' extraction of macroeconomic stabilization parameters:
  - Statutory liquidity ratio (SLR)
  - Bank rate data
  - Cash reserve data
  - Bank credit data
- Fiscal and macroeconomic stabilization features
- Money supply and economic growth model
- Financial development inequality and poverty
- Policy recommendations discussed:
  - The government will use the Public Distribution System to counter inflation and protect the poorer sections of the population from high prices and shortages.
  - Reducing the fiscal deficit and completely eliminating the revenue deficit over the next 5 years.
  - The government decided to begin a phased reduction in the SLR, which would release funds for banks to expand credit to agriculture and industry. This reduction was in line with the proposed reduction in the fiscal deficit.
  - Inflation fighting policy must be transparent, and the Reserve Bank of India must commit to pursuing it.

© The Author(s), under exclusive license to Springer Nature Switzerland AG 2024
C. Vuppalapati, *Assessing Policy Effectiveness using AI and Language Models*,
International Series in Operations Research & Management Science 354,
https://doi.org/10.1007/978-3-031-56097-2_4

This chapter explains statutory liquidity ratio (SLR), bank rate data, cash reserve data, and bank credit data. Then, it discusses ML model with budget deficit, money supply, and inflation linkage and policy implications. Lastly, it helps to derive the policy implications of money supply models.

Economic sustainability means using resources wisely and carefully to achieve long-term economic stability and growth without harming the environment or putting future generations at risk. It requires assessing the environmental effects of economic activity and setting sustainability targets to create a better future. This can vary depending on how an organization changes, such as creating less wasteful systems, finding ways to decrease land use, or improving supply chains. The common idea behind economic sustainability is avoiding wasteful short-term methods and prioritizing the planet's long-term health.

On the other hand, social sustainability includes environmental justice, human health, resource security, and education, among other important social elements of society. It focuses on improving social equality and aims to foster economic and environmental benefits. Promoting sustainable development practices and protecting the environment can help alleviate poverty by ensuring long-term access to natural resources and improving resilience to climate change impacts [1, 2]. For businesses, efforts to generate social sustainability could include focusing company efforts on employee retention instead of economic priorities. For example, investments in the well-being of employees are likely to generate economic benefits for the company by increasing employee motivation. Together, economic sustainability and social sustainability are part of the three pillars of sustainability, which also includes environmental sustainability. True sustainability occurs when these three pillars are balanced (please see Fig. 4.1).

To ensure economic and social sustainability, the poor need to have more economic opportunities, growth, price stability, and rising incomes. Many statistical studies show that fast economic growth is the main way to reduce poverty, using both income and non-income indicators of poverty. Domestic policies also have a significant impact on sustained growth, such as careful macroeconomic management, more open markets, and a stable and predictable environment for private sector activity. Macroeconomic stability is an important condition for higher growth rates and also helps avoid balance of payment crises and high inflation—both of which hurt the poor [3, 4]. High inflation can also slow down economic expansion and limit the chances of poor people getting assets that help them cope with income shocks. The process of getting assets is not only determined by market forces, however. Legal and regulatory structures, as well as political, social, and demographic forces, also affect the ability of poor people to get a variety of financial and human capital assets with high and stable returns. Growth also depends on many factors that are beyond the control of developing countries' governments, such as weather and trade (quotas and terms of trade) and foreign assistance policies in industrial countries [5].

General principles of macroeconomic stability include (please see Fig. 4.2) many factors. The following list would be a healthy measure of items [6]:

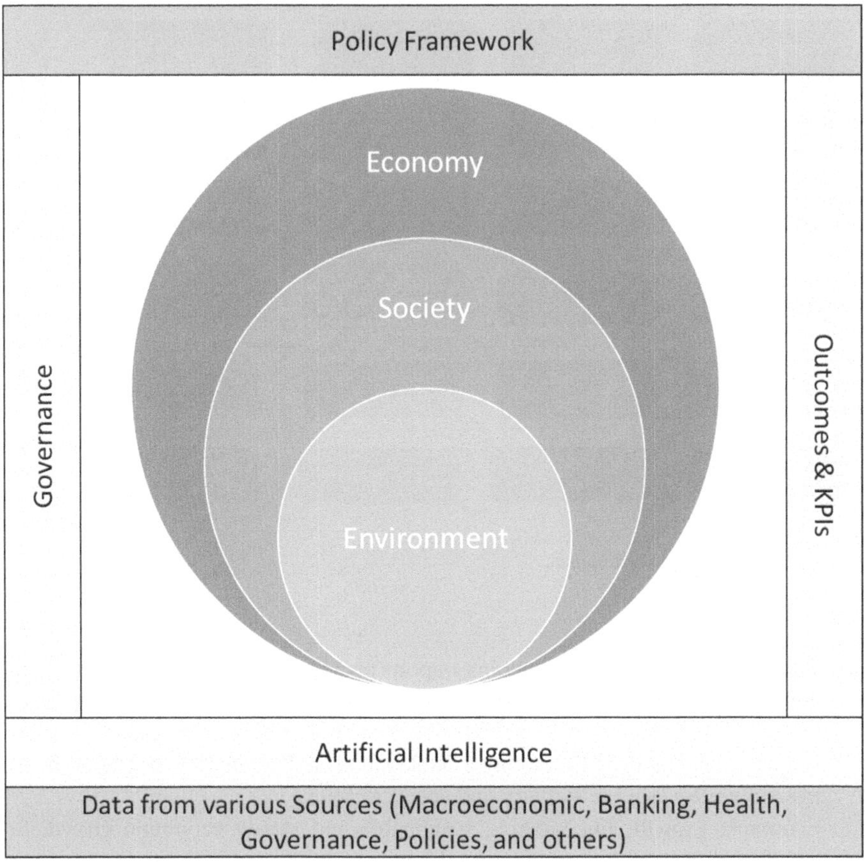

Fig. 4.1 Sustainability

1. Fiscal prudence: Ensuring that government spending and revenue collection are balanced to avoid excessive borrowing and debt accumulation.
2. Monetary stability: Maintaining stable and low inflation through appropriate monetary policy measures, such as controlling money supply and interest rates.
3. Exchange rate stability: Managing the value of the domestic currency to prevent excessive fluctuations and maintain competitiveness in international trade.
4. Price stability: Controlling inflation and preventing excessive price increases to protect the purchasing power of the currency and maintain stable consumer prices.
5. Financial sector stability: Ensuring the stability and soundness of the banking and financial system through regulations and supervision to prevent financial crises and maintain confidence in the system.

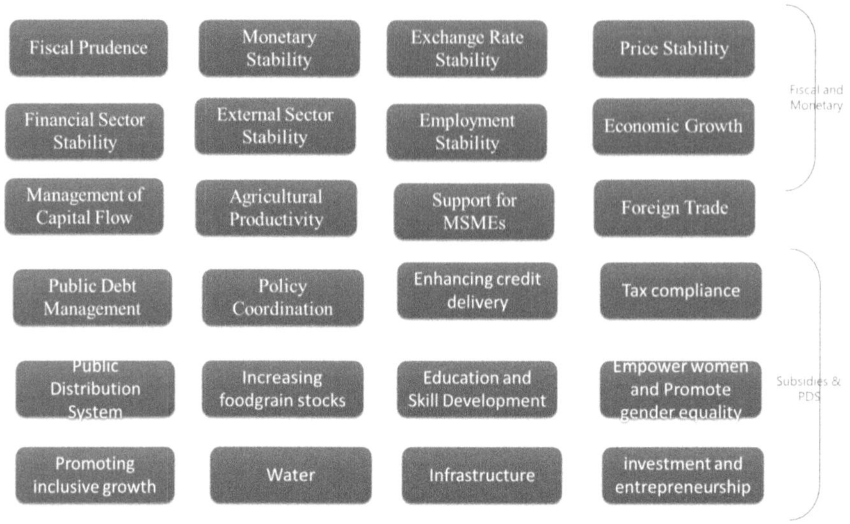

Fig. 4.2 Macroeconomuc strategies

6. External sector stability: Balancing imports and exports to avoid excessive trade deficits or surpluses and managing external debt levels to prevent vulnerability to external shocks.
7. Employment stability: Promoting a stable labor market with low unemployment rates to ensure income stability and social welfare.
8. Economic growth: Encouraging sustainable and steady economic growth to create employment opportunities and improve living standards.
9. Public debt management: Implementing effective debt management strategies to prevent excessive accumulation of public debt and ensure debt sustainability.
10. Policy coordination: Coordinating macroeconomic policies across different sectors and levels of government to ensure coherence and effectiveness in achieving macroeconomic stability.
11. Higher growth: The government aims to achieve sustained growth of a higher threshold that lifts a greater number of people out of poverty within the next 3–4 years.
12. Lower inflation: The government aims to achieve lower levels of inflation.
13. Sustained level of external sector balance: The government aims to maintain a manageable current account deficit.
14. Prudent policy stance: The government aims to introduce fiscal prudence and consolidation to ensure fiscal stability.
15. Expenditure management: The government will establish an Expenditure Management Commission to review and reform government expenditure for maximum efficiency and output.

16. Overhaul of subsidy regime: The government plans to reform the subsidy regime, including food and petroleum subsidies, to make it more targeted and efficient.
17. Increase tax-to-GDP ratio: The government aims to improve the tax-to-GDP ratio and increase non-tax revenues to generate more resources for the economy.
18. Strengthening banking system: The government plans to infuse equity into banks to meet Basel-III norms and raise additional resources. It also aims to increase public ownership in banks through the sale of shares to common citizens.
19. Encouraging investment: The government aims to encourage investment in the economy, including capital investment by PSUs and the creation of a virtuous investment cycle.

> Without price stability, the economy does not work for anyone. In particular, without price stability, we will not achieve a sustained period of strong labor market conditions that benefit all [4].
>
> Jerome H. Powell,
> Federal Chairman

Fiscal and Macroeconomic Stabilization

Economic and social sustainability policies are guided by the long-term vision of the economy and the objective of improving living standards for the people. Economic crisis creates an opportunity to assess, reconsider, and reform policies and thus achieve sustainable development goals. The 1991 economic crisis has transformed and reformed the Indian economy. In 1991, the Finance Minister of India at that time had said his poor nation had started a fundamental shift in priorities that looks, for the first time, to South Korea and wealthy Southeast Asian nations as leading economic examples. "India is now in a new game," said Manmohan Singh, the newly appointed Minister of Finance, a modest international economist who almost admits that he was trying to lead a revolution. "Our country is now ready for big changes, and I think we can start a process of changing the mindsets of our people." With the worst balance-of-payments crisis in the nation's history, *a huge international debt, a plummeting credit rating*, 14% inflation, and shrinking foreign exchange reserves, the mandate was clear—develop policies and incentives to attract high-tech industries and foreign investment, which can provide access to advanced technology and global markets [7, 8]. India was credit rated[1] upper medium grade (A2) at the beginning of 1988. Due to economic crisis, at the time of 1991–1992

[1] India Credit Rating—https://tradingeconomics.com/india/rating

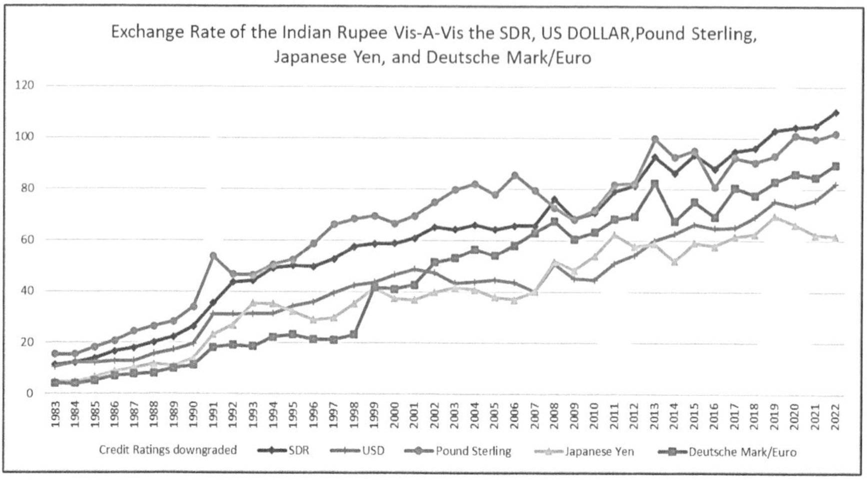

Fig. 4.3 Exchange rate vs. other currencies
†Standard & Poor's (S&P), Moody's, and Fitch: In 1991, S&P downgraded India's long-term foreign currency rating to below investment grade (BB+)
*In 2012, S&P revised India's outlook to negative, but maintained the long-term rating at investment grade (BBB−)
¥In 2016, Fitch affirmed India's long-term foreign currency rating at BBB−, with a stable outlook
‡In 2020, S&P downgraded India's long-term foreign currency rating to the lowest investment grade (BBB−), with a stable outlook

Budget Speech, credit rating[2] was lowered to medium grade (BBB-). It took 3 years, 1994, to move to non-investment grade. Till date, 2023, India's credit rating is lower medium grade. The 1991 economic crisis has had a profound impact on poor and marginalized communities as it increased borrowing costs and increased cost of imports and thus increased balance of payment. Sovereign credit ratings significantly affect bond and stock markets, with average yield spreads increasing by two percentage points and average stock returns decreasing one percentage point after downgrade.[3] They observed that rating changes had stronger effects during crises in both domestic and foreign financial markets (please see Fig. 4.3) [9].

Second, stabilization of economy through measures such as fiscal discipline, inflation control, export promotion, import reduction, financial reform, credit delivery, and better banking services was an acute need of the 1991 economic crisis. The policy outcome was to limit the growth of internal and external debt and reduce the debt servicing burden. These austerity measures were taken with minimal impact on the poor [7].

[2]Credit rating—https://tradingeconomics.com/country-list/rating

[3]Survey—https://www.indiabudget.gov.in/budget202122/economicsurvey/doc/vol1chapter/echap03_vol1.pdf

Policy: Inflation and Public Distribution System (PDS) [7]

The government will use the Public Distribution System to counter inflation and protect the poorer sections of the population from high prices and shortages.

Shri Manmohan Singh
Minister of Finance, 1992–1993[4]

One of the key objectives of economic and social sustainability was to make sure that the adjustment process did not harm the poor by *keeping inflation and prices in check*, *increasing spending on social sectors*, *creating more jobs*, *reducing poverty in rural areas*, and *supporting the weakest and most vulnerable sections of society*. Moreover, allocation of resources for investment in social sectors, especially education, to develop human resources and provide quality education to everyone, including children from disadvantaged backgrounds. Finally, speed up programs for rural development and welfare by fast-tracking programs like Technology Mission for rural development, converting dry latrines quickly, and providing more funds for the rehabilitation and retraining of scavengers [7, 10].

During the 1993–1994 Budget Speech,[5] the government had two major priorities for economic policy: rural development and vital social services such as education and health. The social sustainability measures were clearly directed towards rural empowerment. The government was firmly committed to ensuring that agriculture, agro-processing, and rural development were given top priority in the design of economic policy [11].

Macroeconomic stability ensures sustained growth that helps poor and marginalized communities. Following are some of the important economic stabilizing policies.

Fiscal Discipline

One of the reform policies post-1991 economic crisis was the reduction in fiscal deficit. The objective was to progressively reduce the fiscal deficit of the Central Government, move towards a significant reduction in the revenue deficit, and reduce the current account deficit in the balance of payments [7]. The thrust of the reform process would be to increase efficiency and international competitiveness of industrial production, utilize foreign investment and foreign technology to a greater degree, and improve the performance of the public sector. The government recognized that the fiscal imbalance was the root cause of inflation and other economic problems during the 1991 economic crisis. Efforts were made to reduce the burden of administrative expenditure and consider the effectiveness of subsidies [10]. A

[4] India Budget Speech, 1992–1993—https://www.indiabudget.gov.in/doc/bspeech/bs199293.pdf

[5] Shri Manmohan Singh, Budget Speech 1993–1994—https://www.indiabudget.gov.in/doc/bspeech/bs202021.pdf

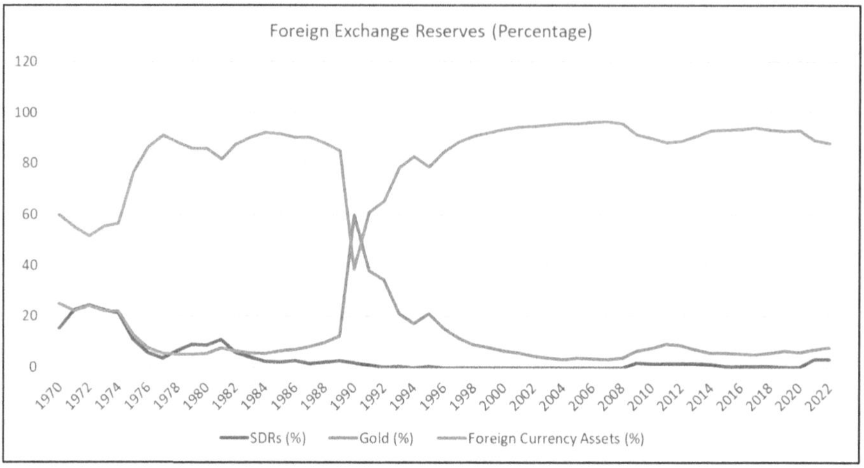

Fig. 4.4 Foreign exchange reserves

decade later, that is in 2001, the reduction in fiscal deficit took central state (please see Fig. 4.4). Reducing the fiscal deficit and completely eliminating the revenue deficit over the next 5 years[6] were budget policies in 2001 [12]. Were the intended results achieved? Except for the surplus year 1981–1982[7] with 988 crores (USD $118 million), all years have an increasing deficit,[8] with the highest deficit reaching 1,820,823 crores (USD $218.427 billion) in 2020–2021 due to COVID-19 (please see Fig. 4.4). The same trend can be applicable to the USA. While revenue increased during the COVID-19 pandemic, from approximately $3.5 trillion in 2019 to $4 trillion in 2021, increased government spending related to widespread unemployment and healthcare caused (please see Fig. 4.5) spikes in the deficit.[9]

Nevertheless, the Central Government deficit to percentage of GDP[10] remains within the range of 46–50%. One reason is growth in GDP along with other macro parameters (please see Fig. 4.6).

[6] Budget Speech 2001–2002—https://www.indiabudget.gov.in/doc/bspeech/bs200102.pdf

[7] Economic situation in 1981–82—https://www.indiabudget.gov.in/budget_archive/es1981-82/1%20Economic%20situation%20in%201981-82.pdf

[8] Combined Deficits of Central and State Governments—https://rbi.org.in/Scripts/PublicationsView.aspx?id=21910

[9] What is the national deficit?—https://fiscaldata.treasury.gov/americas-finance-guide/national-deficit/

[10] Central government debt, total (% of GDP)—https://data.worldbank.org/indicator/GC.DOD.TOTL.GD.ZS

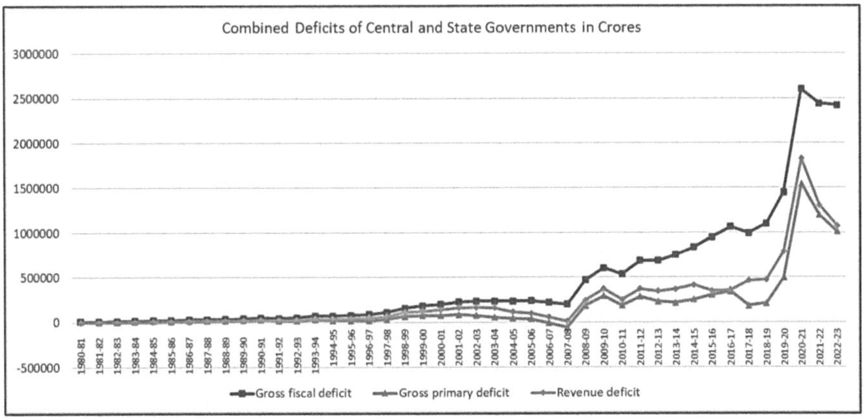

Fig. 4.5 Combined deficit

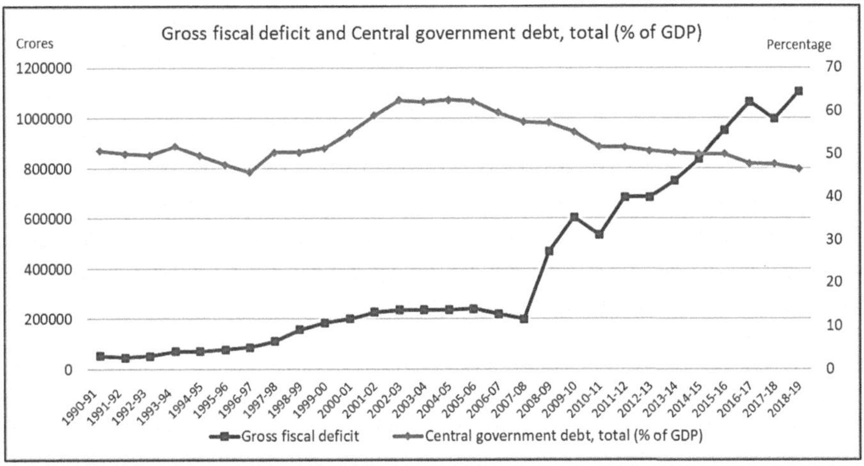

Fig. 4.6 Gross fiscal deficit

Policy: Poverty Reduction Strategy [11]

Reducing the fiscal deficit and ***completely eliminating*** the revenue deficit over the next 5 years.

Shri Yashwant Sinha
Minister of Finance, 2001–2002[11]

[11] Budget Speech 2001–2002—https://www.indiabudget.gov.in/doc/bspeech/bs200102.pdf

Fiscal Correction

To curb inflationary expectations in emerging market economies, it is essential to reduce the fiscal deficit as a percentage of GDP at both the Centre and in the States. This is because persistent debt[12] surprises can raise long term inflation[13]expectations. [14] [13].

Enhancing Credit Delivery

Before the 1991 economic crisis, the Government of India had expressed concern over the low credit-to-deposit ratio in certain regions and urged the Reserve Bank of India to improve credit delivery in those areas while maintaining financial discipline. The credit-to-deposit ratio (CDR), also known as the loan-to-deposit ratio (LDR), is a measure of a bank's liquidity. It is calculated by dividing the bank's total loans disbursed by the total deposits received. In summary, the credit-to-deposit ratio is a crucial indicator of a bank's liquidity and financial health, and it plays a significant role in influencing macroeconomic stability [10].

A traditional signal that a bank may not have enough liquid assets to cover a sudden loss of funding has increased dramatically in rural banks in recent decades. Rural banks' median ratio of the value of their loans outstanding to the value of their deposits has risen from around 62.07% in 1987 to 97.07% in 1990[15] (please see Fig. 4.7) to increase credit available at rural level and prevent rural poverty. In post-1991 economic crisis, the government had attached the highest priority to agriculture and aimed to provide massive thrust to food processing and other agro-based industries to increase the income of farmers, create employment opportunities, diversify the rural economy, and foster rural industrialization, all of which require credit.

For metropolitan areas, on the other hand, liquidity has decreased from 68.07% in 1987 to 58.03% in 1990. The same trend can be observed during the Great Recession of 2008. In hindsight, it can be observed that the increased liquidity at the rural level is a way of ensuring sufficient liquidity and cushioning farmers from the potential economic downturns [14]. Expanding access to financial services, such as savings

[12]Central government debt, total (% of GDP)—India—https://data.worldbank.org/indicator/GC.DOD.TOTL.GD.ZS?locations=IN

[13]Inflation, consumer prices (annual %)—India—https://data.worldbank.org/indicator/FP.CPI.TOTL.ZG?locations=IN

[14]Is High Debt Constraining Monetary Policy? Evidence from Inflation Expectations—https://www.imf.org/en/Publications/WP/Issues/2023/06/30/Is-High-Debt-Constraining-Monetary-Policy-Evidence-from-Inflation-Expectations-534708

[15]Deposits and Credit of Scheduled Commercial Banks—According to Population Group—https://rbi.org.in/Scripts/PublicationsView.aspx?id=21857

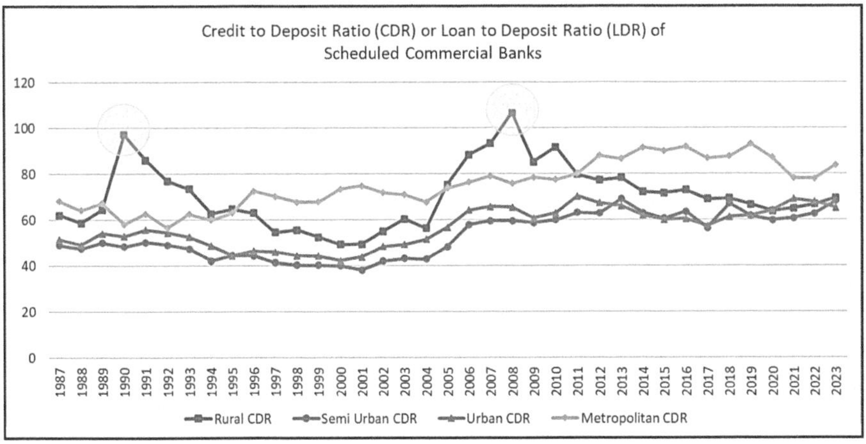

Fig. 4.7 Rural credit

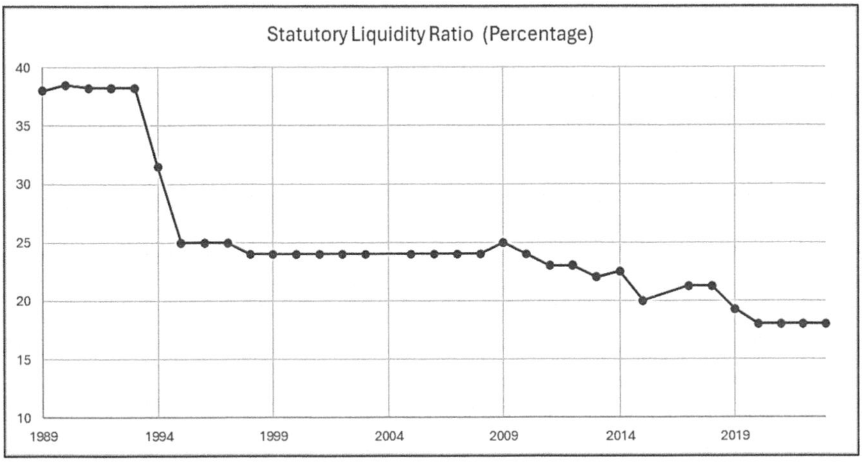

Fig. 4.8 SLR

accounts, microfinance, credit, and insurance, can help individuals and families living in poverty build assets, manage risks, and invest in income-generating activities. The need for fiscal consolidation should be emphasized to eliminate budgetary drags and self-laid traps that hinder economic growth. This can be achieved through efficiently allocating resources, reducing wasteful expenditure, and increasing revenue generation.

Lowering the statutory liquidity ratio (SLR) can enhance credit availability (please see Fig. 4.8). The government in 1993–1994 has decided to begin a phased reduction in the SLR, which would release funds for banks to expand credit to agriculture and industry. This reduction was in line with the proposed reduction in the fiscal deficit. Is the SLR policy effective?

The statutory liquidity ratio data is updated daily, averaging 19.500% from March 1949 to 26 November 2023, with 27,284 observations. The data reached an all-time high of 38.5% in 1989 and a record low of 18.000% in 2023. India's statutory liquidity ratio data[16] is reported by the Reserve Bank of India. Please see SLR data.[17]

> Policy: Reduction in the Statutory Liquidity Ratio (SLR) [11]
>
> The government decided to begin a phased reduction in the SLR, which would release funds for banks to expand credit to agriculture and industry. This reduction was in line with the proposed reduction in the fiscal deficit.
>
> Shri Manmohan Singh
> Minister of Finance, 1993–1994[18]

Machine Learning Model: Money Supply and Economic Growth—SLR, CRR, and BR

In macroeconomics, the money supply (or money stock) refers to the total volume of money held by the public at a particular point in time. Money supply and interest rates have an inverse relationship. A larger money supply lowers market interest rates, making it less expensive for consumers to borrow. Conversely, smaller money supplies tend to raise market interest rates, making it pricier for consumers to take out a loan. As part of the 1991 reforms, the government decided to begin a phased reduction in the SLR, which would release funds for banks to expand credit to agriculture and industry. This reduction was in line with the proposed reduction in the fiscal deficit, in line with J. M. Keynes's thought[19] on the development of modem economic theory—fiscal policy in stabilizing macroeconomics [15]. Having higher interest rates would hold the poor back in poverty. On the other hand, the reduction in SLR and financial deepening contributed to the reduction of rural poverty rates across India by enabling more entrepreneurship in the rural areas and by enticing interstate migration into the tertiary sector [16, 17]. The purpose of the model is to assess policy effectiveness. Before we go deep in data science, let me provide context into monetary policy.

[16] SLR—https://www.rbi.org.in/commonman/English/History/Scripts/BankrateCRRandSLRChanges.aspx

[17] Table 44: Major Monetary Policy Rates and Reserve Requirements—Bank Rate, LAF (Repo, Reverse Repo and MSF) Rates, CRR & SLR—https://rbi.org.in/Scripts/PublicationsView.aspx?id=21850

[18] India Budget Speech, 1993–1994—https://www.indiabudget.gov.in/doc/bspeech/bs199394.pdf

[19] Fiscal Deficits and Balance of Payments Disequilibrium in IMF Adjustment Programs—https://www.elibrary.imf.org/display/book/9780939934287/ch06.xml

The economic growth and the development in India depend a lot on the scheduled commercial banks. Joseph Schumpeter (1911) was the first to propose the link between financial development and economic growth in his work on "Financial Development and Economic Growth." He said that the services that financial institutions offer, like collecting savings, assessing projects, handling risk, overseeing managers, making transactions, etc., are crucial for any economy. It stimulates technological innovation and therefore overall economic growth and development [18, 19].

The Indian banking sector has seen major changes since the early 1990s. The reforms in the sector were triggered by the economic reforms during that period. The banking sector reforms were part of the financial sector reforms that the Government of India started in the early 1990s. The main banking sector reforms were carried out by the Reserve Bank of India and the Government of India in 1991 and 1998 with the guidance of two reports, both chaired by M. Narasimham, who is arguably the most influential banker in post-independence India. The two Committees that he led—the Narasimham Committee on Financial System[20] (1991) and the Narasimham Committee on Banking Sector Reforms (1998)—are still the basic documents for any discussion on banking sector reforms and banking policy. He is also known for laying the groundwork for significant moves such as bank mergers, the emergence of new-generation private banks, and asset reconstruction companies [20–22].

The banking sector reforms aimed to improve operational flexibility, strengthening, competitiveness, and technological and institutional infrastructure. The Indian banking sector has changed significantly and quickly in the post-reform period, and the second committee recommendation was to modernize the banking sector, improve regulation and supervision, and adopt prudential norms. The financial and banking sector reforms have led to fundamental changes in the functioning of commercial banks in India. The reforms have altered the organizational structure, ownership pattern, and operational activities of scheduled commercial banks and exposed them to a competitive environment. The reforms have required all banking sectors to adjust their banking activities to face these challenges [22].

Bank Rate

The bank rate is the rate that the central bank charges for lending money to commercial banks. It affects the interest rates that commercial banks charge their borrowers. When the bank rate is higher, commercial bank interest rates are also higher. The bank rate is one of the ways that the central bank can control the amount of money or inflation in the economy.

Cash Reserve Ratio

The Reserve Bank of India, in order to maintain the monetary stability in the country, has set the cash reserve ratio (CRR) for the scheduled commercial banks according

[20]Report of the Committee on Banking Sector Reforms. Reserve Bank of India, Committee Report, Ministry of Finance, Government of India—https://ibbi.gov.in/uploads/resources/Narasimham%20Committee%20I-min.pdf

to the master circular of the RBI Act,[21] 1934, without any minimum or maximum rate. The cash reserve ratio was very high, around 15%, during the financial and banking sector reforms in India, 1991, but it was lowered gradually based on the committee recommendation. Now, in 2023, the CRR rate is 4.5% of the total demand and time liabilities (DLT) of the scheduled commercial banks. In case of default on daily maintenance, the penalty is 3% above the bank rate applied on the number of days of default multiplied by the amount by which the amount falls short of the prescribed level.

Statutory Liquidity Ratio
As per Section 24 of the Banking Regulations Act, 1949, the scheduled commercial banks must invest in unencumbered government and approved securities a certain minimum amount as statutory liquidity ratio (SLR) every day. Besides investment in unencumbered government and other approved securities, gold, cash, and surplus CRR balance also count as liquid assets for SLR purposes. The SLR was very high when banking sector reforms started in India, which limited the lending ability of the scheduled commercial banks. However, due to the committee recommendation and execution by relevant authorities, the SLR has been lowered gradually. At present, 2023, every scheduled commercial bank has to maintain at least 19% SLR.

The model uses multiple regression models to test the selected banking variables by using the time series data collected from the RBI. The study found that there is an inverse relationship between CRR and bank credit and SLR and bank credit, but there is no association between bank rate by the RBI and bank credit by the scheduled commercial banks in India since the reform period. The purpose of the machine learning model is to codify and explain the association of bank rate (BR), cash reserve ratio (CRR), and statutory liquidity ratio (SLR) with bank credit and its influence on the overall macroeconomic stability.

Financial development, inequality, and poverty
Machine Learning

Model_MonetaryPolicy_StatutoryLiquidityRatio(SLR)_and_BankCredit_India.ipynb

The goal of the model is in line with the proposed financial sector and banking reform policies: implement the recommendations of the Narasimham Committee on Financial System in a phased manner. Focus on improving the financial health of the banking system by addressing issues such as low profitability, poor portfolio quality, and inadequate provisioning for bad debts. Introduce better capital adequacy norms and provisions for bad debts. Encourage a more efficient and competitive banking system, including a larger role for the private sector [7].

[21] Master Circular—Cash Reserve Ratio (CRR) and Statutory Liquidity Ratio (SLR)—https://www.rbi.org.in/commonperson/English/Scripts/Notification.aspx?Id=1476#11

Step 1: Load the Libraries

Install required libraries.

```
import numpy as np
import pandas as pd
from functools import reduce
from sklearn.model_selection import train_test_split
from sklearn.linear_model import LinearRegression
from sklearn.metrics import r2_score
from sklearn import metrics
from sklearn.impute import SimpleImputer
import eli5
import warnings
warnings.filterwarnings("ignore")

import numpy as np
import pandas as pd
import functools
from datetime import date
import plotly.graph_objects as go
import seaborn as sns; sns.set()
import statsmodels.api as sm
from statsmodels.tsa.arima_process import ArmaProcess
from scipy import stats
from scipy.stats import pearsonr
import matplotlib.pyplot as plt
import plotly.express as px
from sklearn.preprocessing import MinMaxScaler
from sklearn.model_selection import train_test_split
from sklearn.linear_model import LinearRegression
from sklearn.metrics import r2_score
from sklearn import metrics
from sklearn.impute import SimpleImputer
import eli5
```

For eli5, if it's not available, pip install eli5.

Step 2: Load Statutory Liquidity Ratio (SLR) Data

Load SLR data.

```
SLR_df= pd.read_csv("SLR.csv")
SLR_df
```

```
SLR_df["Year"] = SLR_df["Year"].astype(int)
```

```
SLR_df.dtypes
```

Output

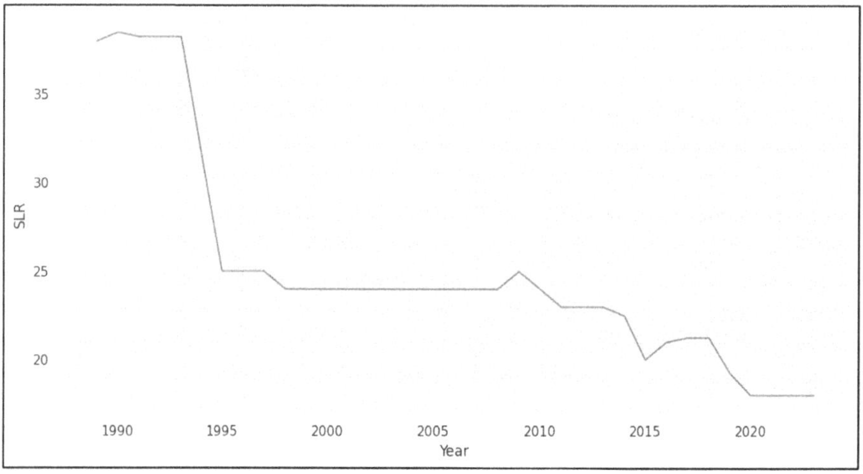

Step 3: Load Bank Rate Data

The Reserve Bank of India maintains bank rate to control monetary policies. For one, factor in maintaining high rates (1995–1998 and 2013–2015) is the resilience of the Indian economy, which continues to grow at a strong clip even as the global economy shows signs of slowing down.[22]

```
BankRate_df= pd.read_csv("BankRate.csv")
BankRate_df
```

```
BankRate_df["Year"] = BankRate_df["Year"].astype(int)
```

```
BankRate_df
```

[22] India Central Bank Stands Pat as Higher Inflation Looms—https://www.wsj.com/articles/india-central-bank-maintains-policy-rate-9e9f71a7

Output

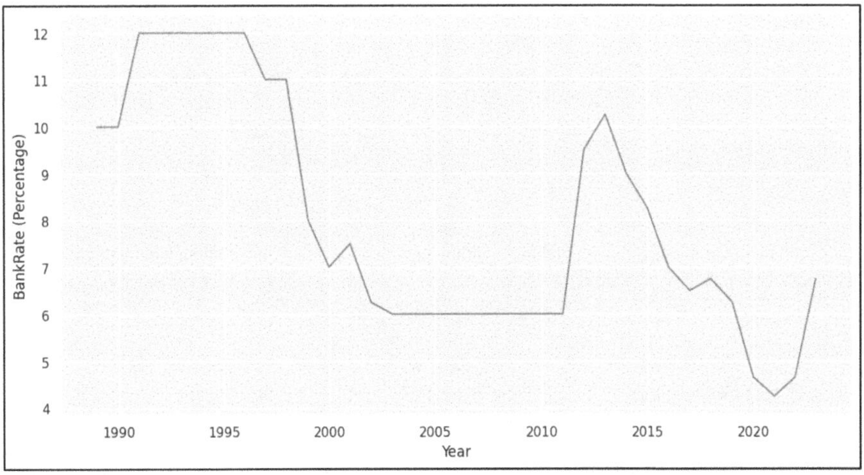

Step 4: Load Bank Credit Data

Bank credit refers to the total amount of money a person or business can borrow from a bank or other financial institution. A rise in bank credit leads to an increase in the money supply, which can lead to inflation. Therefore, by controlling bank credit through the CRR, SLR, and bank rate, the RBI can control inflation and stabilize the economy. Bank credit growth drives robust resource flow to India's commercial sector. Bank credit growth reflects factors such as swings in economic activity, the availability and cost of non-bank funding, the relative cost of bank funding, the banking sector's health, and risk appetite.[23]

```
BankCredit_df= pd.read_csv("BankCredit.csv")
BankCredit_df
```

```
BankCredit_df["Year"] = BankCredit_df["Year"].astype(int)
```

```
BankCredit_df.dtypes
```

[23] RBI report. Bank credit growth drives robust resource flow to India's commercial sector—https://www.thehindubusinessline.com/money-and-banking/bank-credit-growth-drives-robust-resource-flow-to-indias-commercial-sector/article67314951.ece

Output

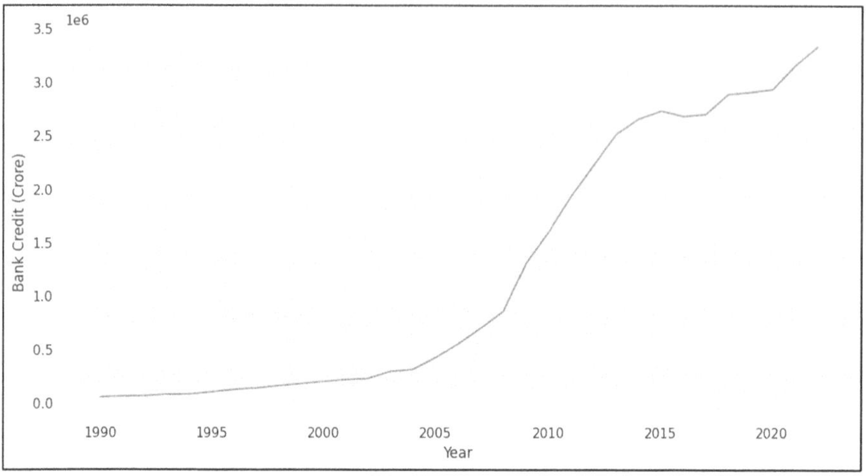

Step 5: Load Cash Reserve Data

Cash reserve ratio (CRR) is a mandatory reserve that every commercial bank has to keep with the RBI, which is a certain percentage of their net demand and time liabilities. The CRR is held in the form of cash balances with the RBI. The main purpose of CRR is to provide some liquid cash against the depositor's money so that the bank can meet the depositor's needs. During high inflation in the economy, the RBI increases the CRR percentage for the commercial banks and lowers the bank's loanable fund that helps to control inflation. Let's load cash reserve ratio data.[24]

```
CashReserveRatio_df= pd.read_csv("CashReserveRatio.csv")
CashReserveRatio_df
```

```
CashReserveRatio_df["Year"] = CashReserveRatio_df["Year"].astype(int)
```

```
CashReserveRatio_df.dtypes
```

[24] RBI report. Bank credit growth drives robust resource flow to India's commercial sector—https://www.thehindubusinessline.com/money-and-banking/bank-credit-growth-drives-robust-resource-flow-to-indias-commercial-sector/article67314951.ece

Output

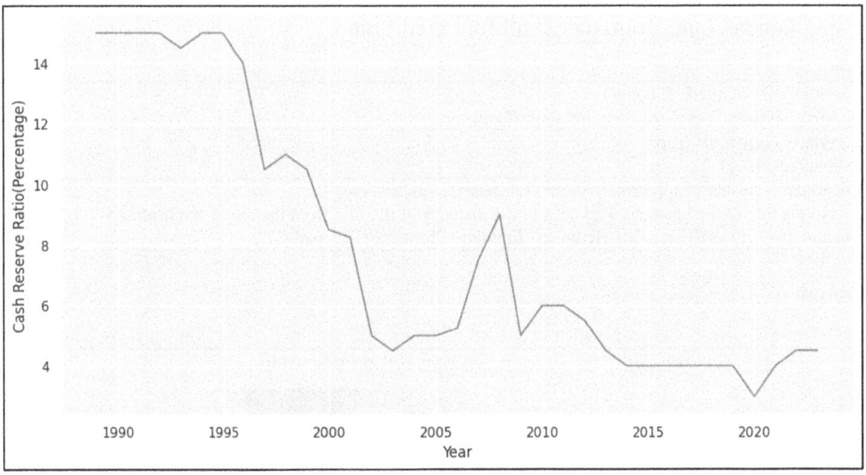

Step 6: Merge CRR, SLR, Bank Rate, and Bank Credit

Using Pandas Inner Join, merge all four credit rates.

```
dfs = [SLR_df,BankRate_df,BankCredit_df,CashReserveRatio_df]

merged_df =reduce(lambda left, right : pd.merge(left, right, on = "Year", how= "inner"),dfs)
merged_df
```

Output

	Year	SLR	BankRate (Percentage)	Bank Credit (Crore)	Cash Reserve Ratio(Percentage)
0	1990	38.50	10.00	53805	15.00
1	1991	38.25	12.00	61689	15.00
2	1992	38.25	12.00	65240	15.00
3	1993	38.25	12.00	78662	14.50
4	1994	31.50	12.00	80482	15.00
5	1995	25.00	12.00	102310	15.00
6	1996	25.00	12.00	124937	14.00
7	1997	25.00	11.00	138548	10.50
8	1998	24.00	11.00	161038	11.00

Step 7: Correlation Analysis

Using Pandas Inner Join, merge all four credit rates.

```
cormat_df = merged_df.copy()
cormat_df.drop('Year', axis =1, inplace = True)
cormat= cormat_df.corr()
round(cormat,2)
heatmap = sns.heatmap(cormat, vmin=-1, vmax=1, annot=True)
# Give a title to the heatmap. Pad defines the distance of the title from the top of the heatmap.
heatmap.set_title('Correlation Heatmap', fontdict={'fontsize':12}, pad=12);
```

Output

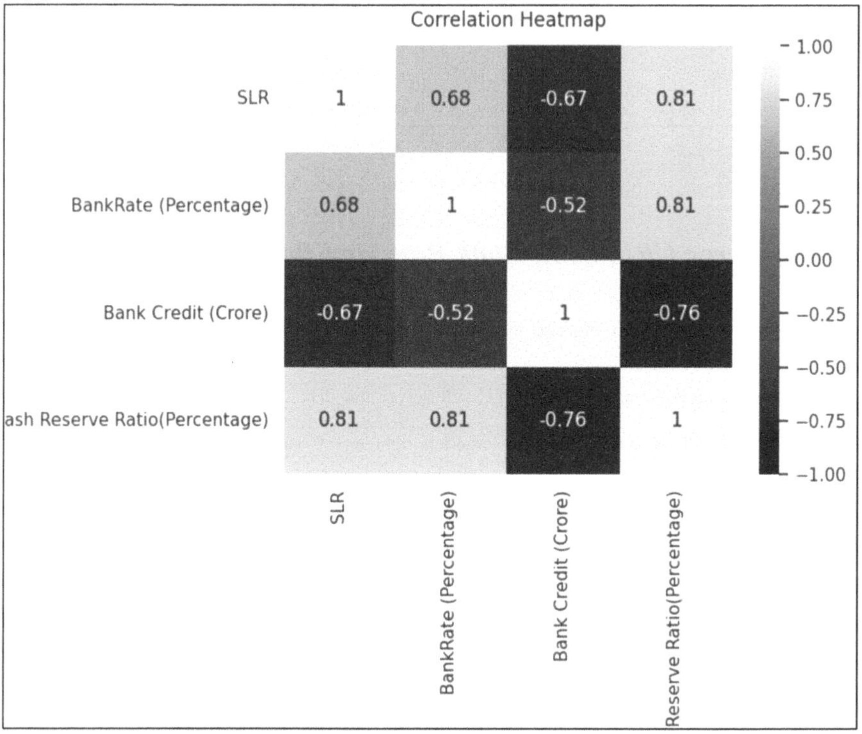

Please note:

- CRR and bank credit: There is an inverse relationship between CRR and bank credit (−76%). This means when the CRR increases, the amount of loans and advances decreases. This is because an increase in CRR reduces the amount of funds available with banks for lending [20, 23].

- SLR and bank credit: Similar to CRR, there is also an inverse relationship (−67%) between SLR and bank credit. When SLR increases, the amount of loans and advances decreases. This is because an increase in SLR reduces the amount of funds available with banks for lending [20, 23].
- Bank rate and bank credit: There is no significant association between the bank rate set by the Reserve Bank of India (RBI) and bank credit by the scheduled commercial banks in India since the reform period. However, this could have changed recently due to COVID-19.
- CRR and SLR: Both CRR and SLR have a significant impact on loans and advances (+82%). If CRR and SLR rates increase, then loans and advances will decrease [23].

Please note that these relationships are based on historical data and trends, and they may not necessarily predict future behaviors accurately. Also, these relationships can be influenced by various other factors such as economic conditions, policy changes, etc.

Step 8: Linear Regressor Model

Let's build a linear regressor model and see its feature variables.

```
numeric_features=[ 'SLR', 'BankRate (Percentage)',
    'Cash Reserve Ratio(Percentage)'
        ]
```

```
X = merged_df[[ 'SLR', 'BankRate (Percentage)',
    'Cash Reserve Ratio(Percentage)'
        ]]

y= merged_df['Bank Credit (Crore)']

X_train,X_test, y_train , y_test = train_test_split(X ,y, test_size = 0.2,random_state=0 )

regressor = LinearRegression()
regressor.fit(X_train,y_train)

y_pred = regressor.predict(X_test)
y_pred

r2_score(y_test,y_pred)
```

Output
0.37826 or 37.826%

Step 8: Evaluate Linear Regressor Model Metrics

Let's see the model metrics.

```
import pandas as pd
from sklearn import metrics
from sklearn.metrics import mean_squared_error, r2_score

mse = mean_squared_error(y_test, y_pred)
print("MSE:", mse)

rmse = np.sqrt(mse)
print("RMSE:", rmse)

r2 = r2_score(y_test, y_pred)
print("R2:", r2)
```

Output

```
MSE: 752956586171.5973
RMSE: 867730.7106306641
R2: 0.3782633215501954
```

Observe a lower accuracy and a higher MSE.

Step 8: Model Explainability

Let's run model explainability analysis. The output clearly points to a higher BIAS, which is unexplained by the model, a fact that can corroborate with higher MSE values.

Weight	Feature
0.8596 ± 0.4783	SLR
0.0929 ± 0.4846	Cash Reserve Ratio(Percentage)
0.0475 ± 0.0879	BankRate (Percentage)

Step 9: Gradient Boosting Regressor Model

Let's build a linear regressor model and see its feature variables.

```
# high accuracy and low bias
from sklearn.ensemble import GradientBoostingRegressor
X = merged_df[numeric_features]

y= merged_df['Bank Credit (Crore)']

X_train,X_test, y_train , y_test = train_test_split(X ,y, test_size = 0.2,random_state=0 )

modelGradientBoostingRegressor = GradientBoostingRegressor(n_estimators=5000,
random_state=0,loss='squared_error')
modelGradientBoostingRegressor.fit(X_train,y_train)

y_pred = modelGradientBoostingRegressor.predict(X_test)
y_pred

r2_score(y_test,y_pred)
```

Output
0.8846 or 88.46%

Step 10: Evaluate Linear Regressor Model Metrics

Let's see the model metrics.

```
import pandas as pd
from sklearn import metrics
from sklearn.metrics import mean_squared_error, r2_score

mse = mean_squared_error(y_test, y_pred)
print("MSE:", mse)

rmse = np.sqrt(mse)
print("RMSE:", rmse)

r2 = r2_score(y_test, y_pred)
print("R2:", r2)
```

Output

```
MSE: 139732392493.31946
RMSE: 373807.96205179935
R2: 0.8846191730357721
```

Observe the higher accuracy of the model.

Step 11: Model Explainability

Let's check model explainability.

```
eli5.show_weights(modelGradientBoostingRegressor)
```

```
eli5.show_weights(modelGradientBoostingRegressor, feature_names =list(X.columns), target_names =
"Poverty Rate in India" )
```

Output

Weight	Feature
0.8944 ± 0.4131	SLR
0.0572 ± 0.3132	BankRate (Percentage)
0.0484 ± 0.4354	Cash Reserve Ratio(Percentage)

Model explainability clearly shows the influence of SLR on the bank credit.

Step 12: Random Forest Regressor

Let's construct random forest regressor.

```
from sklearn.ensemble import RandomForestRegressor
X = merged_df[numeric_features]

y= merged_df['Bank Credit (Crore)']

X_train,X_test, y_train , y_test = train_test_split(X ,y, test_size = 0.2,random_state=0 )

modelRandomForestRegressor = RandomForestRegressor(max_depth=2, random_state=0)
modelRandomForestRegressor.fit(X_train,y_train)

y_pred = modelRandomForestRegressor.predict(X_test)
y_pred

r2_score(y_test,y_pred)
```

```
import pandas as pd
from sklearn import metrics
from sklearn.metrics import mean_squared_error, r2_score

mse = mean_squared_error(y_test, y_pred)
print("MSE:", mse)

rmse = np.sqrt(mse)
print("RMSE:", rmse)

r2 = r2_score(y_test, y_pred)
print("R2:", r2)
```

```
eli5.show_weights(modelRandomForestRegressor)
eli5.show_weights(modelRandomForestRegressor, feature_names =list(X.columns), target_names =
"Poverty Rate in India" )
```

Output

```
0.9595979778746982
MSE: 48929023666.0094
RMSE: 221199.0589175492
R2: 0.9595979778746982
```

Weight	Feature
0.8596 ± 0.4783	SLR
0.0929 ± 0.4846	Cash Reserve Ratio(Percentage)
0.0475 ± 0.0879	BankRate (Percentage)

Conclusions and Policy Implications

The result shows that the model is well fit for data, and it is considered a good regression model. Bank rate, CRR, and SLR have significant influences on inflation during normal as well as crisis periods (1991–1992, 2008–2009, and 2021–2022) [24].

This model validates through data science techniques the inflation and the RBI rate policies from 1991 to 2021 and provides intuition that the RBI rate policies affect inflation during both crisis periods and normal periods [24]. In additional to monetary policies, expanding access to banking services, i.e., enhancing financial inclusion, particularly for marginalized sections of society, through initiatives like opening bank accounts for every household and providing credit facilities would ensure sustained social and economic sustainability [25].

Policy: Reduction in the Statutory Liquidity Ratio (SLR) [25]

The government decided to begin a phased reduction in the SLR, which would release funds for banks to expand credit to agriculture and industry. This reduction was in line with the proposed reduction in the fiscal deficit.

Shri Manmohan Singh
Minister of Finance, 1993–1994[25]

Final_Machine Learning
Modeling_BudgetDeficitsMoneyGrowthAndinflation
MonetaryPloicyLinkageModelDataAll.ipynb

[25] India Budget Speech, 1993–1994—https://www.indiabudget.gov.in/doc/bspeech/bs199394.pdf

Machine Learning Model: Budget Deficit, Money Supply, and Inflation Linkage and Policy Implications

The main goal of the monetary policies of most central banks in the world is to keep price stability. Therefore, it is very important to know what factors affect inflation, and this topic has been widely studied and discussed by researchers and policymakers. Inflation means a rapid and continuous rise in the price level and is driven by a high growth rate of the money supply. A budget deficit can be a source of inflation, but it depends on how long the deficit lasts and how it is financed. On one hand, a short-term budget deficit can cause only a short-term increase in the price level regardless of how the deficit is funded. On the other hand, if budget deficits are long term and financed by money creation, inflation occurs. For example, the central bank and commercial banks buy government bonds, leading to an increasing money supply. Long-term budget deficits do not lead to money expansion if government bonds are exchanged only between firms and households. The aim of the model is to show the relationship among budget deficits, money growth, inflation, and GDP and combined impact on economic and social sustainability.

The main policy implication of our model is as follows. Government deficits and money supply are key causes of the price level as per the model. Both the deficit and changes in the money supply positively impact on and cause inflation. Over the past couple of decades (1990–2020), India's budget deficit has risen sharply, and money supply has also been on the rise, as shown in Fig. 4.9. Given that one of the policy objectives is to maintain low inflation rates so as to achieve social sustainability, our findings suggest that controlling budget deficits is one way of achieving this objective [26].

The correlation between increases in M1—money held by the general public—and inflation becomes apparent only when the time frame is extended beyond a 1–3-year period. Over this longer time frame, the correlation becomes almost perfect. The negative correlation here implies that whenever the M1 growth increases, the CPI

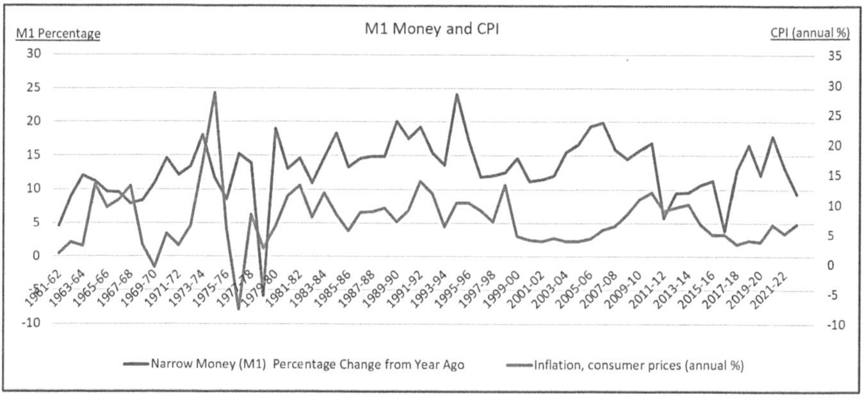

Fig. 4.9 Money supply

growth actually decreases. The reason for this is probably that deflation is counteracted by the Reserve Bank by money printing, and this causes M1 to increase during deflation [12]. The monetary policy initiative shows the results 10 years after the policy initiative. By 2020, the prices of essential commodities, and of manufactured products as a whole, remained stable. Inflation, excluding energy, was around 6% during the year. The economy remained secure with record levels[26] of foreign exchange reserves and public food stocks [12].

The government has benefited from the lower SLR and increased money supply to carry out programs such as a comprehensive policy on agricultural credit, accountability of sponsor banks for regional rural banks (RRBs), and microfinance through self-help groups (SHGs) [1]. The comprehensive policy on agricultural credit aimed to double the credit flow in 3 years, from 2004 to 2007. This policy involved the participation of public and private sector banks, regional rural banks (RRBs), and cooperative banks in executing the policy. The purpose was to enhance the social and economic well-being of rural areas. With the availability of money, the government provided insurance for farming and livestock, ensuring the viability of the agricultural sector.

Data Sources

Source	Time frequency	Source
Inflation rate, average consumer prices, annual percent change	The average consumer price index (CPI) is a measure of a country's average level of prices based on the cost of a typical basket of consumer goods and services in a given period. The rate of inflation is the percentage change in the average CPI[27] Source: World Economic Outlook (October 2023)	
Real GDP growth, annual percent change[28]	Gross domestic product is the most commonly used single measure of a country's overall economic activity. It represents the total value at constant prices of final goods and services produced within a country during a specified time period, such as 1 year Source: World Economic Outlook (October 2023)	
Current account balance, percentage of GDP[29]	The current account is the record of all transactions in the balance of payments covering the exports and imports of goods and services, payments of income, and current transfers	

(continued)

[26] Budget Speech 2001–2002—https://www.indiabudget.gov.in/doc/bspeech/bs200102.pdf

[27] IMF Inflation—https://www.imf.org/external/datamapper/PCPIPCH@WEO/IND

[28] GDP Growth—https://www.imf.org/external/datamapper/NGDP_RPCH@WEO/FR_FCWORLD/IND

[29] Current account balance, percent of GDP—https://www.imf.org/external/datamapper/BCA_NGDPD@WEO/IND

Source	Time frequency	Source
	between residents of a country and nonresidents Source: World Economic Outlook (October 2023)	
Reserve Bank of India	Money stock—currency with the public (% of GDP)[30] Money stock—time deposits (% GDP)[31]	

Step 1: Load Libraries for performing Time Series and ML Modeling

Load scikit-learn and Matplot libraries.

```
import pandas as pd
import numpy as np
import plotly.graph_objects as go
import seaborn as sns; sns.set()
import matplotlib.pyplot as plt
from scipy import stats
from scipy.stats import pearsonr
import matplotlib.pyplot as plt
from sklearn.preprocessing import MinMaxScaler
from sklearn.model_selection import train_test_split
from sklearn.linear_model import LinearRegression
from sklearn.ensemble import RandomForestClassifier
from sklearn.metrics import r2_score
from sklearn import metrics
import eli5
import matplotlib.pyplot as plt
import matplotlib as mpl
import numpy as np
import scipy.stats as spstats
import seaborn as sns
import utils
%matplotlib inline
mpl.style.reload_library()
mpl.style.use('classic')
mpl.rcParams['figure.facecolor'] = (1, 1, 1, 0)
mpl.rcParams['figure.figsize'] = [6.0, 4.0]
mpl.rcParams['figure.dpi'] = 100
```

[30] Money Stock—Currency with the Public—https://rbi.org.in/Scripts/PublicationsView.aspx?id=21846

[31] Money Stock—Currency with the Public—https://rbi.org.in/Scripts/PublicationsView.aspx?id=21846

Step 2: Load Broad Money Growth (Annual %), Money Stock, Inflation, GDP, and Deficit Data for India

Load broad money growth (annual %), money stock, inflation, GDP, and deficit data for India.

```
BudgetDeficitsMoneyGrowthAndinflationDF=
pd.read_csv("FinalBudgetDeficitsMoneyGrowthAndinflationMonetaryPloicyLinkageModelDataAll.csv")
```

Output

	Year	Broad money growth (annual %)	Money Stock - Currency with the Public (Crore)_log	Current account balance, percent of GDP (Percent of GDP)
0	1990	15.066614	4.724669	-2.9
1	1991	18.316952	4.786027	-0.4
2	1992	16.862495	4.834249	-1.7
3	1993	17.007663	4.915405	-0.4
4	1994	20.277728	5.002948	-1.0
5	1995	11.011121	5.072831	-1.6
6	1996	18.736274	5.120860	-1.1
7	1997	17.656754	5.163099	-1.3
8	1998	18.173015	5.227743	-0.9
9	1999	17.149180	5.276650	-1.0

As can be seen, the DataFrame consists Money-Banking, Financial Markets, Public Finances, and Trade-Balance of Payments Linkage Data for India from 1990 to 2022.

The following variables can be found:

Banking and money markets:

- Money stock—currency with the public (% of GDP)
- Money stock—time deposits (% GDP)

Macroeconomic variables:

- Current account balance, percent of GDP (percent of GDP)
- Inflation rate, average consumer prices (annual percent change)
- General government net lending/borrowing (percent of GDP)

Step 3: Check for Nulls in the Data

Check any variables that need imputations by running Is NULL() on the DataFrame.

```
BudgetDeficitsMoneyGrowthAndinflationDF.isnull().sum() ## missing values
```

Output

PDS-Off-Take(Crore)	0
Institutional Credit for Agriculture and Allied Activities - Loans Issued (Crore)	0
Bank Credit to Commercial Sector (Crore)	0
Population growth (annual %)	0
GDP growth (annual %)	0
Money Stock - Currency with the Public (% of GDP)	0
Money Stock - Time Deposits(% GDP)	0
Current account balance, percent of GDP (Percent of GDP)	0
Foreign Trade Non-Oil Exports Crore)	0
Foreign Trade Non-Oil Imports Crore)	0
Total Foreign Investment Inflows (Crore)	0
Net Inflow of Aid (Crore)	0
Inflation rate, average consumer prices (Annual percent change)	0
General government net lending/borrowing (Percent of GDP)	0

No null found in the data.

Step 4: Correlation Analysis

Let's perform correlation analysis to see how the variables interact with each other.

```
cormat=BudgetDeficitsMoneyGrowthAndinflationDF.corr()
plt.figure(figsize=(12, 9))

sns.heatmap(cormat)
heatmap = sns.heatmap(cormat, vmin=-1, vmax=1, annot=True)
# Give a title to the heatmap. Pad defines the distance of the title from the top of the heatmap.
heatmap.set_title('Correlation Heatmap', fontdict={'fontsize':12}, pad=12);
```

Output

The relationships among broad money growth, money stock currency with public, current account balance, inflation rate, GDP, and gross fiscal deficit in India can be summarized as follows:

- Broad money growth: Broad money is a measure of the amount of money circulating in an economy. Central banks often monitor broad money growth to help forecast inflation (42% correlation between money growth and inflation).
- Money stock currency with public: This is part of the money supply, which includes cash and other types of deposits that can be used as easily as cash. The nominal GDP tends to rise with the money supply (6% correlation between GDP and money stock).
- A rise in growth rate, real effective exchange rate, fiscal deficit, trade openness, institutional quality, financial market development, and stage of development generates larger current account deficits. A rise in terms of trade, inflation rate (representing macroeconomic stability), and crude oil export reduces the current account deficits [27].
- Inflation rate: Inflation results when money growth exceeds real GDP growth. The inflation rate is directly proportional to money growth (inflation and money growth +42% correlated).
- GDP: The gross domestic product (GDP) is a measurement of the total value of all the finished goods and services produced within a country's borders within a specified period of time. The nominal GDP tends to rise with the money supply (6% correlated).
- Gross fiscal deficit: A rise in the fiscal deficit can lead to larger current account deficits [27].

These relationships are complex and can be influenced by a variety of factors, including government policies, global economic conditions, and domestic economic factors. It's also important to note that these relationships can vary from country to country and over time.

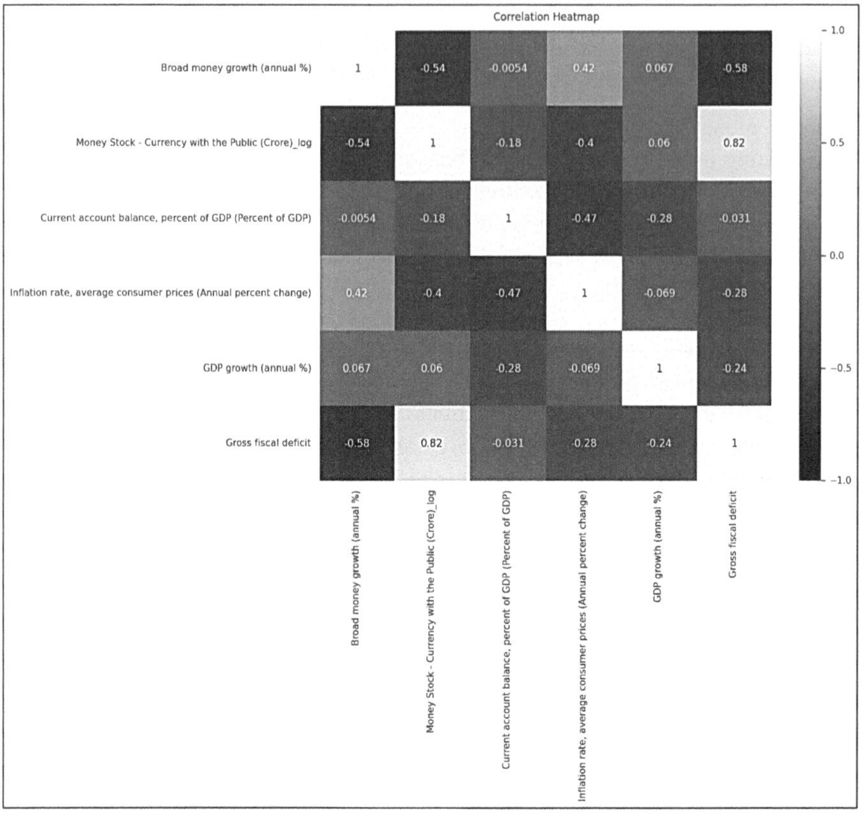

Step 5: Plot Dataset

Plot dataset.

```
# Plot
fig, axes = plt.subplots(nrows=4, ncols=4, dpi=120, figsize=(10,16))
for i, ax in enumerate(axes.flatten()):
 data = dataset[dataset.columns[i]]
 ax.plot(data, color='red', linewidth=1)
 ax.set_title(dataset.columns[i])
 ax.xaxis.set_ticks_position('none')
 ax.yaxis.set_ticks_position('none')
 ax.spines["top"].set_alpha(0)
 ax.tick_params(labelsize=6)
plt.tight_layout();
```

Output

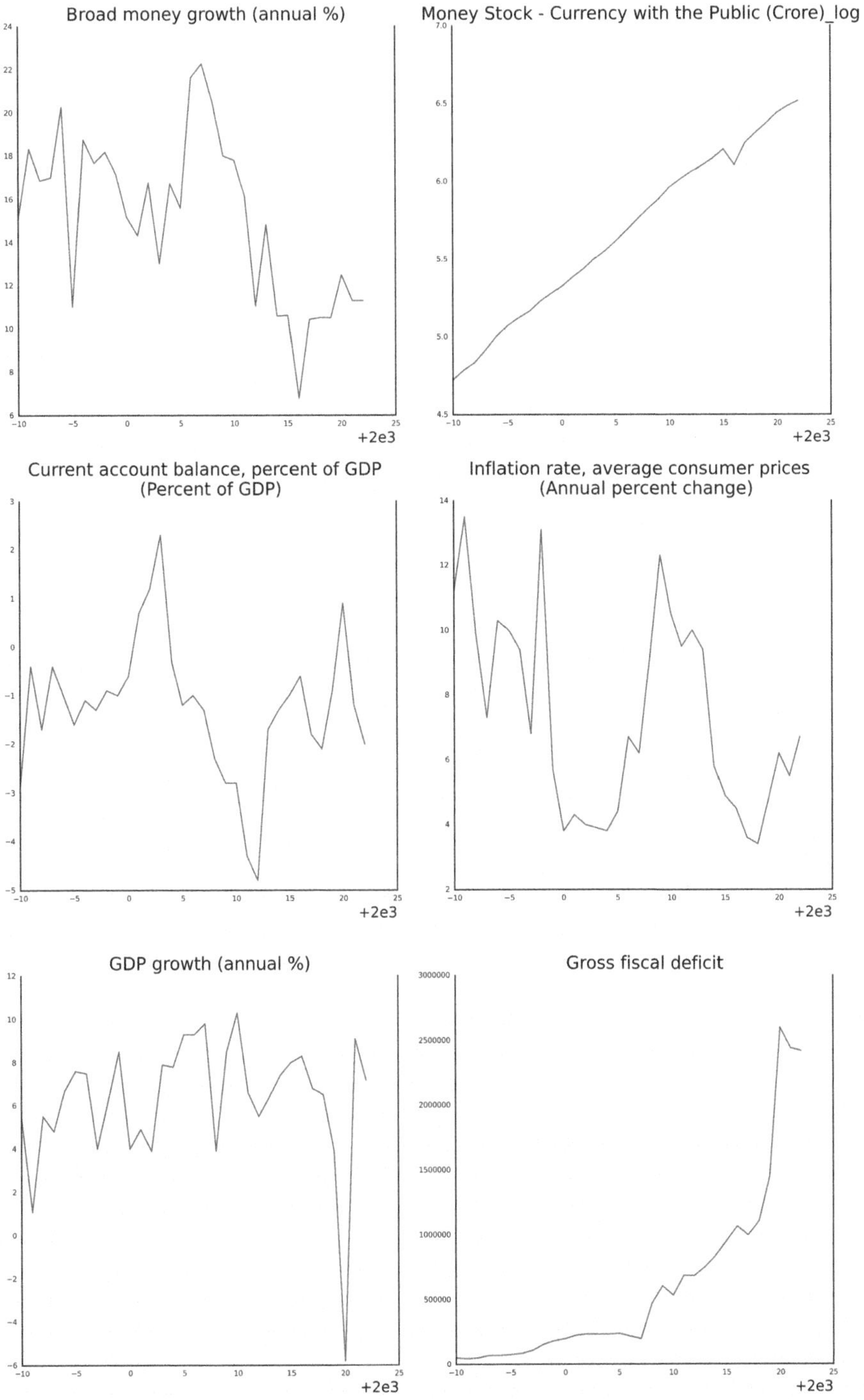
Broad money growth (annual %)
+2e3
Money Stock - Currency with the Public (Crore)_log
+2e3
Current account balance, percent of GDP
(Percent of GDP)
+2e3
Inflation rate, average consumer prices
(Annual percent change)
+2e3
GDP growth (annual %)
+2e3
Gross fiscal deficit
+2e3

Step 6: Transformation

There are several transformations available in Python's NumPy library including logarithms, square roots, and more. I have created a new column for a few of these transformations to validate the stationarity of the data that I will cover later in the chapter. No single transformation method will universally turn all-time series stationary; you will have to test them for yourself.

- Broad money growth (annual %): No transformation applied.
- Money stock: Currency with the public (crore) is transformed using log function.
- Current account balance, percent of GDP (percent of GDP): No transformation applied.
- GDP growth (annual %), inflation rate, and gross fiscal deficit: No transformation applied.

These transformations are required to pass the normality tests.

Step 7: Normality Test

We want to get the most information from our data, so it is good to have a normal or Gaussian distribution of the data. We have done a normality test to see if this is true, based on the idea of null and alternate hypothesis. Augmented Dickey-Fuller Tests assume certain things about your data. They can only tell you how much evidence you have to reject or not reject a null hypothesis. The result has to be interpreted for a specific problem to make sense. However, they can give you a quick check and confirm if your time series is stationary or non-stationary. A unit root test is a kind of statistical test that Augmented Dickey-Fuller Test[32] belongs to. The idea behind a unit root test is that it measures how much a time series follows a trend. There are different kinds of unit root tests, and Augmented Dickey-Fuller Test may be one of the more common ones. It uses a model that is self-regressing and chooses the best information criteria for different lag values.

[32] How to Check if Time Series Data is Stationary with Python—https://machinelearningmastery.com/time-series-data-stationary-python/

```
from statsmodels.tsa.stattools import adfuller

def stationary_test(df,col):
   print("Observations of Dickey-fuller test")
   dftest = adfuller(df[col],autolag='AIC')
   dfoutput=pd.Series(dftest[0:4],index=['Test Statistic','p-value','#lags used','number of observations
used'])
   for key,value in dftest[4].items():
      dfoutput['critical value (%s)'%key]= value
   print(dfoutput)
   if dfoutput[0] < dfoutput[4] or dfoutput[0] < dfoutput[5] or dfoutput[0] < dfoutput[6]:

      print(f"{col}: The series is stationary ")
   else :
      print(f"{col}: The series is not stationary ")
```

```
for i in dataset.columns:
   stationary_test(dataset,i )
```

- p-value >0.05: Fail to reject the null hypothesis (H0); the data has a unit root and is non-stationary.
- p-value ≤0.05: Reject the null hypothesis (H0); the data does not have a unit root and is stationary.

Output

Observations of Dickey-fuller test Test Statistic -1.314549 p-value 0.622490 #lags used 1.000000 number of observations used 31.000000 critical value (1%) -3.661429 critical value (5%) -2.960525 critical value (10%) -2.619319 dtype: float64 Broad money growth (annual %): The series **is not stationary**.	Observations of Dickey-fuller test Test Statistic -1.228633 p-value 0.661159 #lags used 1.000000 number of observations used 31.000000 critical value (1%) -3.661429 critical value (5%) -2.960525 critical value (10%) -2.619319 dtype: float64 Money Stock - Currency with the Public (Crore)_log: The series is **not stationary**

In summary, the following is the stationarity of the variables:

- Current account balance, percent of GDP (percent of GDP): The series is stationary.
- Inflation rate, average consumer prices (annual percent change): The series is stationary.
- GDP growth (annual %): The series is stationary.
- Gross fiscal deficit: The series is not stationary.

Step 8: Check for Autocorrelation

It is essential to check for serial correlation within the historical data.

```
import pandas as pd
import matplotlib.pyplot as plt
import statsmodels.api as sm
MAX_LAGS=1
# plots the autocorrelation plots for each stock's price at 150 lags
for i in dataset:
    sm.graphics.tsa.plot_acf(dataset[i])
    plt.title('ACF for %s' % i)
plt.show()plt.show()
```

Output

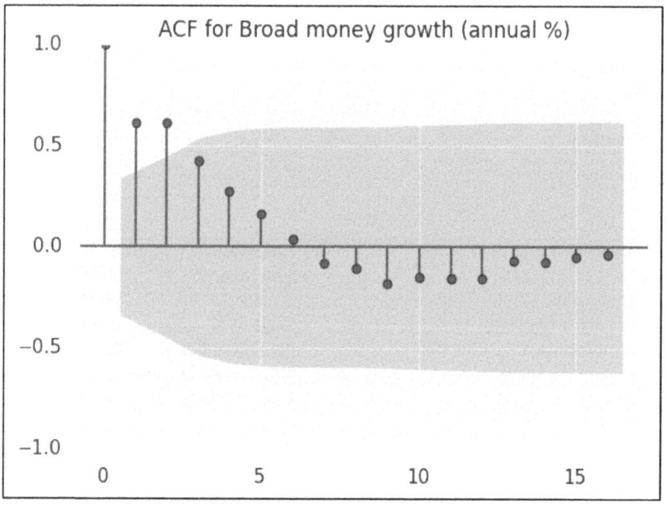

Other series also show autocorrelation of +1 which represents a perfect positive correlation, which means an increase seen in one time series leads to a proportionate increase in the other time series. We need to apply transformation and neutralize this to make the series stationary. It measures linear relationships; even if the autocorrelation is minuscule, there may still be a nonlinear relationship between a time series and a lagged version of itself.

Step 10: Train and Test Data

The VAR model will be fitted on X_train and then used to forecast the next N observations. These forecasts will be compared against the actual present in test data. Observation could vary based on the frequency of the data.

```
n_obs=2
X_train, X_test = dataset[0:-n_obs], dataset[-n_obs:]
print(X_train.shape, X_test.shape)
```

Output

(31,14) (2,14)

Step 11: Transformation

Applying the first difference on training set to make all the series stationary. However, this is an iterative process where, after first differencing, the series may still be non-stationary. We shall have to apply a second difference or log transformation to standardize the series in such cases.

```
transform_data = X_train.diff().dropna()
transform_data.head()
```

Output

	Broad money growth (annual %)	Money Stock - Currency with the Public (Crore)_log	Current account balance, percent of GDP (Percent of GDP)
Year			
1991	3.250339	0.061358	2.5
1992	-1.454457	0.048222	-1.3
1993	0.145168	0.081156	1.3
1994	3.270066	0.087542	-0.6
1995	-9.266607	0.069883	-0.6

Step 12: Stationarity Check

Rerun Augmented Dickey-Fuller Test.

```
limport statsmodels.tsa.stattools as sm
def augmented_dickey_fuller_statistics(time_series):
    result = sm.adfuller(time_series.values, autolag='AIC')
    print('ADF Statistic: %f' % result[0])
    print('p-value: %f' % result[1])
    print('Critical Values:')
    for key, value in result[4].items():
        print('\t%s: %.3f' % (key, value))
```

```
for i in dataset.columns:
    stationary_test(transform_data,i )
```

Output

```
Augmented Dickey-Fuller Test: Broad money growth (annual %)
ADF Statistic: -9.463425
p-value: 0.000000
Critical Values:
        1%: -3.679
        5%: -2.968
        10%: -2.623
Augmented Dickey-Fuller Test: Money Stock - Currency with the Public (Crore)_log
++++++++++++++++++++++++++++++++
ADF Statistic: -3.200493
p-value: 0.019956
Critical Values:
        1%: -3.738
        5%: -2.992
        10%: -2.636
Augmented Dickey-Fuller Test: Current account balance, percent of GDP (Percent of GDP)
```

In summary, the following is the stationarity of the variables:

- Broad money growth (annual %): The series is stationary.
- Money stock—currency with the public (crore)_log: The series is stationary.
- Current account balance, percent of GDP (percent of GDP): The series is stationary.
- Inflation rate, average consumer prices (annual percent change): The series is stationary.
- GDP growth (annual %): The series is stationary.

Step 12: Plot Visualization to See Variables Stationarity

```
fig, axes = plt.subplots(nrows=4, ncols=4, dpi=120, figsize=(10,6))
for i, ax in enumerate(axes.flatten()):
    d = transform_data[transform_data.columns[i]]
    ax.plot(d, color='red', linewidth=1)
    # Decorations
    ax.set_title(dataset.columns[i])
    ax.xaxis.set_ticks_position('none')
    ax.yaxis.set_ticks_position('none')
    ax.spines['top'].set_alpha(0)
    ax.tick_params(labelsize=6)
plt.tight_layout();
```

Output

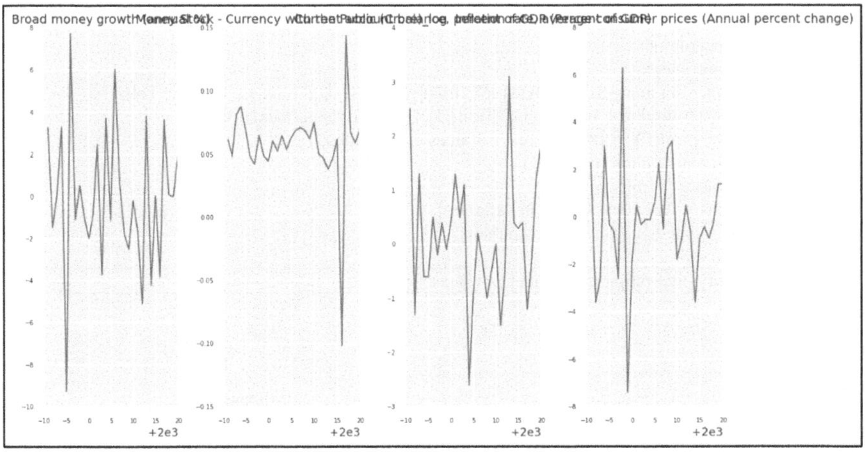

Step 13: Granger's Causality Test

Prof. Clive W.J. Granger, who won the 2003 Nobel Prize in Economics, created the concept of causality to enhance the accuracy of forecasting. It is essentially an econometric speculative test for checking the validity of using one variable to predict another in multivariate time series data with a specific delay.

Before doing the Granger causality test,[33] the data must be stationary, i.e., it should have a fixed mean, fixed variance, and no seasonal component. Make the non-stationary data stationary by subtracting it, either once or twice. Do not continue with the Granger causality test if the data is still not stationary after subtracting it twice.

[33] Granger Causality in Time Series—Explained using Chicken and Egg problem—https://www.analyticsvidhya.com/blog/2021/08/granger-causality-in-time-series-explained-using-chicken-and-egg-problem/

The formal definition of Granger causality can be explained as whether past values of *x* aid in the prediction of *y*(*t*), conditional on having already accounted for the effects on *y*(*t*) of past values of *y* (and perhaps of past values of other variables). If they do, the *x* is said to "Granger cause" *y*. So, the basis behind VAR is that each of the time series in the system influences each other.

Granger's causality tests the null hypothesis that the coefficient of past values in the regression equation is zero. So, if the *p*-value obtained from the test is lesser than the significance level of 0.05, then, you can safely reject the null hypothesis. This has been performed on the original dataset.

```
import statsmodels.tsa.stattools as sm
maxlag=MAX_LAGS
test = 'ssr-chi2test'
def grangers_causality_matrix(X_train, variables, test = 'ssr_chi2test', verbose=False):
  dataset = pd.DataFrame(np.zeros((len(variables), len(variables))), columns=variables,
index=variables)
  for c in dataset.columns:
    for r in dataset.index:
      test_result = sm.grangercausalitytests(X_train[[r,c]], maxlag=maxlag, verbose=False)
      p_values = [round(test_result[i+1][0][test][1],4) for i in range(maxlag)]
      if verbose: print(f'Y = {r}, X = {c}, P Values = {p_values}')
      min_p_value = np.min(p_values)
      dataset.loc[r,c] = min_p_value
  dataset.columns = [var + '_x' for var in variables]
  dataset.index = [var + '_y' for var in variables]
  return dataset
```

```
grangers_causality_matrix(dataset, variables = dataset.columns)
```

Output

	Broad money growth (annual %)_x	Money Stock - Currency with the Public (Crore)_log_x	Current account balance, percent of GDP (Percent of GDP)_x	Inflation rate, average consumer prices (Annual percent change)_x	GDP growth (annual %)_x	Gross fiscal deficit_x
Broad money growth (annual %)_y	1	0.0162	0.8202	0.8836	0.4305	0.02
Money Stock - Currency with the Public (Crore)_log_y	0.954	1	0.9353	0.6189	0.9421	0.9033
Current account balance, percent of GDP (Percent of GDP)_y	0.1858	0.2258	1	0.223	0.3918	0.2954
Inflation rate, average consumer prices (Annual percent change)_y	0.1664	0.468	0.0436	1	0.8093	0.5718
GDP growth (annual %)_y	0.3539	0.7826	0.4372	0.608	1	0.9538
Gross fiscal deficit_y	0.5567	0.0707	0.591	0.3927	0.6763	1

- There is no bidirectional causation among any variables.

Step 14: VAR Model

VAR requires stationarity of the series which means the meaning of the series does not change over time (we can find this out from the plot drawn next to Augmented Dickey-Fuller Test). So, I will fit the VAR model on the training set and then use the fitted model to forecast the next 14 observations. These forecasts will be compared against the actual present in test data. I have taken the maximum lag (1) to identify the required lags for VAR model [27].

```
import numpy as np
import pandas
import statsmodels.api as sm
from statsmodels.tsa.api import VAR

mod = VAR(transform_data)
results = mod.fit(maxlags=1)
print(results.summary())
```

Output

```
Correlation matrix of residuals
                                                                  Broad money growth (annual %)  Money Stock
Broad money growth (annual %)                                                          1.000000
Money Stock - Currency with the Public (Crore)_log                                     0.279495
Current account balance, percent of GDP (Percent of GDP)                               0.104002
Inflation rate, average consumer prices (Annual percent change)                        0.257898
GDP growth (annual %)                                                                 -0.272778
Gross fiscal deficit                                                                   0.301946
```

Step 15: Residual Plot

Plot the residuals.

```
y_fitted = results.fittedvalues
y_fitted

y_fitted = results.fittedvalues
residuals = results.resid
plt.figure(figsize = (15,5))
plt.plot(residuals, label='resid')
plt.plot(y_fitted, label='VAR prediction')
plt.xlabel('Date')
plt.xticks(rotation=45)
plt.ylabel('Residuals')
plt.grid(True)
```

Output

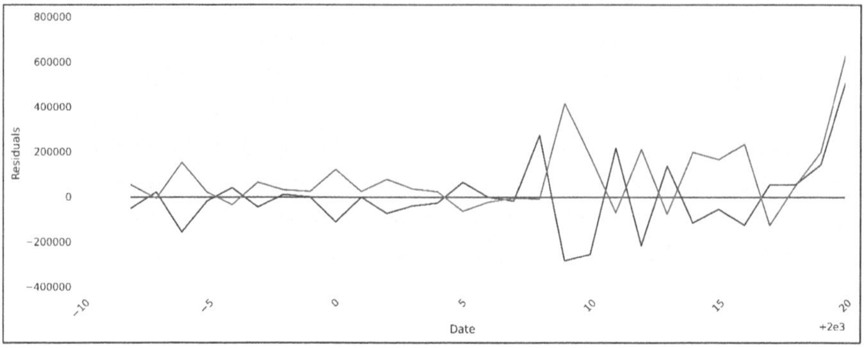

It is evident that there are price spikes during specific economic events.

Conclusion

First, inflation rose from 2008 through 2011 in response to a positive shock to money growth. In particular, the strongest effect of a positive shock to money growth on inflation was observed in this period. Second, budget deficits had no effect on money growth and, therefore, inflation, implying a certain level of independence between the monetary policy and the fiscal policy. Third, money growth decreased in or after 1 year in response to an increase in inflation. This implies that it took double-digit months for the State Bank of India to fully respond to inflation shocks. The Reserve Bank of India aimed to lower money growth to curb inflation in the period[34] 2008–2012. Fourth, budget deficit growth was unaffected by shocks to real GDP growth to the interest rate in the short run because the budget deficit growth was quite persistent. Fifth, the interest rate was not an effective instrument for the Reserve Bank of India to combat inflation because interest rate changes had no effect on inflation. Finally, inflation and interest rates in India were persistent because they were highly expected and retentively memorized. This requires that an inflation-fighting policy must be transparent and that the Reserve Bank of India must commit to pursuing it. As indicated in the 2014[35] Budget Speech, containing fiscal deficit at 4.6% of GDP, containing current account deficit, adding USD to foreign exchange reserves, and focusing on fiscal stability and maintaining a target of 3% fiscal deficit of GDP by 2016–2017 were the government's strategy [28], all of which require inflation fighting strategies at the central bank.

[34]India Central Bank Stands Pat as Higher Inflation Looms—https://www.wsj.com/articles/india-central-bank-maintains-policy-rate-9e9f71a7?tpl=cb

[35]Budget Speech Budget 2013–2014—https://www.indiabudget.gov.in/doc/bspeech/bs201314.pdf

Policy Recommendation

Inflation fighting policy must be transparent and that the Reserve Bank of India must commit to pursuing it.

Author

Impacts of Financial Development and Other Factors on Inequality and Poverty

Many people in the world escape from poverty every year. They benefit from regional development and economic possibilities like new employment, innovations, and enterprises that help them achieve more secure economic lives. However, many people still face a persistent cycle of poverty that is hard to break. We think that financial marginalization is a major cause of this cycle.

Around 1.7 billion people in the world lack access to formal financial services like savings, payments, insurance, and credit. In developing economies, only 63% of adults have an account, and women—almost 1 billion of them—face more barriers to accessing beneficial financial systems.

Many households in poverty mostly use cash for all their transactions. This means their savings are in physical assets, such as animals or jewelry. Cash can be spent, animals can die, and jewelry can be lost or stolen. Also, these forms of savings do not earn any interest and can actually depreciate over time. To transfer money to relatives, those without a bank account have to depend on couriers or friends who carry cash by bus, which is costly, unsafe, and slow. To get money in a crisis, they must go to moneylenders who charge very high interest rates [29].

In India, many farmers buy and sell livestock and cattle. Without reliable digital records, they face the risk of being cheated and losing their money. This leads to many of them becoming poor. Cow Necklace, a sensor for animal health, helps to create digital records and make the transactions successful, thus raising the quality of life at the grassroots level.

When people lack official records of their financial activity, they also lose access to possible benefits like establishing credit or obtaining a loan to launch a business. And it's more challenging to cope with common financial problems, such as severe sickness, a bad crop, or an economic slowdown. Financial exclusion often makes the cost of poverty hard to escape [29].

By connecting people and making it easier to move money and share information, Digital public infrastructure (DPI) is in many ways the modern-day equivalent of the roads and bridges that helped reshape economies in the nineteenth century.

Kanwaljit Singh, Ph.D.
Senior Program Officer
Bill & Melinda Gates Foundation

The main goal of the model is to explore how financial development affects inequality and poverty. There is no agreement in the literature about how financial development influences inequality[36] and poverty, even though this topic is frequently discussed. Some studies say that financial development reduces poverty and makes income distribution fairer, while others say the opposite. The studies that support the positive effect of financial development say that expanding financial services in the economy gives the poor more access to finance opportunities. This leads to more equal income distribution and lower poverty levels. Financial development impacts poverty and inequality in two ways: directly and indirectly. Economic growth is the indirect result of financial development. The argument is that financial development causes economic growth and higher economic growth, which then reduces poverty and improves inequality [30].

Furthermore, it is believed that economic growth, inequality, and poverty are strongly connected [31, 32]. By increasing the average incomes of the poorest 20% of society proportionately, growth causes a change in poverty [33]. It is revealed that policies that promote growth bring about benefits that are mostly in favor of the poor, which is known as pro-poor growth. If policies are pro-poor, their impact on reducing poverty will be much more significant. For instance, the question whether fintech [34] and its sub-measures reduce poverty in China or not and their findings show that fintech complements economic growth and financial development in the country and, therefore, helps to reduce poverty. The same can be established in India with the intervention of fintech many small and marginalized communities who didn't have access to financial markets historically, can access now and have integrated as part of their day-to-day operations [35].

India is a developing country that has faced different economic situations after the 1990s, and financial stability and economic stability are important factors for sustainable development, income distribution, and poverty reduction. Therefore, understanding the role of financial development in these issues can help policymakers. Although the effect of financial development on inequality and poverty is widely studied in the literature, there are few machine learning models that focus on the Indian case. Hence, the results of the ML model not only address this gap in the literature but also offer a chance to compare the results with other models that concentrate on the Indian economy and poverty reduction.

With this respect, several model analyses are employed to demonstrate the relationship between financial development, poverty, and inequality. First, correlation matrices and the Granger causality test are utilized. According to the correlation matrices, financial development, poverty, and inequality are negatively correlated.

Simultaneous equation regression models are used to produce clear results. These equations solve the endogeneity problem and give reliable and unbiased results. The model findings agree with theoretical expectations. Model shows that financial development has a positive effect on poverty and inequality, and thus, for India,

[36]The impact of financial development on income inequality and poverty—https://www.ncbi.nlm.nih.gov/pmc/articles/PMC10602251/

financial development helps to achieve more equal distribution and lower poverty levels. It is inferred that the growth of financial services leads to poverty reduction and betterment in income inequality as it enables improvement for the restricted access of the poor to financial services.

When the findings of this model are compared with the previous literature about India, it is clear that the obtained results are consistent with the previous research. Besides, the model is not only consistent with the theoretical expectations and previous empirical studies about India, but also as it focuses on a long period, it gives an opportunity to examine the relationship between financial development, inequality, and poverty for the last four decades. In addition, the model contributes to the literature by employing a methodology which is not utilized for the Indian economy before. In that sense, compared to the other models, it is the first one that used simultaneous regressive model equations for the Indian case. For further research, to make a comparison with the empirical findings, in the model, different inequality and poverty measures could be utilized to reveal the impact of financial development on inequality and poverty in more detail.

Main macroeconomic and financial indicators:

- GDP growth
- Inflation rate
- Interest rate
- Borrowing requirement (PSBR) as a percentage of total income
- Current account balance (as % of GDP)
- Money supply (M2) as % of GDP
- Private bank credits
- Total TL deposits
- Real effective rate (% change)
- Nominal exchange rate (% change)

Policy: The paper aims to examine the effects of financial technology (Kisan Credit Card and ATM) on India's poverty and empirically analyze the influence of financial technology on poverty scores. We consider data from 22 Indian states for 3 fiscal years, i.e., from 2018 to 2020. The current study uses panel data analysis to test the hypothesis. The results suggest a negative association between financial technology (Kisan Credit Card and ATM) and its influence on the poverty score in India. The current study suggests that the government should pay more attention to expanding financial technology, including increasing economic growth and financial stability and eliminating poverty by investing in infrastructure that would enhance banking services [36].

Share Prices

India's equity market is thriving as investors bet on one of the few bright spots in a weak global economy. The country's stocks are so in demand that India now has the

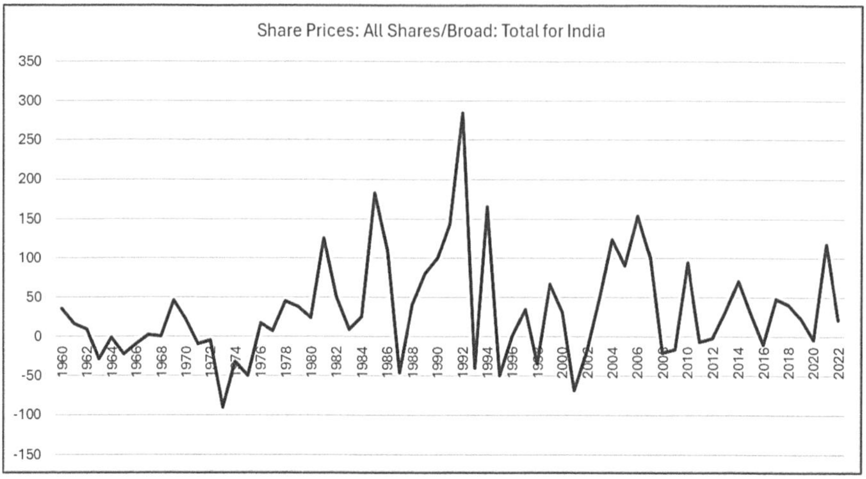

Fig. 4.10 Share prices

world's fourth most valuable stock market, behind only the USA, China, and Japan. India's main Sensex Index, which measures 30 large companies, has risen 10% over the last 3 months, while the wider Nifty 50 Index has increased 11% in that time.[37] The total value of Indian stocks has reached \$3.5 trillion, more than the value of Europe's two biggest equity markets, in the UK and France, according to the data from Refinitiv [37]. The rise in Indian stocks is a sign of the power and potential of the country's economy, according to economists and fund managers. Indian equity market is one of the main channels to attract foreign investment, a proxy for health of economy (please see Fig. 4.10).

Inflation

Containment of inflation was required in the 1990s, just before the economic crisis of 1991. Inflation was around 9% in 1990, and the Government of India formed the Cabinet Committee on Prices and took effective steps to increase the supply of essential commodities, break the inflationary psychology, and contain inflation [10]. However, the policy did not work as deemed! By 1991, inflation shot to 14%.

[37] Share Prices: All Shares/Broad: Total for India (SPASTT01INA657N)—https://fred.stlouisfed.org/series/SPASTT01INA657N#0

Imports

Reduction of import bills: The government aimed to reverse the trend of increasing import bills, particularly for bulk items like oil, which was causing a significant outflow of foreign exchange [38].

Food Grain Stocks

Healthy food grains stocks are necessary to maintain price stability and macroeconomic stability. Food security and PDS, importantly, are necessary social sustainability pragmatic outcomes of food grain stocks [10]. Utilizing the fiscal flexibility obtained through controlling expenses to increase spending on development, specifically for programs aimed at reducing poverty. Control over expenditure: maintaining strict control over expenditure, while making unavoidable increases for subsidies and other necessary provisions for PDP and poverty alleviation [11].

Exchange Rate

A lower credit rating and depreciated rupee have immense impact on the Indian poor and are significantly costlier with respect to terms of borrowings. It's important to note that the impact of a lower credit rating on the Indian poor is complex and multifaceted, as it intersects with various economic factors and policies. Here are a few ways it can affect them:

- Higher borrowing costs: A lower credit rating makes it more expensive for the Indian government and businesses to borrow money. This can lead to higher interest rates on loans, including those for housing, education, and small businesses. The increased borrowing costs can make it harder for the poor to access affordable credit.
- Reduced investment and economic growth: A lower credit rating can deter foreign investors and lenders, leading to reduced investment in the country. This can result in slower economic growth, fewer job opportunities, and limited income growth for the poor.
- Weaker currency: A lower credit rating can put pressure on the country's currency, leading to depreciation. This can increase the cost of imports, including essential goods and commodities, which can directly impact on the purchasing power of the poor.
- Government spending cuts: To address the impact of a lower credit rating, governments may implement austerity measures and reduce public spending. This can result in cuts to social welfare programs, healthcare, education, and other services that are crucial for the well-being of the poor.

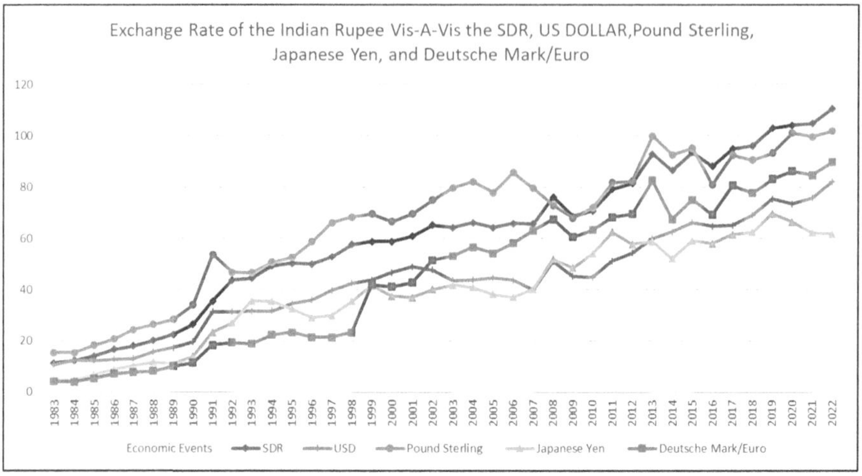

Fig. 4.11 India exchange rate
†Shaded areas: The Kargil War (1999–2000), the 9/11 Attack (2001–2002), the Great Recession (2008–2009), the Growth Crisis (2011–2012), the Great Indian Demonetization (2016–2017), and the COVID-19 pandemic (2019–*)
*Data from 1992 to 1993 onwards are based on FEDAI (Foreign Exchange Dealers' Association of India) indicative rates
‡The data on exchange rate for Japanese yen is in ₹ per 100 yen
°Data from 1983–1984 to 1991–1992 are based on official exchange rates
¥The euro replaced the deutsche mark w.e.f. January 1, 1999

The 1991 economic crisis has had a profound impact on poor and marginalized communities as it increased borrowing costs and increased cost of imports and thus increased balance of payment. Sovereign credit ratings significantly affect bond and stock markets, with average yield spreads increasing by two percentage points and average stock returns decreasing one percentage point after downgrade.[38] They observed that rating changes had stronger effects during crises in both domestic and foreign financial markets (please see Figs. 4.11 and 4.12) [9].

Removal of Bureaucratic Interference

Removing unnecessary bureaucratic interference in economic activity to create an environment conducive to innovation, production, and growth.

[38] Survey—https://www.indiabudget.gov.in/budget202122/economicsurvey/doc/vol1chapter/echap03_vol1.pdf

Fig. 4.12 Central government debt

Credit is when banks lend money to borrowers. Credits are what the bank owns. Deposits are the amount that customers put in the bank as savings. Deposits are what the bank owes. So, the credit-deposit ratio generally means the ratio of what the bank owns and owes. The credit-to-deposit (CTD) or loan-to-deposit ratio (LTD) is a measure of a bank's liquidity by dividing the bank's total loans given out by the total deposits taken in. It shows how much of a bank's main funds are used for lending, which is the main banking activity. CD ratio helps in assessing a bank's liquidity and indicates its financial health. A higher ratio indicates that the loans disbursed are more than the deposits and vice versa.

Manufacturing

Manufacturing refers to industries belonging to International Standard Industrial Classification (ISIC) divisions 15–37. Value added is the net output of a sector after adding up all outputs and subtracting intermediate inputs. It is calculated without making deductions for depreciation of fabricated assets or depletion and degradation of natural resources. The origin of value added is determined by the International Standard Industrial Classification (ISIC), revision 3. Please find manufacturing, value added (annual % growth) for India.[39]

[39]Manufacturing, value added (annual % growth)—India—https://data.worldbank.org/indicator/NV.IND.MANF.KD.ZG?locations=IN

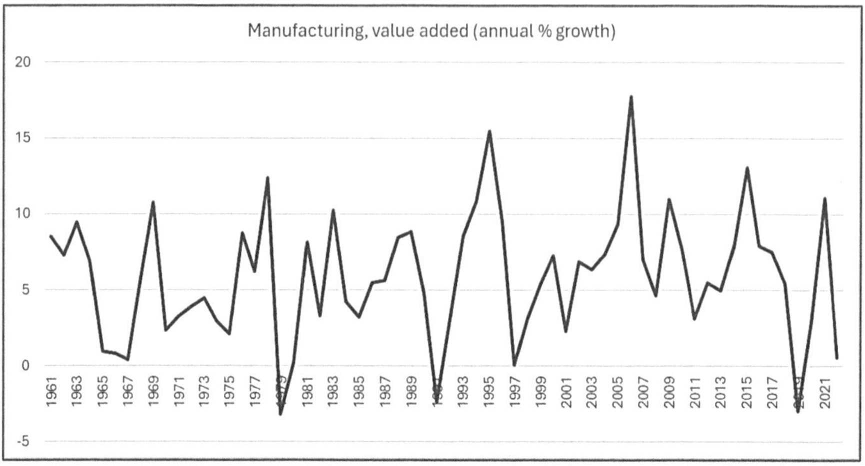

Fig. 4.13 Manufacturing India

- India manufacturing output for 2022 was $450.86B, a 1.11% decline from 2021.
- India manufacturing output for 2021 was $455.91B, a 20.82% increase from 2020.
- India manufacturing output for 2020 was $377.35B, a 1.1% decline from 2019.
- India manufacturing output for 2019 was $381.55B, a 5.14% decline from 2018.

Manufacturing, value added (current US$) for India is as follows (please see Fig. 4.13):

Figures 4.13 and 4.14 illustrate the different macroeconomic connections and limitations within a country and show the main choices that policymakers have to make. The starting point is the initial formulation of the country's poverty reduction strategy, based on consultations with representatives of the government, stakeholders, and development partners. Ideally, these consultations will have led to the creation of a comprehensive action plan that specifies priority sectoral policies to support poverty reduction, such as in education, health, and rural infrastructure. Since poverty has many dimensions, the action plan will also likely include priority actions for governance, structural reform, and other relevant areas, each of which may affect the budget [4].

The initial step is to estimate the total cost of the planned poverty reduction strategy. A thorough system for budget planning of poverty reduction strategies needs the creation of medium-term expenditure frameworks (MTEFs) [4].

The next step is to assess the government's spending program based on priority spending, nondiscretionary spending, and discretionary nonpriority spending. In this process, policymakers should consider the option of moving existing government spending from ineffective nonpriority areas, or areas where public intervention is not required, to priority areas [4].

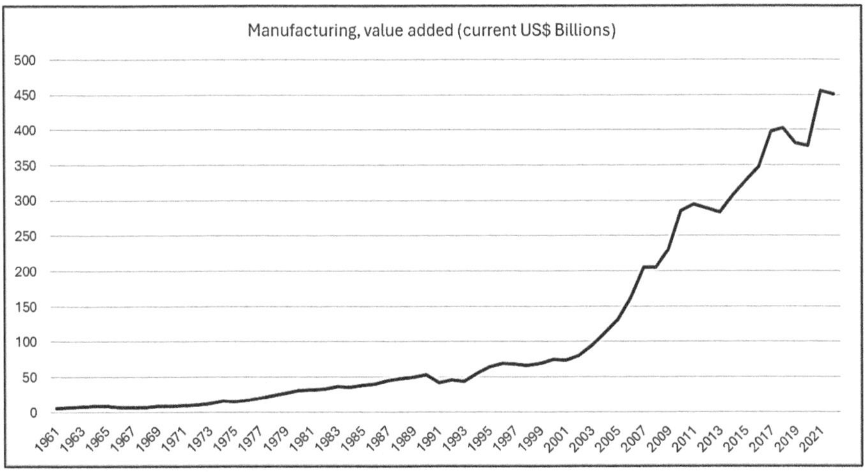

Fig. 4.14 Manufacturing value-added service

The next step (third as shown in Fig. 4.15) is to assess the internal and external sources of budget financing. This would involve looking at (1) the current base of tax and non-tax revenues, including the likely impact of any changes in the tax system planned under the poverty reduction strategy; (2) the possibility of funding public spending through net domestic borrowing in light of the need to maintain macro-economic stability and to ensure enough availability of credit to support private sector growth and economic growth; and (3) the possibility of external financing (e.g., grants, net external borrowing, and debt relief) that is realistic and sustainable under existing conditions [4].

For policymakers who want to design effective poverty reduction strategies, they need a quantitative framework that can help them measure how different macroeconomic policy choices affect the distribution of income among the population. The effects of macroeconomic policies on poverty are complex, and a quantitative framework that reveals the key factors that determine the outcome can help in assessing these trade-offs. Such a framework should be able to show some of the important trade-offs in macroeconomic policies that reduce poverty. For example, how do the benefits (in terms of poverty) of higher spending and higher fiscal deficits compare with the costs of targeting that spending to the poor? Second, the framework should be consistent with both economic theory and basic data sources, such as national accounts and household income and expenditure surveys, to be able to facilitate a dialogue between different parties on these issues. Third, and most importantly, the framework should be simple enough that government officials can use it on their desktop computers. It should not require too much data, and it should be based on commonly available spreadsheet software [4]. For instance, using a machine learning model through a simple web interface or calling from Excel macro is essential to glean insights.

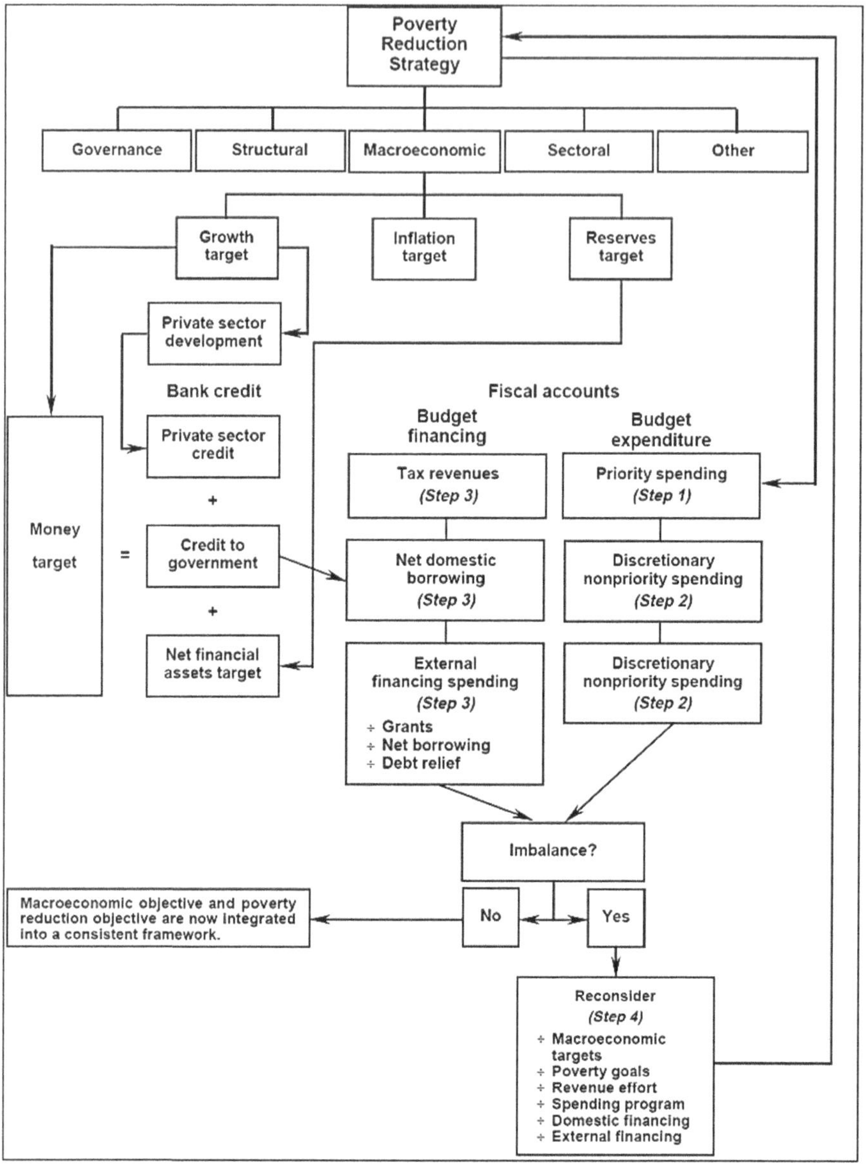

Fig. 4.15 Sustainable economic policies for the poor [4]

In India, many farmers buy and sell livestock and cattle. Without reliable digital records, they face the risk of being cheated and losing their money. This leads to many of them becoming poor. Cow Necklace,[40] a sensor for animal health, helps to create digital records and make the transactions successful, thus raising the quality of life at the grassroots level [39, 40].

Eliminating obstacles to accessing new products, technology, and investment possibilities (through trade, investment, and financial openness) has typically been linked to economic growth. Policies to enhance the performance of markets are therefore essential. Likewise, good governance[41] is vital to speeding up private investment and thus economic growth. Some of the main sources of economic and social sustainability are listed below (please see Table 4.1) [5].

The availability of data is vital to achieve any measurable progress on the achievement of social and economic sustainability. The lack of timely available data means an opportunity to create a sustainable environment to grow for the most vulnerable and marginalized communities.

Insecurity can be understood as vulnerability to decline in well-being. The shock triggering the decline can occur at the microeconomic or household level (for instance, illness or death); at the meso or community level (pollution or riots); or at the national or international level (national calamities or macroeconomic shocks). In poor rural areas, the most important risks are those affecting the harvest. Agriculture is the main source of livelihood for approximately 64% of rural areas in India, and crop losses often lead to poverty for small and subsistence farmers. Vulnerability need not be unexpected and could be seasonal. Everywhere, the risk of illness is a prime concern of the poor [5].

The poor suffer more from income losses than the rich because the poor have less assets—or less access to insurance or credit—to cope with income fluctuations. Households can mitigate some of the microeconomic risks, but public actions are needed to lower the risk of the shock or to help alleviate its adverse effects at the meso- and macroeconomic levels [5].

The poor use different methods to reduce and deal with risks, such as saving for emergencies and relying on informal group-based risk sharing through family and community networks. However, consumption variability is often high among the poor, partly because the shared networks may experience simultaneous shocks, such as the impact of a poor harvest.

The extent and nature of the country's vulnerability to exogenous shocks, and the impact of such shocks on the poor, could be assessed. These sources of vulnerability can reduce the likelihood of successful pursuit of poverty reduction strategy (PRS). At the same time, a good understanding of sources of vulnerability may lead to policies to reduce risk. For example, poverty analysis could be linked to information

[40] CLASS 10 Veterinary Sensor—Hanumayamma Innovations and Technologies, Inc.

[41] A sourcebook for Poverty Reduction Strategies (Vol. 1): Core Techniques and Cross-Cutting Issues—https://documents1.worldbank.org/curated/en/156931468138883186/text/298000182131497813.txt

Table 4.1 Sustainability parameters

	Micro	Meso	Macro
Natural		Rainfall Landslides Volcanic eruption Pests	Earthquakes Floods Drought High winds
Environmental	Indoor air pollution	Pollution Deforestation Soil degradation Desertification Extreme weather	Climate change
Health	Illness Injury Disability Death	AIDS	Pandemic
Social	Crime Domestic violence	Terrorism Gangs	Civil strife War
Economic	Unemployment Harvest failure		High inflation Balance of payments, financial crisis Terms of trade shock Growth collapse

on food shortages and relative price changes in order to identify specific social protection strategies needed to reduce the risks faced by vulnerable groups. Public investment in effective safety net programs may well be an important element of a long-term strategy for growth and poverty reduction [39].

Please see Table 4.2 for the examples of poverty reduction indicators and targets to achieve economic and social sustainability [5].

Implementation of stabilization measures:

- The government implemented measures to reduce the fiscal deficit and restore macroeconomic balance. This helped to stabilize the economy and restore credibility.
- Structural and policy reforms: India embarked on a medium-term program of structural reforms, including initiatives in trade policy and industrial policy. These reforms aimed to improve the efficiency, productivity, and international competitiveness of Indian industry.
- Diversified industrial structure: India had developed a well-diversified industrial structure, which served as an asset during the implementation of structural reforms. This diversification helped in mitigating risks and promoting growth.
- Improvement in infrastructure sectors: The infrastructure sectors, which are crucial for industrial growth, performed well during this period. This provided a strong foundation for future growth.

Table 4.2 Poverty reduction indicators and targets

	Intermediate indicator (inputs and outputs)	Final outcome indicators (outcomes and impact)
Poverty and inequality	Percentage of roads in good and fair condition Productive asset ownership (land, cattle, or other physical capital)	Poverty headcount Poverty gap Average income Gini coefficient Quintile ratio
Macroeconomic stability	Inflation Exchange rate fluctuations Unemployment Fiscal deficit	Per capita economic growth rate Unemployment
Security	Unemployment rate Variability in production of chief staples Expenditure on and number of beneficiaries of public works programs	Food consumption variability Income variability Wasting among children Malnutrition prevalence Death rate due to violence

- Increase in foreign exchange reserves: There was a welcome increase in foreign exchange reserves, which helped to stabilize the economy and improve India's ability to deal with external shocks.
- Credibility restoration: The government successfully restored India's credibility, which helped in rebuilding foreign exchange reserves and attracting investments.
- Support from multilateral financing institutions: India obtained quick disbursing funds from international institutions like the IMF, World Bank, and Asian Development Bank. This support helped in managing the balance of payments.
- Focus on reducing inflation: The government prioritized reducing inflation, as it had a significant impact on the poor and fixed-income earners. Efforts were made to control inflation and maintain price stability.
- Emphasis on equitable growth: The government aimed to achieve a pattern of production that was labor-intensive and generated more employment opportunities. The objective was to reduce income and wealth disparities between rural and urban areas.
- Opportunities for growth: The economic conditions at the time provided opportunities for growth, and the government emphasized the need for adequate developmental outlays to seize these opportunities. It is important to note that these factors contributed to the economic sustainability of India post-1991, but the process of reforms and sustainable growth is ongoing and requires continuous efforts.

Modeling Key Macroeconomic Key Indicators

For data science and machine learning process, most of the work centers on the data, data quality, fixing of missing information, imputation strategies, and transformation of the data for data mining purposes, a design process standardized by the knowledge discovery in databases (KDD). In essence, much of the discussion on the reliability of statistics centers[42] on issues of "data quality." Without going into too much detail on what is meant by the term, it is sufficient to say that it embodies several dimensions including relevance, accuracy, credibility, timeliness, punctuality, accessibility, interpretability, coherence, cost-efficiency, and, in the international context, comparability. These are true characteristics of the data that is useful for both academic and business purposes. Nevertheless, when predicting economies that involve lives in terms of economic progress, the transparency is of utmost importance.

There are several macroeconomic key indicators data collected by international and national agencies. As part of the book, the following data sources are studied to model: The United Nations Statistics Department[43] (UNSD), the World Bank Open data,[44] the International Monetary Fund (IMF) data to macroeconomic and financial data,[45] Food and Agriculture Organization statistics data,[46] *OCED Main Economic Indicators* data,[47] the US Bureau of Labor Statistics,[48] Central Statistics Office (CSO) India,[49] and the Reserve Bank of India *Handbook of Statistics on Indian Economy*.[50]

UNSD Statistical Yearbook

The Statistical Yearbook[51] is a one-volume compilation of internationally available statistics on the social and economic conditions and activities in the world, regions,

---

[42] MEI: COMPARATIVE METHODOLOGICAL ANALYSIS: CONSUMER AND PRODUCER PRICE INDICES—https://www.oecd.org/sdd/prices-ppp/1947731.pdf

[43] The UNSD—https://unstats.un.org/home/

[44] The World Bank Open Data—https://data.worldbank.org/

[45] The IMF Data Access to Macroeconomic & Financial Data—https://data.imf.org/?sk=7A51304B-6426-40C0-83DD-CA473CA1FD52

[46] The FAOSTAT Data—https://www.fao.org/faostat/en/#data

[47] The OCED MEI—https://www.oecd-ilibrary.org/economics/data/main-economic-indicators/main-economic-indicators-complete-database_data-00052-en

[48] The U.S. Bureau of Labor Statistics—https://www.bls.gov/cpi/tables/

[49] The CSO India—http://mospi.nic.in/data

[50] The RBI—https://www.rbi.org.in/scripts/AnnualPublications.aspx?head=Handbook%20of%20Statistics%20on%20Indian%20Economy

[51] The UNSD Statistical Yearbook—https://unstats.un.org/unsd/publications/statistical-yearbook/

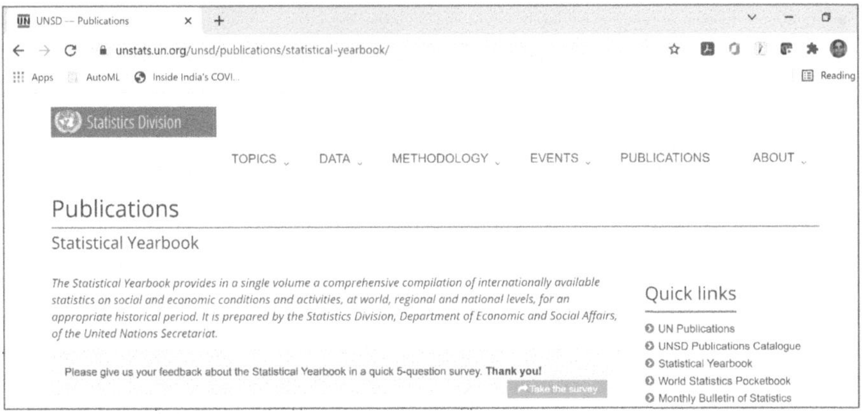

Fig. 4.16 UNSD website

and countries, for an appropriate historical period. The Statistics Division, Department of Economic and Social Affairs, of the United Nations Secretariat is the producer of it. The UNSD Statistical Yearbook provides data on major economic activities.

The Yearbook[52] mainly uses data from more specific, specialized databases that are created by the Statistics Division and many other international statistical services. These specialized databases focus on monitoring topics and trends in social, economic, and environmental fields, while the Statistical Yearbook tables aim to provide data for a comprehensive, overall description of social, economic, and environmental structures, conditions, changes, and activities. The goal has been to collect, organize, coordinate, and present in a consistent way the most important components of comparable statistical information that can show a broad picture of social, economic, and environmental processes (please see Fig. 4.16).

OCED Main Economic Indicators

The OCED dataset[53] has main economic indicators that show different aspects of the economy, such as quarterly national accounts, business surveys, retail sales, industrial production, construction, consumer prices, total employment, interest rates,

[52] Statistical Yearbook—https://unstats.un.org/unsd/publications/statistical-yearbook/files/syb63/SYB_5_Introduction.pdf

[53] Main Economic Indicators—complete database—https://www.oecd-ilibrary.org/economics/data/main-economic-indicators/main-economic-indicators-complete-database_data-00052-en

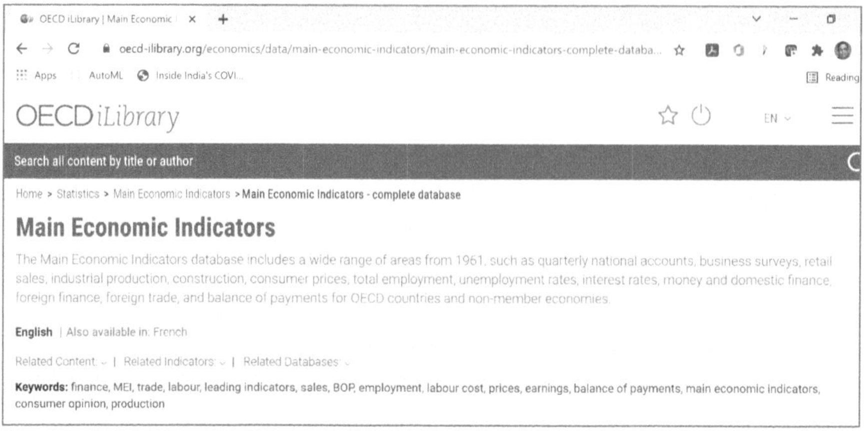

Fig. 4.17 OECD indicators

money and domestic finance, foreign trade, and balance of payments. Data is available from 1955 to now. This database (please see Fig. 4.17) contains financial indicators, balance of payments, foreign trade, sales, national accounts, composite leading indicator, statistics, production, price indices, foreign trade indicators, national accounts of nonmember economies, business tendency, consumer prices, and labor indicators (please see Fig. 4.17).

Statistical Methodological Information for International Comparisons: Dissemination Model

For any data science and machine learning, transparency is crucial. The dissemination model below aims to give clarity to the data so that the average end user can assess their suitability and importance for his or her objective. It is specific to OCED MEI, but it applies to all the macroeconomic statistical process. In the model shown above, for any particular statistical series (e.g., CPI, PPI, industrial production index, unemployment rate, etc.), the information explaining the methods behind the data gets more in depth as one goes down from the top of the pyramid (please see Fig. 4.18).

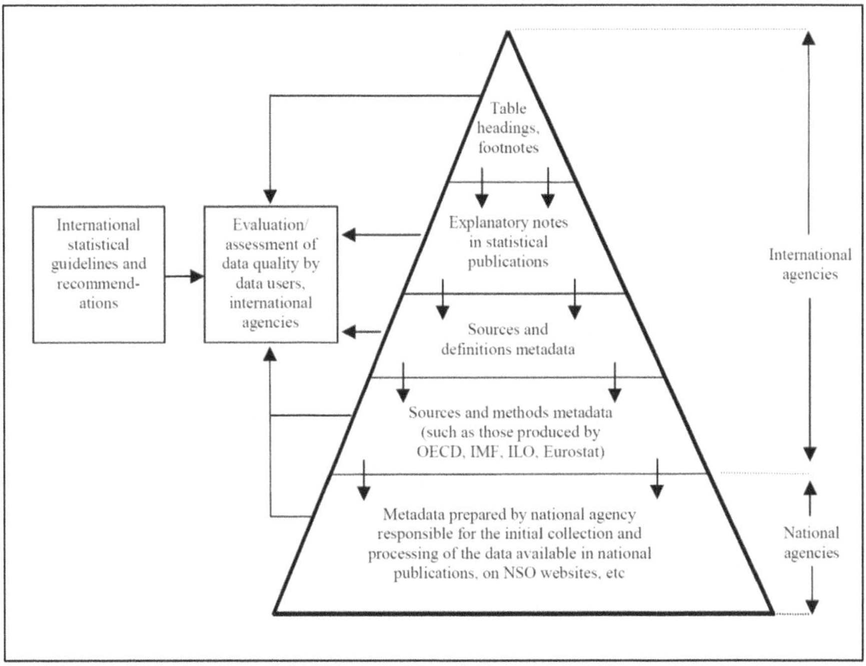

Fig. 4.18 Main economic indicators

Main Economic Indicators (MEI) Short-Term Aggregates

The main economic indicators (MEI) include a wide range of specific short-term indicators. The table below shows a list of "target" indicators that the OECD aims to include in the monthly publication. The main goal of the list is to guide OECD requests to member country agencies and other international organizations for MEI data and methodological information. The short-term aggregates are divided into the major aggregates: national accounts, production, commodity output, manufacturing sales, manufacturing new orders, composite indicators, construction, business tendency surveys, labor, wages, producer prices, consumer prices, retail sales, international trade, domestic finance, balance of payments, interest rates, and foreign finances.

National accounts 1. GDP (value) 2. GDP (volume) 3. Implicit price value **Production** 4. Industry excluding construction. 5. Manufacturing 6. – Consumer goods: total 7. – Consumer nondurable goods 8. – consumer durable goods 9. – Investment goods 10. Intermediate goods including energy 11. Intermediate goods excluding energy 12. Energy 13. Construction 14. Services 15. Rate of capacity utilization **Commodity output** 16. Cement 17. Crude steel 18. Crude petroleum 19. Natural gas 20. Commercial vehicles 21. Passenger cars **Manufacturing—sales (volume)** 22. Total 23. – Domestic 24. – Export 25. Consumer goods 26. – Consumer nondurable goods 27. – Consumer durable goods 28. Investment goods 29. Intermediate goods including energy **Manufacturing—new orders (volume)** 30. Total 31. – Domestic 32. – Export 33. Consumer goods 34. – Consumer nondurable goods 35. – Consumer durable	**OCED composite leading indicator** 42. Trend restored 43. 6-month rate of change (annual rate) **Construction** 44. Orders/permits: total construction 45. Orders/permits: residuals 46. Work put in place: total construction 47. Work put in place: residential **Business tendency surveys** 48 Industrial business climate 49. Industry production: future tendency 50. Industrial orders inflow: tendency 51. Industrial order books: level 52. Industrial finished goods stocks: level 53. Industrial export order books or demand: level 54. Industrial rate of capacity utilization 55. Industrial employment: future tendency 56. Industrial selling prices: future tendency 57. Construction order inflows: future tendency 58. Construction employment: future tendency 59. Retail/wholesale: present business situation 60. Retail/wholesale business situation: future tendency 61. Retail/wholesale stocks: level 62. Other services: present business situation 63. Other services' business situation: future tendency 64. Other services employment: future tendency **Consumer tendency surveys** 65. Consumer confidence indicator 66. Consumer expected inflation 67. Consumer expected economic situation **Retail sales** 68. Total retail sales (value) 69. – Total retail sales (volume)	**Labor** 78. Employment: total 79. – Employment: agriculture 80. – Employment: industry 81. – Employment: services 82. Total employees 83. – Part-time employees 84. – Temporary employees 85. Total unemployment (level) 86. Total unemployment (rate) 87. Unemployment: short-term index 88. Worked hours 89. Job vacancies **Wages** 90. Hourly earnings: all activities 91. Hourly earnings: manufacturing 92. United labor costs: manufacturing **Producer prices** 93. Total 94. Manufacturing 95. – Consumer goods 96. – Investment goods 97. – Intermediate goods including energy 98. – Intermediate goods excluding energy 99. – Energy 100. Food 101. Services **Consumer prices** 102. Total 103. Food 104. All items less food and energy 105. Energy 106. All services less rent 107. Rent 108. National core inflation **Domestic finance** 109. Narrow money 110. Broad money 111. Domestic credit to total economy 112. New capital issues 113. Fiscal balance

(continued)

goods 36. Investment goods 37. Intermediate goods including energy **Manufacturing—stocks (volume)** 38. Total 39. Finished goods 40. Work in progress 41. Intermediate goods	70. New passenger car registration (level) **International trade** 71. Imports c.i.f or f.o.b (value) 72. Exports c.i.f or f.o.b (value) 73. Net trade (value) 74. Imports c.i.f or f.o.b (volume) 75. Exports c.i.f or f.o.b (volume) 76. Import prices 77. Export prices	114. Public debt **Balance of payments** 115. Current account balance 116. – Balance on goods 117. – Balance on services 118. – Balance on income 119. – Balance on current transfer 120. Capital and financial account balance 121. – Reserve assets 122. Net errors and omission **Interest rates—share price** 123. 3-month interest rate 124. Prime interest rate 125. Long-term interest rate 126. All shares price index **Foreign finances** 127. US dollar exchange rate: spot 128. Euro exchange rate: spot 129. Reserve assets excluding gold

The World Bank Open Data

The World Development Indicators[54] is a trustworthy source of statistics on worldwide development and poverty alleviation. The database has 1400 indicators for 217 economies and more than 40 country groups, with data for many indicators going back over 50 years. Data theme[55] under the World Bank Open data covers poverty and inequality, people, environment, economy, states and markets, and global links. The indicators in the economy section help us learn about different aspects of national and global economic activity. The signal mining of these indicators shows how countries produce and consume goods and services and trade them within or across borders; economic indicators measure levels and changes in the size and structure of various economies and indicate growth and decline. For

[54] The World Development Indicators—https://datatopics.worldbank.org/world-development-indicators/

[55] Data Theme—https://datatopics.worldbank.org/world-development-indicators/

Table 4.3 World Bank data themes

Data themes		
Poverty and inequality		Poverty, prosperity, consumption, income distribution
People		Population dynamics, education, labor, health, gender
Environment		Agriculture, climate change, energy, biodiversity, water, sanitation
Economy		Growth, economic structure, income and savings, trade, labor productivity
States and markets		Business, stock markets, military, communications, transport, technology
Global links	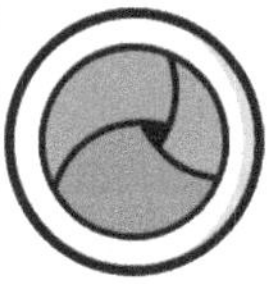	Debt, trade, aid dependency, refugee, tourism, migration

macroeconomic signal modeling, we will focus on economy. We will also focus on poverty and inequality and macroeconomic aggregates (please see Table 4.3).

Poverty and Inequality

A shared aim[56] for national governments and their development partners is to lower poverty. Poverty is more than just a lack of material resources but also impacts many

[56] Reducing Poverty and inequality—https://datatopics.worldbank.org/world-development-indicators/themes/poverty-and-inequality.html

aspects of a person's life: the world's poor are more likely to suffer from hunger; they have less access to services like education, electricity, sanitation, and healthcare; and they are more vulnerable to conflict and climate change. Therefore, it is important to comprehend poverty to comprehend how societies can progress.

The UN's Sustainable Development Goals (SDGs) and the World Bank Group's twin goals for 2030 have a common objective of lowering poverty and inequality in a sustainable manner. They aim to eliminate extreme poverty and ensure everyone's prosperity in all countries. To track these global goals, national statistical systems, household surveys, and methods for measuring poverty are crucial.

- Poverty rates at national poverty lines.[57]
  - Poverty headcount ratio at national poverty lines (% of population)
  - Urban poverty headcount ratio at national poverty lines (% of urban population)
  - Rural poverty headcount ratio at national poverty lines (% of rural population)
  - Poverty gap at national poverty lines (%)
  - Urban poverty gap at national poverty lines (%)
  - Rural poverty gap at national poverty lines (%)
- Poverty rates at international poverty lines
  - Poverty headcount ratio at $1.90 a day (2011 PPP) (% of population)
  - Poverty headcount ratio at $3.20 a day (2011 PPP) (% of population)
  - Poverty headcount ratio at $5.50 a day (2011 PPP) (% of population)
  - Poverty gap at $1.90 a day (2011 PPP) (%)
  - Poverty gap at $3.20 a day (2011 PPP) (%)
  - Poverty gap at $5.50 a day (2011 PPP) (%)
- Distribution of income or consumption
  - GINI index (World Bank estimate)

Economic Indicators

Economic indicators include measures of macroeconomic performance (please see Fig. 4.19) (gross domestic product [GDP], consumption, investment, and international trade) and stability (central government budgets, prices, the money supply, and the balance of payments). It also includes broader measures of income and savings adjusted for pollution, depreciation, and depletion of resources. Many economic indicators from WDI are used in tracking progress towards SDG Goal

[57] World Development Indicators—https://wdi.worldbank.org/tables

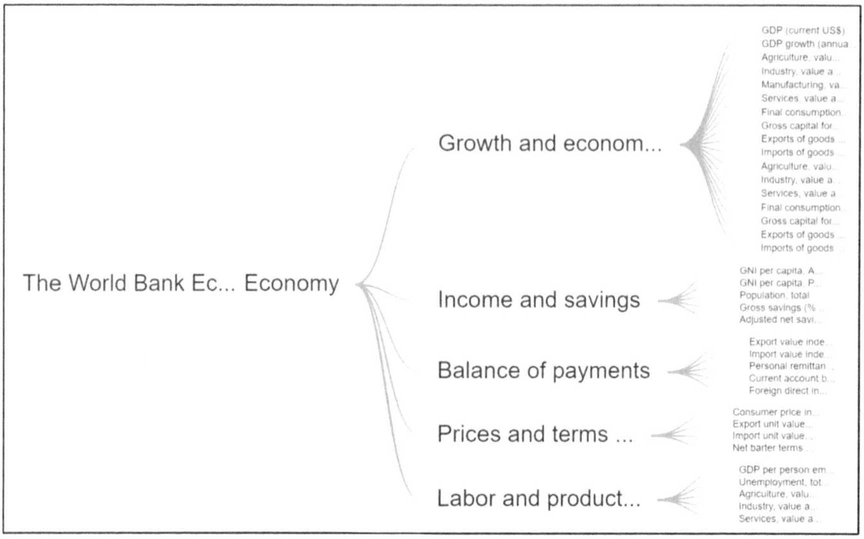

Fig. 4.19 Budget speech indicators

8, promoting decent work and economic growth, and Goal 2, which encourages sustainable consumption and production. The growth indicators include (a) growth and economic structure, (b) income and savings, (c) balance of payments, (d) prices and terms of trade, and (e) labor and productivity.

Gross Savings (% of GDP)

Gross savings[58] measure the gap between disposable income and consumption. In the face of so much economic instability caused by the coronavirus outbreak, people are keeping their money—the household savings rate reached 33%, the highest since 1975.[59] The world[60] saw 26.683%, the highest since 1979 and fell to 24.97% in 2019. During the downturns of 1991 and 2009, the savings rate dropped to 22.023% and 23.268%, respectively.

[58] Gross Savings—https://databank.worldbank.org/metadataglossary/world-development-indicators/series/NY.GNS.ICTR.ZS

[59] Saving vs. spending during a recession—https://www.marketplace.org/2020/06/04/saving-vs-spending-during-a-recession/

[60] The World Bank—https://data.worldbank.org/indicator/NY.GNS.ICTR.ZS

Macroeconomic Signal: Gross Savings (% of GDP)
A higher percentage of GDP that is saved indicates a robust national macroeconomic condition. A higher savings rate means the country is ready for any unforeseen economic challenges. As mentioned in the Indian Union Budget Speech for the year 1960–1961:[61] The House will recall that in the previous session, we enacted laws to allow depositors in the Post Office Savings Bank and holders of Savings Certificates to nominate beneficiaries. The Small Savings movement is not just a regular method for collecting savings. It also has a strong emotional appeal in giving a chance for the common people to join the national effort for development. I would, therefore, urge every family in this country to save more and do their part in making the movement more successful.

The National Saving Certificate from the Post office was a major support to my family in 1990 when my father died suddenly from a heart attack. We escaped poverty because of the assured returns from the National Saving certificates and we are very thankful to the government of India. Both National Saving and Kissan Vikas Patra were life savers and we've sustained quality of life due to guidance and investments into savings by late father.

Author

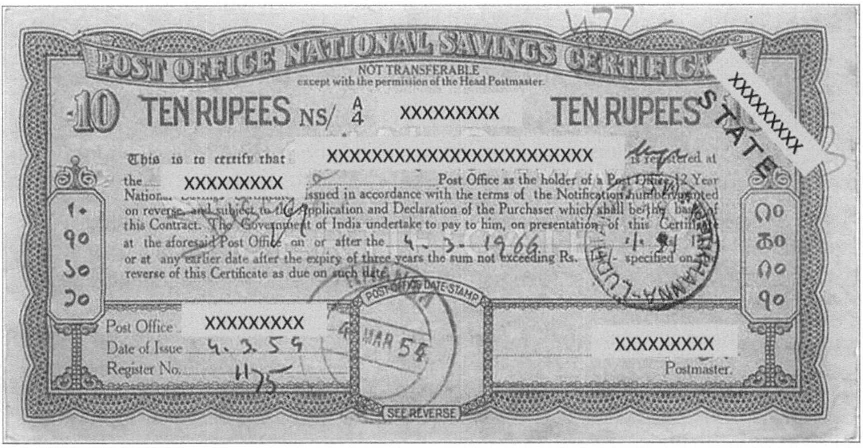

(continued)

[61] Budget Speech for the year 1960–1961—https://www.indiabudget.gov.in/doc/bspeech/bs196061.pdf

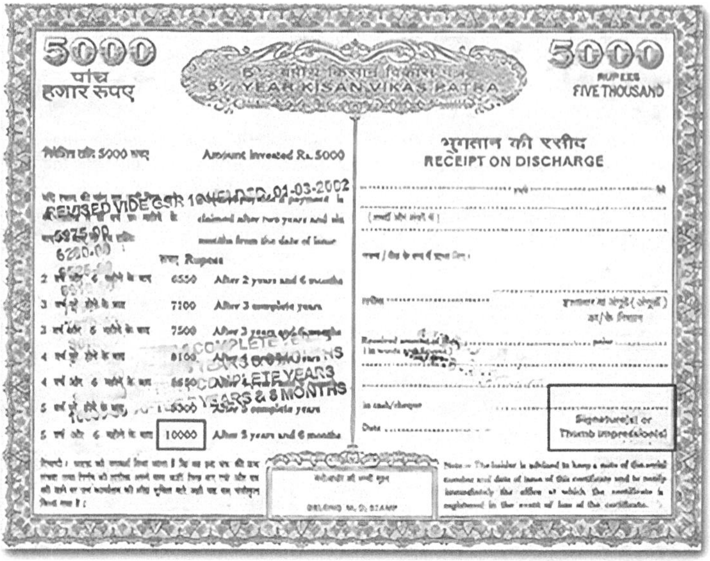

If the standard of living of our people is to be substantially raised by undertaking large schemes of development, both the rich and the middle classes should come forward to place their savings at the disposal of the Government. The willing help and co-operation of all sections of the community is required in maintaining peace and order, [41] in increasing production and in avoiding internecine quarrels whether between communities or between capital and labour.[62]

Building a mutually beneficial ecosystem where government policies and financial inclusion for disadvantaged communities create a network of opportunities to generate middle class and helps to foster trust at the grassroots level. This strengthens democracies and eradicates poverty in a sustainable way.

Author

Chapter Machine Learning Experiments

ML Experiment 1 The policy connection for machine learning model using budget deficit, money supply, and inflation was that the Reserve Bank of India must be clear about its inflation-combating policy and stick to it. How does the model change if we include market index numbers and gold and silver prices from the RBI data?

[62] India Budget Speech, 1947–1948—https://www.indiabudget.gov.in/doc/bspeech/bs194748.pdf

ML Experiment 2 If we include gas prices and energy prices, how does it affect the machine learning model that links budget deficit, money supply, and inflation?

This chapter will help you learn about statutory liquidity ratio (SLR), bank rate data, cash reserve data, and bank credit data. Then, you will be able to use budget deficit, money supply, and inflation variables and build a linkage model.

The policy recommendations that should help to achieve economic and social sustainability are as follows:

- The government will use the Public Distribution System to counter inflation and protect the poorer sections of the population from high prices and shortages.
- Reducing the fiscal deficit and completely eliminating the revenue deficit over the next 5 years.
- The government decided to begin a phased reduction in the SLR, which would release funds for banks to expand credit to agriculture and industry. This reduction was in line with the proposed reduction in the fiscal deficit.
- Inflation-fighting policy must be transparent, and the Reserve Bank of India must commit to pursuing it.

References

1. P. Chidambaram, Budget 2004-2005, July 8, 2004, https://www.indiabudget.gov.in/doc/bspeech/bs200405.pdf, Access Date: November 30, 2023
2. Chandrasekar Vuppalapati, Specialty Crops for Climate Change Adaptation, Publisher Springer Cham, Hardcover ISBN 978-3-031-38398-4 Published: 15 October 2023
3. Aparna Jayashankar and Anthony Murphy, High inflation disproportionately hurts low-income households, January 10 2023, https://www.dallasfed.org/research/economics/2023/0110, Access Date: 11/22/2023
4. Chair Jerome H. Powell, Monetary Policy and Price Stability, August 26 2022, https://www.federalreserve.gov/newsevents/speech/powell20220826a.htm, Access Date: 11/22/2023
5. Achikbache, Bahjat; Alkire, Sabina; Ames, Brian; Aoki, Aya; Baharoglu, Deniz; Bamberger, J. Michael; Bebbington, Anthony J.; Belkindas, Misha V.; Blackden, C. Mark; Bojö, Jan; Bosch, Christophe E.; Brown, Ward; Bruns, Barbara; Bucknall, Julia; Christiaensen, Luc; Claeson, Mariam; Cord, Louise J.; Coudouel, Aline; Devarajan, Shantayanan; Dinc, Mustafa; Domelen, Julie Van; Dongier, Philippe; Donnelly-Roark, Paula; Drabble, Michael; Eele, Graham; Esmail, Talib B. K.; Ezemenari, Kenechukwu Maria; Fort, Lucia; Fozzard, Adrian; Gannon, Colin A.; Girishankar, Navin; Griffin, Charles C.; Grosh, Margaret Ellen; Gwilliam, Kenneth M.; Hamilton, Kirk E.; Hammergren, Linn A.; Hentschel, Jesko S.; Hoekman, Bernard M.; Holmes, Malcolm R. G.; Hommann, Kirsten; Izquierdo, Alejandro; Johnston, Timothy A.; Kenny, Charles J.; Kessides, C. Fallert; Kishor, Nalin M.; Klugman, Jeni [editor]; Knack, Stephen; Kraus, Christiane; Lamech, Ranjit J.; Levy, Brian David; Litvack, Jennie I.; Liu, Zhi;

Malmberg Calvo, Christina; Manning, Nicholas Paul; Manoukian, Violeta; Marope, Mmantsetsa; Mclachlan, Mildred; Messick, Richard E.; Michalopoulos, Constantine; Mingat, Alain Jean Louis; Moock, Peter Russell; Mozammel, Masud; Murphy, Patrick D.; Navas-Sabater, Juan; O'Sullivan, Kyran; Ostrom, Elinor; Paci, Pierella; Patrinos, Harry Anthony; Pillai, Poonam; Polski, Margaret M.; Prennushi, Giovanna; Rinne, Jeffrey James; Ryan Rizvi, Andrea C.; Rubio Soto, Gloria Martha; Sadoff, Claudia W.; Schiff, Maurice; Scott, Christopher; Shah, Parmesh; Sherburne-Benz, Lynne D.; Soucat, Agnes L. B.; Subbarao, Kalanidhi; Sutch, M. Helen; Swanson, Eric Valdeman; Tan, Jee-Peng; Tarr, David; Thomas, Christopher J.; Tikare, Seema; Travers, Sumter Lee; Wagstaff, Adam; Wakeman, Wendy E.; Winter, Carolyn; Withers, Kate; Wodon, Quentin T.; Yang, Hongyu; Yazbeck, Abdo S.; Yitzaki, Shlomo; Youssef, Deborah; Zatlokal, Barbara. A sourcebook for Poverty Reduction Strategies (Vol. 1): Core Techniques and Cross-Cutting Issues (English). Washington, D.C.: World Bank Group. https://documents1.worldbank.org/curated/en/156931468138883186/text/2980000182131497813.txt, Access Date: 11/21/2023

6. P. Chidambaram, Budget 2007-2008, February 2007, https://www.indiabudget.gov.in/doc/bspeech/bs200708.pdf, Access Date: December 04, 2023
7. Shri Manmohan Singh, Budget 1992-93, PUBLISHED TUE, 29th February 1992, https://www.indiabudget.gov.in/doc/bspeech/bs199293.pdf, Access Date: 11/23/2023
8. Bernard Weinraub, 'India Is Now in a New Ball Game', July 8 1991, https://www.nytimes.com/1991/07/08/business/india-is-now-in-a-new-ball-game.html, Access Date: 11/23/2023
9. Survey, Does India's Sovereign Credit Rating reflect its fundamentals No!, 2021, https://www.indiabudget.gov.in/budget2021-22/economicsurvey/doc/vol1chapter/echap03_vol1.pdf, Access Date: 11/24/2023
10. Prof. Madhu Dandavate, Introducing the Budget for the Year 1990-91, Published Tue, 1990, https://www.indiabudget.gov.in/doc/bspeech/bs199091.pdf, Access Date: 11/23/2023
11. Shri Manmohan Singh, Budget Speech 1993-1994, 27 February 1993, https://www.indiabudget.gov.in/doc/bspeech/bs199394.pdf, Access Date: 11/22/2023
12. Shri Yashwant Sinha, Budget 2001-2002, February 2001, https://www.indiabudget.gov.in/doc/bspeech/bs200102.pdf, Access Date: November 28, 2023
13. Luis Brandão-Marques, Marco Casiraghi, Gaston Gelos, Olamide Harrison, and Gunes Kamber, June 30 2023, Is High Debt Constraining Monetary Policy? Evidence from Inflation Expectations, https://www.imf.org/en/Publications/WP/Issues/2023/06/30/Is-High-Debt-Constraining-Monetary-Policy-Evidence-from-Inflation-Expectations-534708, Access Date: 11/22/2023
14. Jim DiSalvo and Ryan Johnston, Banking Trends: The Rise in Loan-to-Deposit Ratios: Is 80 the New 60?, Q3 2017, https://www.philadelphiafed.org/the-economy/banking-and-financial-markets/banking-trends-the-rise-in-loan-to-deposit-ratios-is-80-the-new-60, Access Date: 11/23/2023
15. Mr. Vito Tanzi and Mr. Mario I. Bléjer, Fiscal Deficits and Balance of Payments Disequilibrium in IMF Adjustment Programs*, 15 Sep 1984, https://www.elibrary.imf.org/display/book/9780939934287/ch06.xml, Access Date: November 27, 2023
16. Meghana Ayyagari, Mohammad Hoseini, and Thorsten Beck, Finance and Poverty: Evidence from India, JUNE 17 2013, https://blogs.worldbank.org/allaboutfinance/finance-and-poverty-evidence-india, Access Date: 11/20/2023
17. Pierre Wunsch, Two sides of the same coin: money supply, budget deficits and inflation, 19 April 2023, https://www.nbb.be/en/blog/two-sides-same-coin-money-supply-budget-deficits-and-inflation, Access Date: November 27, 2023

18. Joseph A Schumpeter (Author), Theory of Economic Development (Social Science Classics Series), Publisher: Routledge; New edition (December 31, 1980), ISBN-13: 978-0878556984, Access Date: November 27, 2023
19. Schumpeter, J.A. 1911. The Theory of Economic Development. Cambridge, MA: Harvard University Press
20. Mahendra R. Mishra and Sunil S. Narwade, Impact of Monetary Policy on Bank Credit since Reforms Period, December 2018, https://svv-research-data.s3.ap-south-1.amazonaws.com/paper_120053_1627101011.pdf, Access Date: November 27, 2023
21. G. Sampath, Understanding banking reforms in India after the Narasimham era | The Hindu In Focus Podcast, May 10, 2021 11:00 pm | Updated May 12, 2021 02:13 pm IST, https://www.thehindu.com/podcast/understanding-banking-reforms-in-india-after-the-narasimham-era-the-hindu-in-focus-podcast/article34530264.ece, Access Date: November 27, 2023
22. Narasimham, M. 1991. Report of the Committee on Banking Sector Reforms. Reserve Bank of India, Committee Report, Ministry of Finance, Government of India, Mumbai, https://ibbi.gov.in/uploads/resources/Narasimham%20Committee%20I-min.pdf, Access Date: November 27, 2023
23. Taranjeet KaurGrewa, SakshiSachdeva, Anchit Jhamb, IMPACT OF CRR ON RETURN OF COMMERCIAL BANKS IN INDIA, JETIR May 2019, Volume 6, Issue 5, https://www.jetir.org/papers/JETIRCW06001.pdf, Access Date: 10 December 2023
24. Dr. Latha M T, Time Series Analysis of Inflation and RBI Policy Rates, October 2018, https://www.jetir.org/papers/JETIRG006015.pdf, Access Date: November 27, 2023
25. Shri Arun Jaitley, Budget 2014-2015, July 2014, https://www.indiabudget.gov.in/doc/bspeech/bs201415.pdf, Access Date: December 07, 2023
26. Paresh Kumar Narayan, Seema Narayan and Arti Devi Prasad, Modeling the relationship between budget deficits, money supply and inflation in Fiji, Number 2 2006, https://openresearch-repository.anu.edu.au/bitstream/1885/157821/1/212_modeling.pdf, Access Date: November 27, 2023
27. Altayligil, Y.B., Çetrez, M. Macroeconomic, institutional and financial determinants of current account balances: a panel data assessment. Economic Structures 9, 49 (2020). https://doi.org/10.1186/s40008-020-00225-1, Access Date: 10 December 2023
28. Pranab Mukherjee, Budget 2014-2015, Feb 2014, https://www.indiabudget.gov.in/doc/bspeech/bs.pdf, Access Date: December 07, 2023
29. Bill & Melinda Gates Foundation, Financial Services for the Poor, 2023, https://www.gatesfoundation.org/our-work/programs/global-growth-and-opportunity/financial-services-for-the-poor, Access Date: November 26, 2023
30. Ayşe Aylin Baya, The impact of financial development on income inequality and poverty, Oct 2023, https://www.ncbi.nlm.nih.gov/pmc/articles/PMC10602251/, Access Date: November 25, 2023
31. King R.J. and Levine R. (1993). Finance, Entrepreneurship, and Growth: Theory and Evidence, Journal of Monetary Economics, 32 (3), 513–542, Access Date: November 26, 2023
32. Beck T., Demirguc-Kunt A., and Levine R. (2004). Finance, Inequality, and Poverty: Cross-Country Evidence. World Bank Policy Research Working Paper, WP 3338, World Bank, Washington D.C., Access Date: November 26, 2023
33. Chakroun M. (2020). Threshold Effects in the Relationship Between Financial Development and Income Inequality. International Journal of Finance and Economics, 1–23. https://doi.org/10.1002/ijfe.1757, Access Date: November 26, 2023
34. Appiah-Otoo I. and Song N. (2021). The Impact of Fintech on Poverty Reduction: Evidence from China. Sustainability, 13, 5225. https://doi.org/10.3390/su13095225, Access Date: November 26, 2023
35. Alexander, Jonathan; Shi, Lin; Solomon, Bensam Biniam How Fintech is reaching the poor in Africa and Asia: a start-up perspective (English). EMCompass, no. 34 Washington, D.C.: World Bank Group. http://documents.worldbank.org/curated/en/823341492592362552/How-

Fintech-is-reaching-the-poor-in-Africa-and-Asia-a-start-up-perspective, Access Date: November 26, 2023
36. Rahul Singh Gautam Jagjeevan Kanoujiya Venkata Mrudula Bhimavarapu Shailesh Rastogi, Financial Technology and Its Impact on Poverty in India, November 2021, https://www.researchgate.net/publication/356253195_Financial_Technology_and_Its_Impact_on_Poverty_in_India, Access Date: November 26, 2023
37. Anna Cooban, India's stock market is hitting record highs. Here's why, Published 1:50 AM EDT, Wed June 28, 2023, https://www.cnn.com/2023/06/28/investing/india-stock-market-boom/index.html, Access Date: November 25, 2023
38. Valerie Cerra and Sweta Chaman Saxena, What Caused the 1991 Currency Crisis in India?, March 2002, https://www.imf.org/external/pubs/ft/staffp/2002/03/pdf/cerra.pdf, Access Date: 11/23/2023
39. J. S. Vuppalapati, S. Kedari, A. Ilapakurthy, A. Ilapakurti and C. Vuppalapati, "Smart Dairies — Enablement of Smart City at Gross Root Level," 2017 IEEE Third International Conference on Big Data Computing Service and Applications (BigDataService), Redwood City, CA, USA, 2017, pp. 118–123, https://doi.org/10.1109/BigDataService.2017.35
40. C. Vuppalapati et al., "Artificial Intelligence (AI) Infused Cow Necklace – For Diagnosis Of Bovine Respiratory Diseases," 2018 International Conference on Machine Learning and Cybernetics (ICMLC), Chengdu, China, 2018, pp. 222–229, https://doi.org/10.1109/ICMLC.2018.8526937
41. Shri R.K. Shanmukham Chetty, The Budget for the Year 1947-1948, Year 1947-1948, https://www.indiabudget.gov.in/doc/bspeech/bs194748.pdf, Access Date: 11/5/2021

Chapter 5
Economic Sustainability

This chapter covers the following:

- The public distribution system (PDS).
- Inflation, prices, and shortages.
- PDS-off-take is a true indicator of social sustainability?
- Agriculture and climate change.
- The impact of monetary polices.
- Agricultural productivity and impact on PDS.
- Policy implications discussed.
- Any reform on social protection should come in tandem with carefully crafted reforms on the agricultural and food price risk management sides. This is not a call for inaction, but for concerted cross-sectoral policy. Put it simply, economic sustainability pillar depends on three things: (1) increasing farmers' income and promoting agriculture and farmers' welfare; (2) enhancing rural employment and infrastructure development, providing healthcare services, and implementing health protection schemes; and (3) improving education, skills, and job creation to create a productive society.

The public distribution system was studied in this chapter. The PDS-off-take and its role in economic sustainability indicator model were developed. The impact of climate change on agriculture and its role in food security in India were discussed. Finally, the role of agricultural productivity and impact of climate offset are discussed.

In order to ensure economic sustainability, it is crucial for countries like India, where agriculture is the main source of income, to prioritize the empowerment of the poor and rural communities. The government should recognize the important role of

© The Author(s), under exclusive license to Springer Nature Switzerland AG 2024
C. Vuppalapati, *Assessing Policy Effectiveness using AI and Language Models*,
International Series in Operations Research & Management Science 354,
https://doi.org/10.1007/978-3-031-56097-2_5

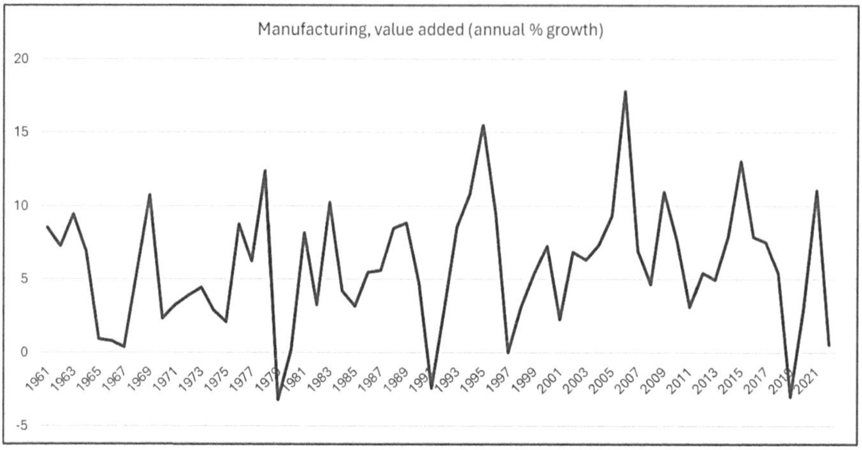

Fig. 5.1 Manufacturing value added

agriculture in promoting inclusive and fair development and make it a key driver of rural and national prosperity. Economic policies should be formulated with a strong commitment to supporting farming, agro-processing [1], and rural development, without compromising on helping the poor and disadvantaged. This has been a priority in India's economic strategy, even during times of economic crisis and restructuring (1991).

To address poverty, various strategies can be implemented, including tax reform and modernization, improving fiscal balance, promoting industrial growth, and changing development policies to tackle issues of poverty, unemployment, and social exclusion. It is also important to focus on strengthening the banking system, reducing dependence on foreign loans, and restructuring regional rural banks to make them more efficient and sustainable [1].

Additionally, prioritizing industrial growth in a competitive and non-monopolistic environment is crucial for sustained economic growth. Reviewing and simplifying industrial licensing policies can encourage competition and prevent monopolies. Additionally, maintaining healthy food stocks is important for controlling inflation, as there is an inverse relationship between inflation and industrial production, especially in developing countries like India. When inflation is low, macroeconomic stability is achieved, and industrial production can thrive. Please see Fig. 5.1[1] 1991 with −2.398% and 2019 with −2.974%.

To achieve economic stability and sustainability, it is crucial to prioritize employment generation, especially considering that India has the largest youth[2] population. Developing a growth strategy that focuses on creating more job opportunities is

[1]Manufacturing, value added (annual % growth) (https://data.worldbank.org/indicator/NV.IND.MANF.KD.ZG)

[2]UNFPA (https://india.unfpa.org/en/topics/adolescents-and-youth-8)

essential. This can be achieved by investing in vocational training programs to enhance the skills of the youth and make them more employable. Additionally, providing credit support for self-employment initiatives will encourage entrepreneurship and further contribute to employment generation [2].

In order to strengthen the rural economy, it is important to focus on improving agriculture and allied activities. This can be done by investing in infrastructure and technology and providing support to farmers.[3] The objective of planning in India is to double real national income in less than 20 years and double per capita income in 25 years[4] [3]. It is crucial to ensure that the benefits of economic reforms are shared by all sections of society, particularly those living in rural areas and marginalized communities. By prioritizing the development of rural areas and marginalized communities, we can ensure the sustainable and inclusive growth of the economy [4]. Here are other strategic priorities:

> No economic strategy can succeed in our country, which does not recognize the central role of agriculture in supporting broad-based and equitable development. The government is firmly committed to ensuring that agriculture, agro-processing, and rural development are given top priority in the design of economic policy.
>
> Shri Manmohan Singh, Minister of Finance
> February 27, 1993

Industrial Production

In order to return to a path of high sustained growth with reasonable price stability and greater social equity post 1991 economic crisis, the task for the planners of Indian economy has become imperative to be an internationally competitive economy. This is when opening of economy to international investment took place. Furthermore, before 1991, there was insufficient focus on reducing costs, upgrading technology, and enhancing living standards. The lack of such competitive measures made the local companies noncompetitive to the world economy, and knowledge training of Indian workforce has become a necessity [5]. Promoting the competitive strength of Indian industry can lead to increased employment and them by increased macroeconomic stability. Focus on improving technology levels and integrating industries into the global economy. Encourage collaboration and cooperation between domestic and foreign investors to leverage their respective strengths.

[3] India Budget Speech, 2000–2001 (https://www.indiabudget.gov.in/doc/bspeech/bs200001.pdf)

[4] Economic Development in India: The First and the Second Five Year Plans (https://www.elibrary.imf.org/view/journals/024/1958/001/article-A002-en.xml)

Foreign Investment

Encouraging foreign investment in critical sectors helps to boost economy and create a tighter integration with global economy. Actively promote foreign investment in critical infrastructure sectors where capacities are inadequate and investment needs are significant. Develop policies to attract private and foreign investment in sectors such as power and hydrocarbon. Maximize the growth potential of these sectors while ensuring the management of the balance of payments [5].

The implementation of policies on the financial markets results in a significant capital inflow, which leads to an increase in the volume of domestic credit. Meanwhile, some studies stress that financial development improves inequality and combats poverty by increasing the ability of advantage of new investment opportunities and by improving the allocation of capital. The widening of the financial sector leads to more equal income distribution and poverty alleviation [6].

Consider contracyclical policies:
In times of economic slowdown, consider implementing contracyclical policies, such as increasing expenditure in infrastructure and rural development, to counter the negative impact of the global recession.

Shri Pranab Mukherjee
Minister of Finance, 2009–2010[5]

Exports

The policy initiatives we have taken do not in any way reduce our commitment to take care of the poor and the disadvantaged. We have taken steps to minimize the burden of adjustment on the poor and working classes.

The Public Distribution System (PDS)

Protecting the poor from a rise in food prices remains a global policy priority. In developing countries like India, households spend 45% of their income on food, and many poor households rely on government subsidized grains for core cereal intake. Thus, programs to mitigate losses from food price shocks are an important part of India's social protection system. The PDS as inflation buster to protect poor and marginalized communities is the de facto practice of the Indian government,

[5] Budget Speech 2009–2010 (https://www.indiabudget.gov.in/budget_archive/ub2009-10(I)/bs/speecha.htm)

a strategy that was applied during the 1991 economic crisis and post economic reforms [1, 5].

India's food subsidy program continues to be one of the largest in the world. The public distribution system (PDS) is a major income transfer program used in India. The PDS has existed since the 1970s and has been expanded over time. The aim of the program is to improve food security. Beneficiaries are offered wheat, rice, sugar, edible oils, kerosene, and soft cake below market prices to ensure their necessary caloric intake is met. In 2007, the program was expanded to include two additional items: pulses like lentils, chickpeas, and salt [7]. The PDS does not operate flawlessly. It sometimes forces the poor to pay prices higher than the market rate, and its provisions only aid the poor marginally [8–10]. Today, after concerted efforts from civil society, administrators, politicians, and the judiciary, many states have seen improvements in the PDS. The program is reaching many more of India's poor, and the diversion of food grains has reduced.

Despite these issues, in a 1998 study, many impoverished Indians stated that the PDS was essential to their well-being [9, 10]. While not a monetary exchange, the PDS represents poverty reduction via income transfers because of the basic and uncomprehensive service it offers. This program follows the fairness approach to ethical standards by attempting to level the playing field for its beneficiaries to obtain necessary food products. Integrated poverty reduction programs are much more common in both nations than income transfers. There is a large debate regarding the effectiveness of income transfer programs because of beneficiaries' abilities to squander their aid.

Policy: inflation and public distribution system (PDS) [1, 5]

The government will use the public distribution system to counter inflation and protect the poorer sections of the population from high prices and shortages. They aimed to ensure that food grains and essential commodities reach the poor and underprivileged in adequate quantities and at affordable prices.

Shri Manmohan Singh
Minister of Finance, 1992–1993[6]

A major integrated poverty reduction program instituted in India in recent years was the Mahatma Gandhi National Rural Employment Guarantee Scheme[7] (NREGS). Implemented in 2005, the program guaranteed everyone at least 100 days of work per year. Provisions of the program include public employment and a statutory wage rate.30 The program has employed 45.5 million households in India, making it the world's largest public works scheme. NREGS has a budget of over $7.5 billion, over 1% of India's GDP.

Despite the size and efforts of the program, the benefits are often lost through corruption at the local levels and failure to reach all areas of the nation [10]. This program can be considered an integrated poverty reduction program through its

[6] India Budget Speech, 1992–1993 (https://www.indiabudget.gov.in/doc/bspeech/bs199293.pdf)

[7] NREGS (https://nrega.nic.in/MGNREGA_new/Nrega_home.aspx)

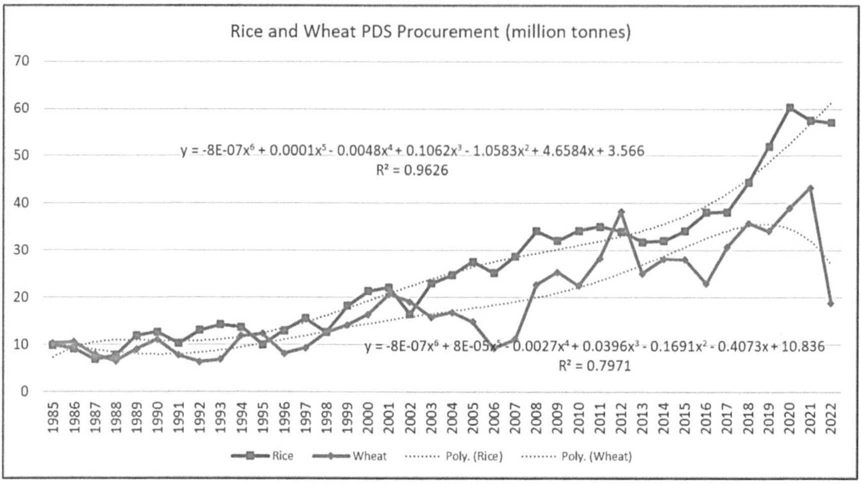

Fig. 5.2 Rice and wheat PDS

attempt to address the societal failure to provide access to employment. One could interpret NREGS as embodying the fair approach of ethical standards through its guarantee to at least 100 days of employment, attempting to make employment universal.

Wheat plays an important role, both in agriculture and food security. Agriculture and wheat were one of the important topics in 2021[8] budget speech. Dwelling on agriculture, the Finance Minister said that the government is committed to the welfare of farmers. In case of wheat, the total amount paid to farmers in 2013–2014 was Rs. 33,874 crores (USD $ 40.734 billion). In 2019–2020, it was Rs. 62,802 crores (USD $ 75.520 billion), and even better, in 2020–2021, this amount, paid to farmers, was Rs. 75,060 crores (USD $ 90.261 billion). The number of farmers growing wheat that benefitted increased in 2020–2021 to 43.36 lakhs (4.3 million) as compared to 35.57 lakhs (3.5 million) in 2019–2020 [11].

Sharp increases (please see Fig. 5.2) in rice and wheat procurements can be seen in 2019 due to COVID-19. As part of its response to the economic impacts of the COVID-19 pandemic, India has sharply increased its distribution of wheat and rice to the 800 million Indian citizens (about 58% of the population) eligible to receive subsidized rations. The public distribution system (PDS) provides nearly 800 million people with subsidized grain through a network of over half a million fair price shops in 2017 [12]. Facing major shocks to employment and incomes associated with nationwide measures to control the virus, India announced a relief program in March 2020 worth $22.3 billion. In 2017, the government spent some $16 billion dollars on the program, or almost 1% of GDP [12]. The program supplements the highly subsidized, standard monthly ration of 5 kilograms (11 lb) per person of wheat or

[8] Summary of the Budget 2021–2022 (https://pib.gov.in/PressReleasePage.aspx?PRID=1693908)

rice with an additional free allocation of 5 kilograms (11 lb). Implementation of the program led to a 75% increase in India's total wheat and rice distribution from April to September 2020 (please see in Fig. 5.2) compared with earlier years, with the average monthly distribution of rice more than doubling. India is a major global holder of food security stocks of both rice and wheat, as well as the world's largest rice exporter—with 2021 exports forecast at 12.5 million tons [13].

Year	Indian households reported buying food grains from fair price shops [12]
2004–2005	22.4%
2011–2012	44.5%
2019–2020	58%

The public distribution system (PDS) of India plays a crucial role in reducing food insecurity by acting as a safety net by distributing essentials at a subsidized rate. While the PDS forms a cornerstone of government food and nutrition policy, India continues to be home to a large population of hungry and malnourished people—one reason is inadequate access to food grains has resulted in the poor of India often not having enough to eat despite the fact that the country is producing enough food grains to sustain its entire population [14]. The address of distribution inefficiencies is beyond the scope of the book. Nevertheless, the aim of the book is to develop a resilient framework that looks into future and provide guidance to the policymakers and honorable civil services officers.

The purpose of the machine learning model is to develop a forecast model that takes into advanced time series and is able to provide procurement forecasts as an effective policy input to the government. Our goal is simple. Helping to make an efficient PDS reduces overall hunger and poverty levels in India.

Machine learning Model_PDSProcurementModel_ForPoorerSectionsOftheSociety_India.ipynb

Machine Learning Linkage Model: Public Distribution System and Inflation—Protect the Poorer Sections of the Population from High Prices and Shortages

The public distribution of grains in India originated from the "rationing" system that the British implemented during the World War II. Since the rationing system and its successor, the public distribution system (PDS) has been important for achieving higher levels of household food security and preventing famines from occurring in the country, and it is appropriate to study its development, functioning, and effectiveness in-depth [15].

The world war forced the British government[9] at that time to start the first organized public distribution of cereals in India through the rationing system—selling a fixed amount of ration (rice or wheat) to eligible families (ration card holders) in certain cities/towns. The system began in 1939 in Bombay and later expanded to other places. By the end of 1943, 13 cities had been included in the rationing system, and by 1946, the number rose to 771 cities/towns. Some rural areas, facing chronic scarcity, were also covered. The Government of India created the Department of Food in 1942, which helped in giving food matters the serious attention of the government. When the war ended, India, like many other countries, decided to end the rationing system. This was in 1943. However, after gaining independence, *India had to reintroduce it in 1950 due to renewed inflationary pressures in the economy right after independence* "which were worsened by the already high global prices of food grains at the end of the war, which were about four times higher than the prewar prices" [15].

India chose to keep public distribution of food grains as a conscious social policy when it started on the way of a planned economic development in 1951. It was actually a key part of the policy of growth with justice. In the first 5-year plan, the system, which was mainly urban focused till then, was expanded to all those rural areas which faced persistent food shortages. The Public Distribution System is implemented to combat inflation and safeguard economically vulnerable sections of society from shortages and steep prices.

The regression model is developed to forecast the procurement for PDS in India. The FCI (Food Corporation of India) procured various food grain to ensure food security to all Indians. Indian PDS (public distribution system) is one of the largest supply chains which covers more than 800 million beneficiaries and has FPS (fair price shops) across the country [15–17]. India[10] operates 5.38 lakh (538 K) fair price shops. The National Food Security Act 2013 (NFSA) provides for distribution of only food grains (rice, wheat, and coarse grains) at highly subsidized prices through targeted public distribution system (TPDS).

Step 1: Load the Libraries

Install required libraries.

[9] Public distribution system in India-evolution, efficacy and need for reforms (https://www.fao.org/3/x0172e/x0172e06.htm)

[10] 5.38 lakh Fair Price Shops (FPS) operational in the country (https://pib.gov.in/PressReleaseIframePage.aspx?PRID=1740687)

```
import numpy as np
import pandas as pd
from functools import reduce
from sklearn.model_selection import train_test_split
from sklearn.linear_model import LinearRegression
from sklearn.metrics import r2_score
from sklearn import metrics
from sklearn.impute import SimpleImputer
import eli5
import warnings
warnings.filterwarnings("ignore")

import numpy as np
import pandas as pd
import functools
from datetime import date
import plotly.graph_objects as go
import seaborn as sns; sns.set()
import statsmodels.api as sm
from statsmodels.tsa.arima_process import ArmaProcess
from scipy import stats
from scipy.stats import pearsonr
import matplotlib.pyplot as plt
import plotly.express as px
from sklearn.preprocessing import MinMaxScaler
from sklearn.model_selection import train_test_split
from sklearn.linear_model import LinearRegression
from sklearn.metrics import r2_score
from sklearn import metrics
from sklearn.impute import SimpleImputer
import eli5
```

For eli5, if it's not available, pip install eli5.

Step 2: Load Public Distribution System Procurement, Off-Take, and Stocks Data

Download public distribution system[11] procurement, off-take, and stocks from the Reserve Bank of India.

```
PDSProcurement_df= pd.read_csv("PDS-Procurement.csv")
PDSProcurement_df
```

```
PDSProcurement_df["Year"] = PDSProcurement_df["Year"].astype(int)
```

```
PDSProcurement_df.dtypes
```

Output

[11] Public Distribution System - Procurement, Off-Take and Stocks (https://rbi.org.in/Scripts/PublicationsView.aspx?id=21833)

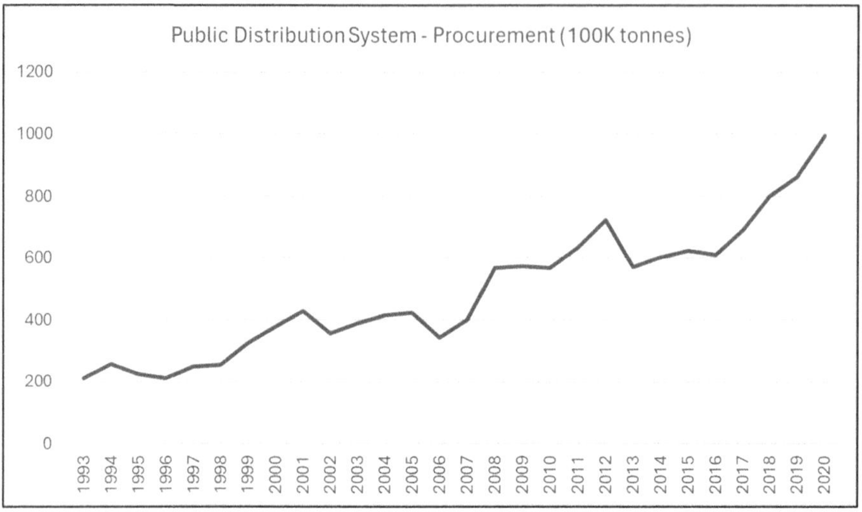

Load PDS-off-take:

```
PDSOffTake_df= pd.read_csv("/content/drive/My Drive/GDP-PPP/PDS-Off-take.csv")
PDSOffTake_df
```

```
PDSOffTake _df["Year"] = PDSOffTake _df["Year"].astype(int)
```

```
PDSOffTake _df.dtypes
```

Output

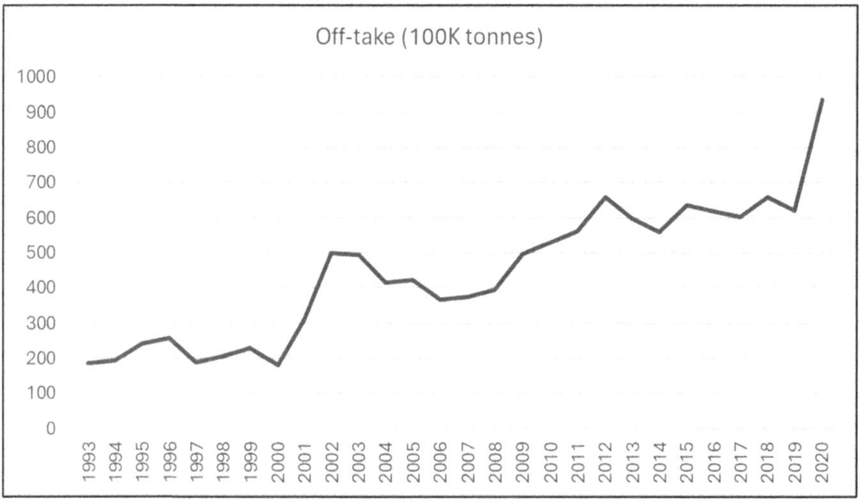

Off-take increased drastically in 2002 dot-com recession. Next, 2019 COVID-19 has increased off-take. This compendium aims to highlight joint efforts of the Indian government and UNWFP India in ensuring food security during the COVID-19 crisis in India by swift and efficient response through India's targeted public distribution system (TPDS), ensuring inter- and intrastate portability of ration cards in TPDS through One Nation One Ration Card scheme, reducing transportation cost through supply chain optimization, conducting social audit and awareness generation, bringing PDS operations on to a single window through a central dashboard, and introducing rice fortification.[12]

Load PDS stocks:

```
PDSStocks_df= pd.read_csv("/content/drive/My Drive/GDP-PPP/PDS-Off-take.csv")
PDSStocks_df
```

```
PDSStocks_df["Year"] = PDSStocks_df["Year"].astype(int)
```

```
PDSStocks_df.dtypes
```

Output

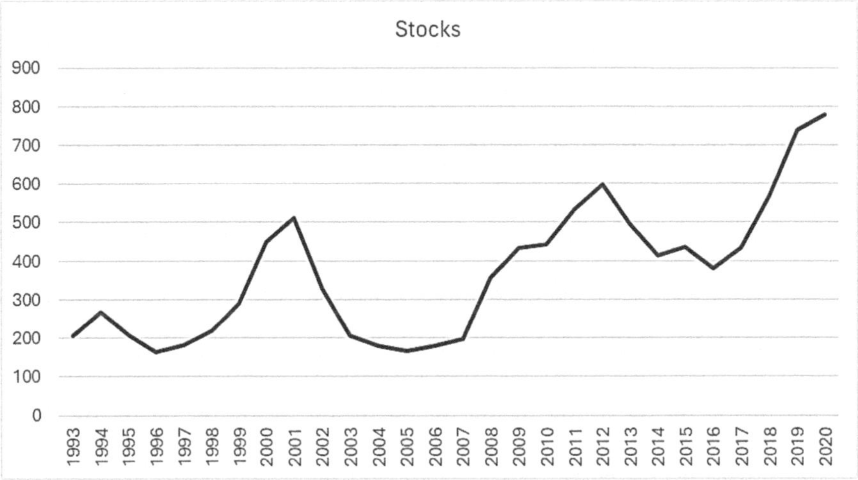

To smooth volatile grain prices,[13] the Indian government procures additional grain quantities, creating agricultural commodity buffers, often redistributed to poor households. This policy can produce a demand shock, triggering increased inflation. Inflation reduces procurement quantities [18].

[12]COVID-19 Lockdown: How India Can Provide Food Grains to Stranded Migrant Labourers (https://thewire.in/rights/covid-19-lockdown-india-food-grains-stranded-migrants)

[13]The inflationary impact of the grain procurement policy in India (https://www.theigc.org/blogs/inflationary-impact-grain-procurement-policy-india)

Step 3: Load Inflation, Consumer Prices (Annual Percentage)

Load inflation[14] data for India.

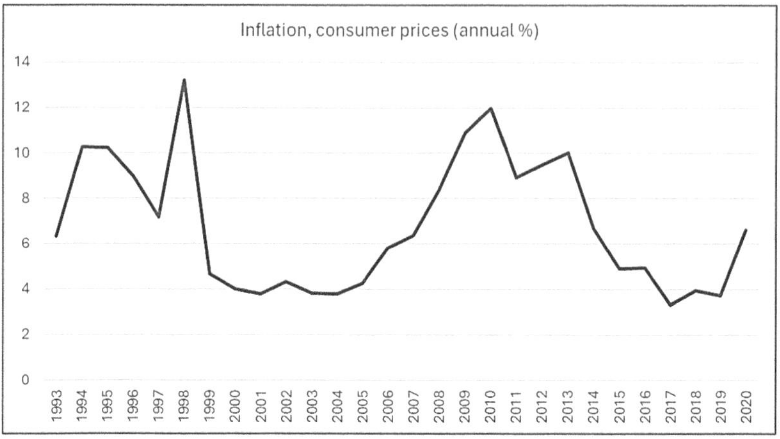

Step 4: Load Population Growth (Annual Percentage) Data

Load population growth[15] (annual %) data from the World Bank for India.

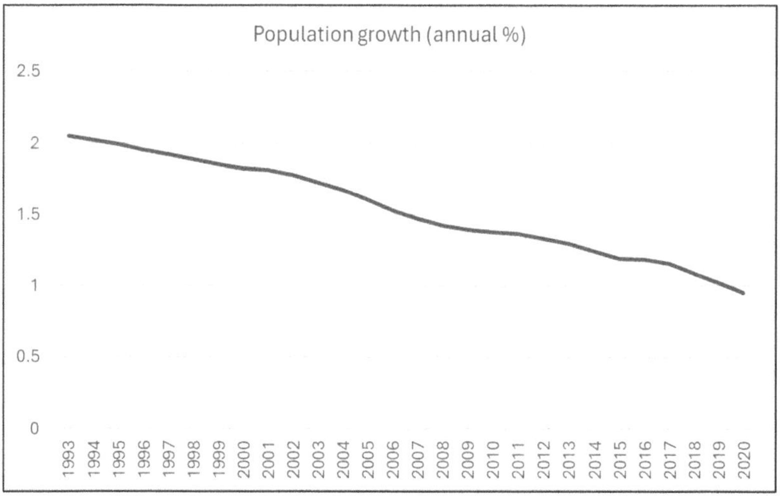

[14]Inflation, consumer prices (annual %)—India (https://data.worldbank.org/indicator/FP.CPI.TOTL.ZG?locations=IN)

[15]Population growth (annual %)—India (https://data.worldbank.org/indicator/SP.POP.GROW?locations=IN)

Step 5: Load Poverty Headcount (<$1.9) Data

Poverty headcount (<$1.9) data.

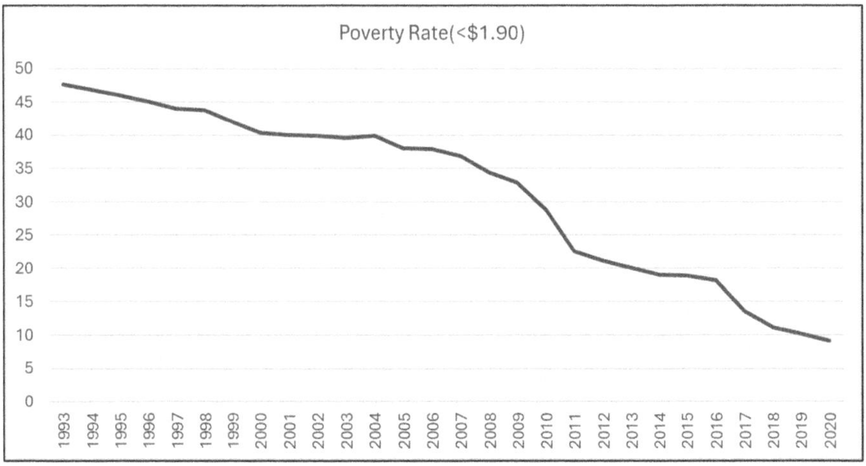

More calculations on the headcount I will cover in the next chapter.

Step 6: Load Economic and Drought Years Binary Data

Please create two synthetic variables: one for economic events and second one for droughts. India has faced the following economic events: notable years of economic slowdown include the East Asian Crisis (1997–1998), the Kargil War (1999–2000), 9–11 attack (2001–2002), the Great Recession (2008–2009), the Growth Crisis (2011–2012), the Great Indian Demonetization (2016–2017), and the COVID-19 pandemic (2019–*).

Drought years include 2000, 2002, 2009, 2012, 2014–2016, and 2019.

```
DroughtsInIndia_df= pd.read_csv("/content/drive/My Drive/GDP-PPP/DroughtsInIndia.csv")
DroughtsInIndia_df
DroughtsInIndia_df ["Year"] = DroughtsInIndia_df ["Year"].astype(int)
DroughtsInIndia_df.dtypes
```

Load of economic events data. Inflation and recession impact poor and marginalized communities. I would like to assess the impact of it.

```
EconomicEvents_df= pd.read_csv("/content/drive/My Drive/GDP-PPP/ EconomicEvents.csv")
EconomicEvents_df
EconomicEvents_df ["Year"] = EconomicEvents_df ["Year"].astype(int)
EconomicEvents_df .dtypes
```

Step 7: Merge Public Distribution System (PDS), Inflation, Poverty Rates, Population Growth, and Recession/Drought Data

Merge all data frames for developing machine learning models.

```
dfs =
[PRate_df,PDSStocks_df,PDSOffTake_df,PDSProcurement_df,InflationCPI_df,Populationgrowth_df,EconomicEvents_df,DroughtsInIndia_df]

merged_df =reduce(lambda left, right : pd.merge(left, right, on = "Year", how= "inner"),dfs)
merged_df
```

Output:

	Year	Poverty Rate(<$1.90)	Stocks	Off-take	Procurement	Inflation, consumer prices (annual %)	Population growth (annual %)	EconomicEvents	DroughtYear
0	1993	47.60	205.4	186.0	211.5	6.326890	2.047837	0	0
1	1994	46.80	268.0	194.4	255.8	10.247936	2.020856	0	0
2	1995	46.01	208.2	243.5	223.8	10.224886	1.991871	0	0
3	1996	45.01	164.1	256.3	211.2	8.977152	1.951436	0	0
4	1997	44.02	181.2	189.6	248.5	7.164252	1.919263	1	0
5	1998	43.80	218.2	207.3	252.5	13.230839	1.887558	0	0
6	1999	42.03	289.1	230.5	323.7	4.669820	1.849333	1	0
7	2000	40.39	449.8	182.1	376.4	4.009436	1.822184	0	1
8	2001	40.01	510.2	313.1	427.6	3.779293	1.808446	1	0
9	2002	39.89	328.1	498.4	354.4	4.297152	1.776768	1	1
10	2003	39.57	206.5	493.3	387.0	3.805859	1.724269	0	0
11	2004	39.87	179.7	414.7	414.7	3.767252	1.672811	0	0
12	2005	38.01	166.2	422.5	423.7	4.246344	1.604129	1	0

```
merged_df.columns
```

Output

```
Index(['Year', 'Poverty Rate(<$1.90) ', 'Stocks', 'Off-take', 'Procurement',
       'Inflation, consumer prices (annual %)', 'Population growth (annual %)',
       'EconomicEvents', 'DroughtYear'],
      dtype='object')
```

Step 8: Linear Regressive Model

To see the impact of independent variables on the PDS procurement.

```
X = merged_df[['Poverty Rate(<$1.90) ', 'Stocks', 'Off-take',
    'Inflation, consumer prices (annual %)',
    'Population growth (annual %)','EconomicEvents', 'DroughtYear'
        ]]

y= merged_df['Procurement']

X_train,X_test, y_train , y_test = train_test_split(X ,y, test_size = 0.2,random_state=0 )

regressor = LinearRegression()
regressor.fit(X_train,y_train)

y_pred = regressor.predict(X_test)
y_pred

r2_score(y_test,y_pred)
```

Output
0.9486 or 94.86%.

Step 9: Model Metrics

Develop model metrics to see the effectiveness of the model.

```
import pandas as pd
from sklearn import metrics
from sklearn.metrics import mean_squared_error, r2_score

mse = mean_squared_error(y_test, y_pred)
print("MSE:", mse)

rmse = np.sqrt(mse)
print("RMSE:", rmse)

r2 = r2_score(y_test, y_pred)
print("R2:", r2)
```

Output

```
MSE: 1108.7837447389632
RMSE: 33.298404537439374
R2: 0.9486882006357653
```

Step 10: Model Equation (Y = MX + C)

Develop model equation and see the impact of the variables on the PDS procurement.

```
mx=""
for ifeature in range(len(X.columns)):
  if regressor.coef_[ifeature] <0:
    # format & beautify the equation
    mx += " - " + "{:.3f}".format(abs(regressor.coef_[ifeature])) + " * " + X.columns[ifeature]
  else:
    if ifeature == 0:
      mx += "{:.3f}".format(regressor.coef_[ifeature]) + " * " + X.columns[ifeature]
    else:
      mx += " + " + "{:.3f}".format(regressor.coef_[ifeature]) + " * " + X.columns[ifeature]

print(mx)
```

Output

```
3.184 * Poverty Rate(<$1.90) + 0.528 * Stocks + 0.179 * Off-take - 3.229 * Inflation, consumer prices
(annual %) - 431.105 * Population growth (annual %) - 11.073 * EconomicEvents - 27.331 *
DroughtYear
```

The $Y = MX + C$

```
# y=mx+c
if(regressor.intercept_ <0):
  print("The formula for the " + y.name + " linear regression line is = " + " -
{:.3f}".format(abs(regressor.intercept_)) + mx )
else:
  print("The formula for the " + y.name + " linear regression line is = " + " +
{:.3f}".format(regressor.intercept_) + mx )
```

Output

```
The formula for the Procurement linear regression line is = + 812.2493.184 * Poverty Rate(<$1.90) +
0.528 * Stocks + 0.179 * Off-take - 3.229 * Inflation, consumer prices (annual %) - 431.105 *
Population growth (annual %) - 11.073 * EconomicEvents - 27.331 * DroughtYear
```

As it can be seen, poverty rate, PDS stocks, and PDS-off-take have positive influence on PDS procurement. That is, the increase in poverty rate mandates higher quantity of PDS [14, 19].

Step 11: Model Explainability

To see the most influenced variables, run model Explainability.

```
eli5.explain_weights(regressor, feature_names = list(X.columns))
```

Output

y top features

Weight?	Feature
+812.249	<BIAS>
+3.184	Poverty Rate(<$1.90)
+0.528	Stocks
+0.179	Off-take
-3.229	Inflation, consumer prices (annual %)
-11.073	EconomicEvents
-27.331	DroughtYear
-431.105	Population growth (annual %)

Step 12: Gradient Boosting Regressor Model

Let's run the gradient boosting regressor.[16] Given the rice procurement is less explained via polynomial regressor, it's required to apply a more robust loss function. Application of gradient boosting technique, function estimation/approximation is viewed from the perspective of numerical optimization in function space, rather than parameter space, could make a difference. A general gradient descent "boosting" paradigm is developed for additive expansions based on any fitting criterion [20].

[16] GradientBoostingRegressor (https://scikit-learn.org/stable/modules/generated/sklearn.ensemble.GradientBoostingRegressor.html)

```
# high accuracy and low bias
from sklearn.ensemble import GradientBoostingRegressor
X = merged_df[numeric_features]

y= merged_df['Procurement']

X_train,X_test, y_train , y_test = train_test_split(X ,y, test_size = 0.2,random_state=0 )

modelGradientBoostingRegressor = GradientBoostingRegressor(n_estimators=5000,
random_state=0,loss='squared_error')
modelGradientBoostingRegressor.fit(X_train,y_train)

y_pred = modelGradientBoostingRegressor.predict(X_test)
y_pred

r2_score(y_test,y_pred)
```

Output

0.957985 or 95.7985%

Step 13: Gradient Boosting Regressor Model Metrics

Let's evaluate model metrics.

```
import pandas as pd
from sklearn import metrics
from sklearn.metrics import mean_squared_error, r2_score

mse = mean_squared_error(y_test, y_pred)
print("MSE:", mse)

rmse = np.sqrt(mse)
print("RMSE:", rmse)

r2 = r2_score(y_test, y_pred)
print("R2:", r2)
```

Output

```
MSE: 907.8731258735608
RMSE: 30.13093304020904
R2: 0.9579858526028698
```

Step 14: Model Explainability

Model explainability helps to identify the dominate parameters.

```
eli5.show_weights(modelGradientBoostingRegressor)
eli5.show_weights(modelGradientBoostingRegressor, feature_names =list(X.columns), target_names =
"PDS" )
```

Output

Weight	Feature
0.4948 ± 0.1414	Population growth (annual %)
0.3968 ± 0.1326	Poverty Rate(<$1.90)
0.0582 ± 0.1784	Off-take
0.0356 ± 0.2228	Inflation, consumer prices (annual %)
0.0131 ± 0.2507	Stocks
0.0014 ± 0.0316	EconomicEvents
0.0001 ± 0.0756	DroughtYear

Population growth (annual percentage), poverty rate (<$1.90), off-take, inflation, and stocks are major drivers of procurement quantity for the PDS. Among the variables, population growth and poverty rate combined would influence 60% on the procurement quantity.

Step 15: Random Forest Model

Random forest[17] is a meta estimator that fits a number of classifying decision trees on various subsamples of the dataset and uses averaging to improve the predictive accuracy and control overfitting. The subsample size is controlled with the max_samples parameter if bootstrap = True (default); otherwise, the whole dataset is used to build each tree.

[17] Random Forest Regressor (https://scikit-learn.org/stable/modules/generated/sklearn.ensemble.RandomForestRegressor.html)

```
# high accuracy and low bias
from sklearn.ensemble import RandomForestRegressor
X = merged_df[numeric_features]

y= merged_df['Procurement']

X_train,X_test, y_train , y_test = train_test_split(X ,y, test_size = 0.2,random_state=0 )

modelRandomForestRegressor = RandomForestRegressor(max_depth=2, random_state=0)
modelRandomForestRegressor.fit(X_train,y_train)

y_pred = modelRandomForestRegressor.predict(X_test)
y_pred
r2_score(y_test,y_pred)
```

Output

0.94382 or 94.382%

Step 16: Random Forest Regressor Model Metrics

Let's evaluate model metrics.

```
import pandas as pd
from sklearn import metrics
from sklearn.metrics import mean_squared_error, r2_score

mse = mean_squared_error(y_test, y_pred)
print("MSE:", mse)

rmse = np.sqrt(mse)
print("RMSE:", rmse)

r2 = r2_score(y_test, y_pred)
print("R2:", r2)
```

Output

```
MSE: 1213.8738730145617
RMSE: 34.840692774607135
R2: 0.943824886574006
```

Step 17: Model Explainability

Model explainability helps to identify the dominate parameters.

```
eli5.show_weights(modelRandomForestRegressor)
eli5.show_weights(modelRandomForestRegressor, feature_names =list(X.columns), target_names =
"PDS" )
```

Output

Weight	Feature
0.3739 ± 0.7924	Population growth (annual %)
0.3424 ± 0.7866	Poverty Rate(<$1.90)
0.1609 ± 0.5980	Off-take
0.1009 ± 0.4482	Stocks
0.0204 ± 0.0715	Inflation, consumer prices (annual %)
0.0015 ± 0.0224	EconomicEvents
0 ± 0.0000	DroughtYear

Population growth (annual percentage), poverty rate (<$1.90), off-take, stocks, and inflation are major drivers of procurement quantity for the PDS. Among the variables, population growth and poverty rate combined would influence 70% on the procurement quantity.

The procurement of PDS is highly influenced by population growth and poverty rates and off-take quantities. Based on these high influence variables, PDS should invest in procuring necessary quantities to serve poor and marginalized communities.

Model Comparison and Summary

The following shows the model summary. Gradient boost regressors (95.79%) outperforms among all the regressors, followed by linear regressor (94.86%) and random forest regressor (94.38%), a narrow higher than linear regression (please see Fig. 5.3).

The "best" metric for a regression model's accuracy depends on the specific problem and the desired outcome. Both mean squared error (MSE) and root mean squared error (RMSE) are commonly used metrics for regression models, but they have different characteristics:

MSE is sensitive to outliers because it squares the errors before they are averaged. It measures the average of the squares of the errors, that is, the average squared difference between the estimated values and the actual value (please see Fig. 5.4).

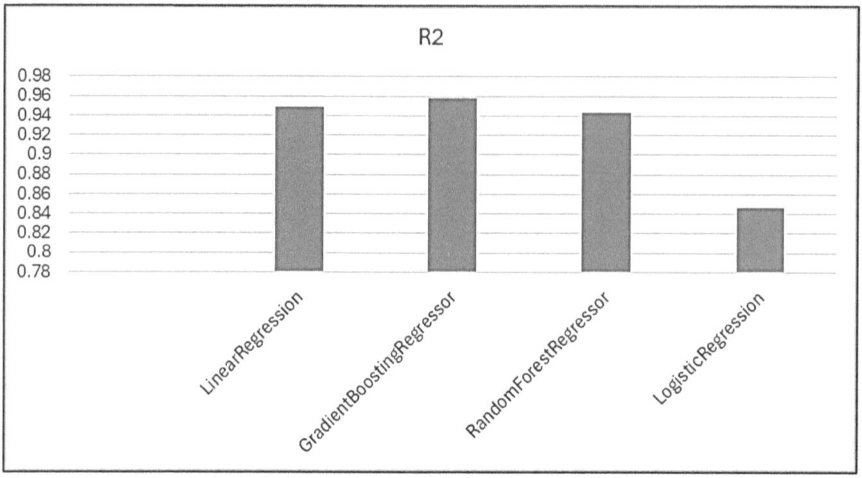

Fig. 5.3 R2 Score

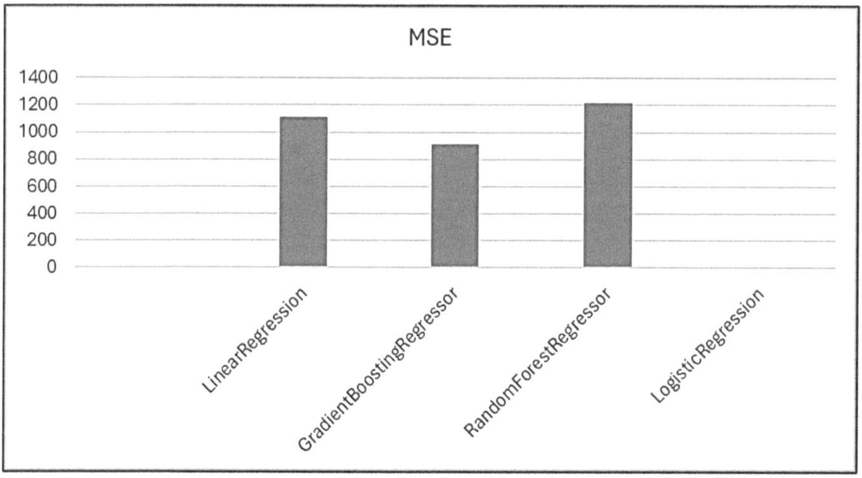

Fig. 5.4 MSE

RMSE, which is the square root of MSE, combines the advantages of MSE while having the same scale as the target variable. It penalizes large errors more severely, making it sensitive to outliers (please see Fig. 5.5).

Gradient boost model performs well with respect to MSE and RMSE. Nevertheless, neither of these metrics alone can be considered the "best" for all situations. For example, during COVID-19 [21], the PDS data has experienced an exponential growth. In order to future forecast with higher accuracies under similar scenarios of COVID, RMSE would be a good candidate because it penalizes large errors more.

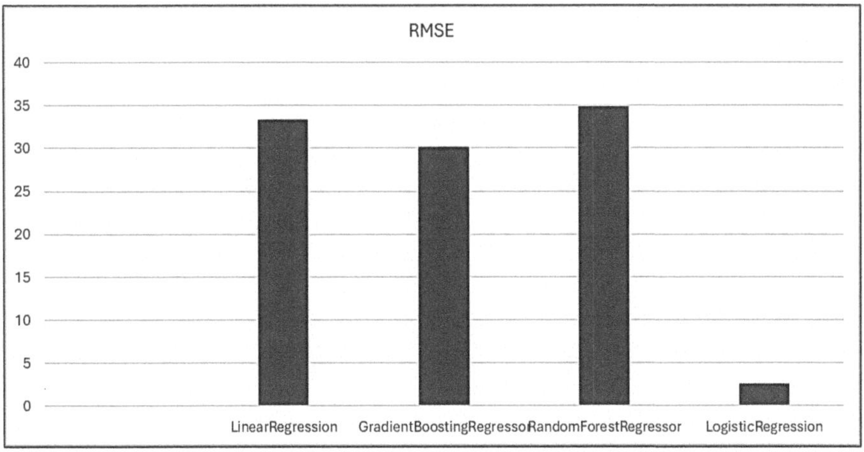

Fig. 5.5 RMSE

On the other hand, if you're more interested in the average error, MSE might be more appropriate. It's also worth noting that these metrics don't tell you anything about the model's bias or variance. A model with a low MSE or RMSE might still be overfitting or underfitting the data. Therefore, it's important to use these metrics in conjunction with other diagnostic tools when evaluating your model's performance.

Policy Recommendations and Future Consideration

Consider climate change and extreme weather to include adaptive procurement strategy. Crop losses and extreme weather put pressure on the low and marginalized communities in terms of earnings and social sustainability. It would be pragmatic to add safety cushion extra procurement percentage (5% of calculated procurement) to increase procurement.

Policy: inflation and public distribution system (PDS) [12]

Any reform on social protection should come in tandem with carefully crafted reforms on the agricultural and food price risk management sides. This is not a call for inaction, but for concerted cross-sectoral policy. Put it simply, economic sustainability pillar depends on three things: (1) increasing farmers' income and promoting agriculture and farmers' welfare; (2) enhancing rural employment and infrastructure development, providing healthcare services, and implementing health protection schemes; and (3) improving education and skills and creating jobs to create a productive society.

Fig. 5.6 The fair price shops. (Image is distorted to respect civility of fellow citizens.) (Source: https://pmjandhanyojana.co.in/fair-price-shops/)

Machine Learning Modeling—Public Distribution System—Off-Take, a True Indicator of Economic and Social Sustainability (1990:2021)

The public distribution system (PDS) is a food security system in India that distributes subsidized food and nonfood items to India's poor. The term "off-take" in the context of PDS refers to the quantity of food grains lifted by the beneficiaries from the fair price shops (please see Fig. 5.6).

Public distribution system means the system for distribution of essential commodities to the ration cardholders through the fair price shops sells items such as rice, wheat, sugar, edible oils, kerosene, and such other commodities as are notified by the central government under clause (a) of Section 2 of the Act; ration card means a document issued under an order or authority of the state government for the purchase of essential commodities under the public distribution system from the fair price shop. The state government shall issue distinctive ration cards to above poverty line, below poverty line, and Antyodaya families and shall conduct periodical review and checking of the ration cards as per paragraph 2 of the Annexe to this Order.[18]

[18] PDS (Control) Order 2001 (https://dfpd.gov.in/pds-control-order-1.htm)

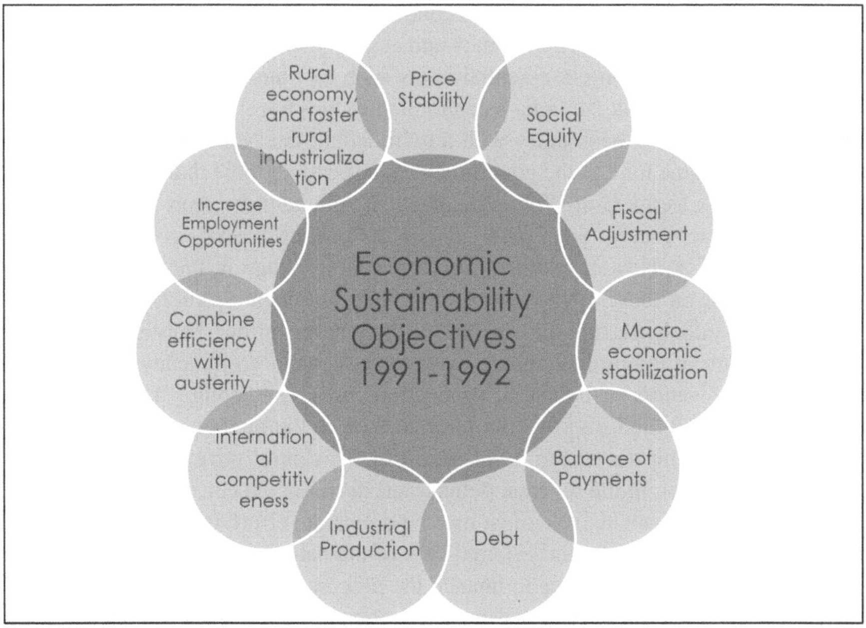

Fig. 5.7 Economic Sustainability

The lines at fair price shops, locally known as ration shops, are a true indicator of wellness of real economy, the economy for poor and marginalized communities. During economic contractions, one tends to observe longer and busy lines. Growing up in India during the 1970s, I have personally witnessed the importance of ration shops in maintaining social and economic order in the country. Rice, oil, and pickle were survival mantras of many of my friends who were school buddies of mine in the 1970s. Milk based products at that time were luxury, meaning out of reach due affordability for many of my school friends and masses in general [22].[19]

The linkage between macroeconomic indicators and real economy, i.e., PDS-off-take, is analyzed as part of the book and machine learning applied to assess relationship between economic cycles and macroeconomic/financial indicators on economic sustainability of marginalized communities. As one would expect, most macroeconomic and financial variables exhibit procyclical behavior during recessions. In addition, recessions are characterized by sharp declines in (residential) investment, industrial production, imports, and housing and equity prices; modest declines in consumption and exports; and some decrease in employment rates. Two key policy related variables, short-term interest rates and fiscal expenditures, often behave countercyclical during recessions (please Fig. 5.7).

[19] Milk Production in India (https://pib.gov.in/FeaturesDeatils.aspx?NoteId=151137)

For example, during the 1990–1991 economic slump, the main economic stability goals were based on balancing efficiency and austerity, concentrating on the needs of the majority, and investing in essential items such as water, education, health, and housing. Please note that the India's economic troubles began to escalate in 1985 as imports grew, leaving the country with a twin deficit: the Indian trade balance was negative at the same time as the government was spending more than its income. By the end of 1990, as the Gulf War approached, the situation became critical. India's foreign exchange reserves were barely enough to cover 3 weeks' worth of imports. Moreover, the Iraq-Kuwait conflict caused a major change in the trade deficit as India depended on these countries for crude oil. The increase in crude oil prices worsened the imbalance in India's balance of payments further. During 1991–1992 budget, the aim of the reform process would be to enhance efficiency and international competitiveness of industrial production, use more foreign investment and foreign technology, and improve the performance of the public sector. Lastly, it's our goal to gradually lower the fiscal deficit of the central government, achieve a significant reduction in the revenue deficit, and decrease the current account deficit in the balance of payments. We must make a deliberate effort to reduce the internal debt of the government and the external debt of the nation, so that we depend more and more on our own resources to finance the process of development [23].

India was affected by the global financial crisis of 2008–2009, even though its financial sector had little connection to global markets and no exposure to mortgage-backed securities (please see Fig. 5.8). The global crisis hit India through a "sudden

Fig. 5.8 Economic sustainability

stop" of capital inflows and a drop in both external and domestic demand [24]. The economy grew by 6.7% in 2008–2009 (April–March), down from 9.0% in the previous year. The initial impact on India was mild: GDP growth fell from 9% in 2007–2008 to 7.8% in April–September 2008. After Wall Street crashed in September, India's growth sank to 5.8%, 5.8%, and 6.1% in the next three quarters. India's trade shrank along with global trade, although its decline started sooner because of a deliberate effort by the Reserve Bank of India to slow down the economy in 2008. Falling foreign trade, capital flows, and exchange rate movements all transmitted negative effects to the Indian economy. Despite these difficulties, India was able to cope with the crisis due to its strong, well-capitalized, and well-regulated financial sector, gradual and careful opening up of the capital account, and the large amount of foreign reserves [25, 26].

The goals were much like those of economic stabilization factors of 1991–1992, except for spending cuts and balance of payments measures. There was more focus on fair distribution of wealth, public distribution system, and rural employment. For instance, institutionalization of the national program such as midday meals in schools has helped to increase school attendance and decrease hunger in classrooms.

India experienced its first recession in decades during 2020–2021 [27]. The COVID-19 pandemic had a significant impact, causing an unprecedented contraction in the economy [28, 29]. In budget speech 2020–2021 [30], pandemic initiatives and health infrastructure were critical measures introduced (please see Fig. 5.9).

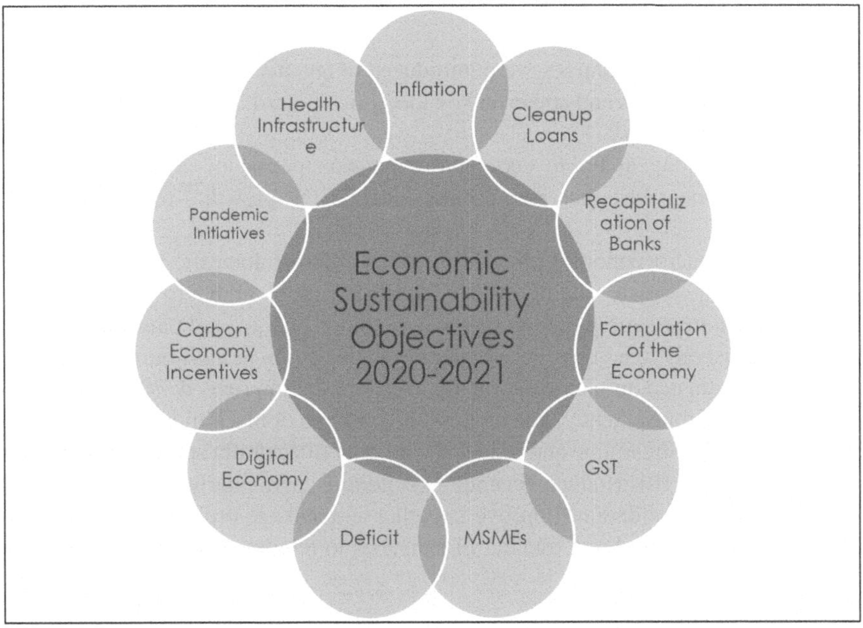

Fig. 5.9 Economic sustainability

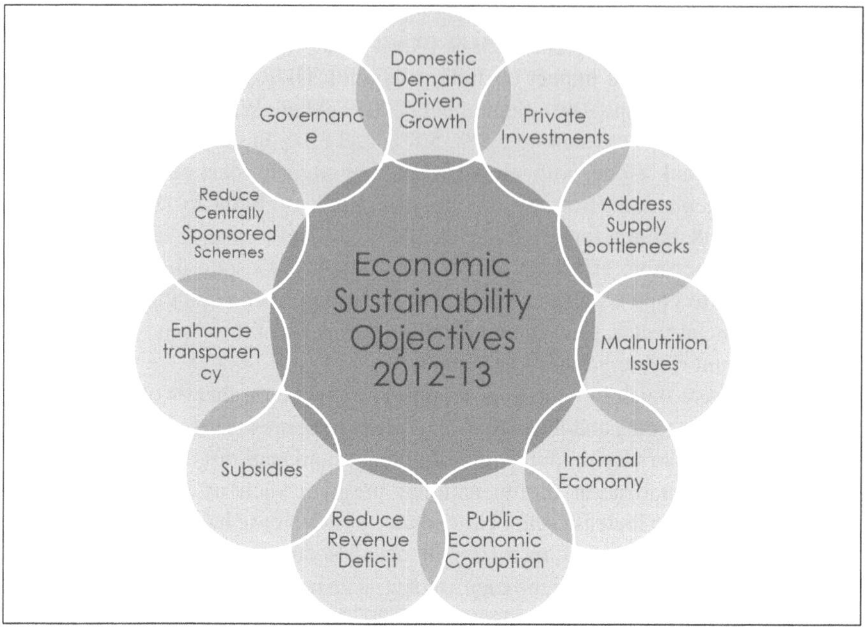

Fig. 5.10 Economic sustainability 2012–2013

Additionally, MSME initiatives were introduced to counter pandemic, which led to supply chain disruptions and spikes in key food prices, causing elevated inflation for most of the year.[20]

Let's take a look at boom economic expansion period of 2012–2013 when balance of payment becomes surplus and Indian GDP growth[21] registered 2.7% annually.

The major macroeconomic policies include focus on domestic demand-driven growth recovery. Create conditions for rapid revival of high growth in private investment. Address supply bottlenecks in agriculture, energy, and transport sectors. Intervene decisively to address the problem of malnutrition in high-burden districts. Expedite coordinated implementation of decisions to improve delivery systems, governance, and transparency, and address the problem of black money and corruption in public life. Streamline and reduce the number of centrally sponsored schemes (please see Fig. 5.10). Reduce revenue expenditure, particularly on subsidies, to improve fiscal consolidation. Improve targeting and leakage-proof delivery of subsidies. And implement direct transfer of subsidies to retailers and farmers using IT strategies [31].

[20] Economic Review (https://www.rbi.org.in/scripts/AnnualReportPublications.aspx?Id=1315)

[21] GDP growth (annual %) (https://data.worldbank.org/indicator/NY.GDP.MKTP.KD.ZG)

Despite recessionary periods or massive economic growth times, the linkage[22] between macroeconomic indicators and real economy is always forefront of economic policymakers. As part of the chapter analyzed, the causative relationships and machine learning are applied to assess the relationship between economic cycles and macroeconomic/financial indicators [32]. As one would expect, most macroeconomic and financial variables exhibit procyclical behavior during recessions. In addition, recessions are characterized by sharp declines in (residential) investment, industrial production, imports, and housing and equity prices, modest declines in consumption and exports, and some decrease in employment rates. Two key policy related variables, short-term interest rates and fiscal expenditures, often behave countercyclical during recessions.

The purpose of ML is to study the interrelations between recessions and the effects of it spread across the globe. Recessions in many advanced countries [32] were bunched in four periods over the past 30 years—the early 1990s, 2000s, 2010s, and the early 2020s—and often coincided with global shocks. Just as many countries experience synchronized recessions, countries also go through simultaneous episodes of credit contractions. Moreover, declines in house and equity prices tend to occur at the same time. The goal is to uncover macroeconomic signals so as to predict future prognostics.

Economic cycle	Average duration
Recessions	4 quarters
Credit crunch	10 quarters
Equity price bursts	18 quarters

Coincidence of Recessions with Credit Crunch and Bursts

In general, it was observed that in about one out of six recessions, there is also a credit crunch underway and, in about one out of four recessions, also a house price bust. Equity price busts overlap for about one-third of recession episodes. A recession, if one occurs, can start as late as four to five quarters after the onset of a credit crunch or an asset bust.

Machine learning model—PDS-off-take and influence of macroeconomic and financial variables. ipynb (Jupyter Notebook Code)

[22] What Happens During Recessions, Crunches and Busts? (https://www.imf.org/external/pubs/ft/wp/2008/wp08274.pdf)

Data Sources

Source	Time frequency	Source
World Bank data	Population growth (annual %)[23] India unemployment rate 1991–2023[24]	THE WORLD BANK IBRD · IDA
Inflation rate, average consumer prices, annual percent change	The average consumer price index (CPI) is a measure of a country's average level of prices based on the cost of a typical basket of consumer goods and services in a given period. The rate of inflation is the percentage change in the average CPI[25] Source: World Economic Outlook (October 2023)	INTERNATIONAL MONETARY FUND
Real GDP growth, annual percent change[26]	Gross domestic product is the most commonly used single measure of a country's overall economic activity. It represents the total value at constant prices of final goods and services produced within a country during a specified time period, such as 1 year Source: World Economic Outlook (October 2023)	
General government net lending/borrowing (percent of GDP)	Net lending (+)/borrowing (−) is calculated as revenue minus total expenditure. This is a core GFS balance that measures the extent to which general government is either putting financial resources at the disposal of other sectors in the economy and nonresidents (net lending) or utilizing the financial resources generated by other sectors and nonresidents (net borrowing). This balance may be viewed as an indicator of the financial impact of general government activity on the rest of the economy and nonresidents. Note: net lending (+)/borrowing (−) is also equal to	

(continued)

[23] Population growth (annual %) India (https://data.worldbank.org/indicator/SP.POP.GROW?locations=IN)

[24] India Unemployment Rate 1991–2023 (https://www.macrotrends.net/countries/IND/india/unemployment-rate)

[25] IMF Inflation (https://www.imf.org/external/datamapper/PCPIPCH@WEO/IND)

[26] GDP Growth (https://www.imf.org/external/datamapper/NGDP_RPCH@WEO/FR_FCWORLD/IND)

Source	Time frequency	Source
	net acquisition of financial assets minus net incurrence of liabilities[27] Source: World Economic Outlook (October 2023)	
Current account balance, percentage of GDP[28]	The current account is the record of all transactions in the balance of payments covering the exports and imports of goods and services, payments of income, and current transfers between residents of a country and nonresidents Source: World Economic Outlook (October 2023)	
Reserve Bank of India	Public distribution system—procurement, off-take, and stocks[29]	
	Institutional credit for agriculture and allied activities[30]	
	Industry-wise deployment of bank credit[31]	
	Money stock—currency with the public (% of GDP)[32]	
	Money stock—time deposits (% GDP)[33]	
	Foreign trade non-oil exports (crore)[34] Foreign trade non-oil imports (crore) Total foreign investment inflows (crore)	
	Net inflow of aid (crore)	

[27] General government net lending/borrowing (https://www.imf.org/external/datamapper/GGXCNL_NGDP@WEO/OEMDC/IND)

[28] Current account balance, percent of GDP (https://www.imf.org/external/datamapper/BCA_NGDPD@WEO/IND)

[29] Public Distribution System - Procurement, Off-Take and Stocks (https://rbi.org.in/Scripts/PublicationsView.aspx?id=21833)

[30] Institutional Credit for Agriculture and Allied Activities (https://rbi.org.in/Scripts/PublicationsView.aspx?id=21858)

[31] Industry-Wise Deployment of Bank Credit (https://rbi.org.in/Scripts/PublicationsView.aspx?id=21977)

[32] Money Stock-Currency with the Public (https://rbi.org.in/Scripts/PublicationsView.aspx?id=21846)

[33] Money Stock - Currency with the Public (https://rbi.org.in/Scripts/PublicationsView.aspx?id=21846)

[34] Foreign Trade Non-Oil Exports Crore) (https://rbi.org.in/Scripts/PublicationsView.aspx?id=21924)

Step 1: Load Libraries for Performing Time Series and ML Modeling

Load Scikit Learn and Matplotlib libraries.

```
import pandas as pd
import numpy as np
import plotly.graph_objects as go
import seaborn as sns; sns.set()
import matplotlib.pyplot as plt
from scipy import stats
from scipy.stats import pearsonr
import matplotlib.pyplot as plt
from sklearn.preprocessing import MinMaxScaler
from sklearn.model_selection import train_test_split
from sklearn.linear_model import LinearRegression
from sklearn.ensemble import RandomForestClassifier
from sklearn.metrics import r2_score
from sklearn import metrics
import eli5
import matplotlib.pyplot as plt
import matplotlib as mpl
import numpy as np
import scipy.stats as spstats
import seaborn as sns
import utils

%matplotlib inline
mpl.style.reload_library()
mpl.style.use('classic')
mpl.rcParams['figure.facecolor'] = (1, 1, 1, 0)
mpl.rcParams['figure.figsize'] = [6.0, 4.0]
mpl.rcParams['figure.dpi'] = 100
```

Step 2: Load Money-Banking, Financial Markets, Public Finances, Trade Balance of Payments Linkage Data for India

Load money-banking, financial markets, public finances, trade balance of payments linkage data for India.

```
MacroFinMonetrayAgriLinkage = pd.read_csv("LinkageModelDataAll.csv")
```

Output

Year	PDS-Off-Take(Crore)	Institutional Credit for Agriculture and Allied Activities - Loans Issued (Crore)	Bank Credit to Commercial Sector (Crore)	Population growth (annual %)	GDP growth (annual %)	Money Stock - Currency with the Public (% of GDP)	Money Stock - Time Deposits(% GDP)	Current account balance, percent of GDP (Percent of GDP)	Foreign Trade Non-Oil Exports Crore)	Foreign Trade Non-Oil Imports Crore)	Total Foreign Investment Inflows (Crore)	Net Inflow of Aid (Crore)
1990	164.9	9829	158790	2.141141	5.5	9.207976	30.017920	-2.9	31620	32377	185	2421
1991	207.4	11199	175716	2.101891	1.1	9.225675	30.598685	-0.4	43020	34724	326	4959
1992	179.5	12141	204164	2.074344	5.5	8.969176	31.522764	-1.7	52309	46233	1713	2233
1993	186.0	14636	226473	2.047837	4.8	9.395173	31.998679	-0.4	68504	55055	13026	2230
1994	194.4	18366	256410	2.020856	6.7	9.797969	32.634077	-1.0	81365	71358	16133	455
1995	243.5	23138	309803	1.991871	7.6	9.809199	31.881347	-1.6	104836	97505	16364	-1625
1996	256.3	25677	349493	1.951436	7.5	9.469851	32.649253	-1.1	117107	103291	21773	39
1997	189.6	27798	391029	1.919263	4.0	9.420797	35.817653	-1.3	128790	123835	19916	-601
1998	207.3	32277	449640	1.887558	6.2	9.532488	37.910801	-0.9	139377	151413	10101	-833

As can be seen, the DataFrame consists of money-banking, financial markets, public finances, trade balance of payments linkage data for India from 1990 to 2022.

The following variable can be found:

Food security

- PDS-off-take (crore)
- Natural gas (USA), spot price at Henry Hub, Louisiana

Banking and money markets

- Institutional credit for agriculture and allied activities—loans issued (crore)
- Bank credit to commercial sector (crore)
- Money stock—currency with the public (% of GDP)
- Money stock—time deposits (% GDP)

Macroeconomic variables

- Current account balance, percent of GDP (percent of GDP)
- Foreign trade non-oil exports (crore)
- Foreign trade non-oil imports (crore)
- Total foreign investment inflows (crore)
- Net inflow of aid (crore)
- Inflation rate, average consumer prices (annual percent change)
- General government net lending/borrowing (percent of GDP)

Others

- Population growth (annual percentage), GDP growth (annual percentage)

Step 3: Check for Nulls in the Data

Check any variables that need imputations by running Is NULL() on the dataframe.

```
MacroFinMonetrayAgriLinkage.isnull().sum() ## missing values
```

Output

PDS-Off-Take(Crore)	0
Institutional Credit for Agriculture and Allied Activities - Loans Issued (Crore)	0
Bank Credit to Commercial Sector (Crore)	0
Population growth (annual %)	0
GDP growth (annual %)	0
Money Stock - Currency with the Public (% of GDP)	0
Money Stock - Time Deposits(% GDP)	0
Current account balance, percent of GDP (Percent of GDP)	0
Foreign Trade Non-Oil Exports Crore)	0
Foreign Trade Non-Oil Imports Crore)	0
Total Foreign Investment Inflows (Crore)	0
Net Inflow of Aid (Crore)	0
Inflation rate, average consumer prices (Annual percent change)	0
General government net lending/borrowing (Percent of GDP)	0

No null found in the data

Step 4: Correlation Analysis

Let's perform correlation analysis to see how the variables interact with each other.

```
MacroFinMonetrayAgriLinkage.corr()
```

Output

The amount of food grains taken by the beneficiaries from the fair price shops (PDF-off-take) is highly correlated with loans issued as part of institutional credit for agriculture and allied activities, unemployment rate, currency with the public, time deposits, non-oil exports, non-oil imports, and total foreign investment inflows and net inflow of aid.

While both PDS-off-take and unemployment rate are important indicators of the socioeconomic status of the population, the specific relationship between the two is complex and can be influenced by a variety of factors. For instance, during times of economic hardship or increased unemployment, there may be a higher reliance on PDS, leading to an increase in PDS-off-take. Conversely, improvements in the economy and reductions in the unemployment rate might lead to a decrease in PDS-off-take as more people are able to afford food items without subsidies. As per correlation analysis, a strong influence can be observed.

The PDS-off-take exhibits a stronger negative relationship with broad money growth, inflation rate, and general government net lending/borrowing. A weaker relationship could be observed with current account balance (percentage of GDP) and GDP growth.

Broad money growth refers to the expansion of the money supply in an economy, which includes currency in circulation and demand deposits [33]. It's a key indicator of economic activity and is closely monitored by central banks as part of their monetary policy [33]. Increased purchasing power, inflation, and government spending all could influence off-take.

Correlation	PDS-Off-Take(Crore)
Broad money growth (annual %)	- 58.89%
Inflation rate, average consumer prices (Annual percent change)	- 34.17%
General government net lending/borrowing (Percent of GDP)	- 23.00%

The economy may have more liquidity when broad money growth is high. This could potentially raise the buying power of individuals, making them less dependent on subsidized goods from the PDS and reducing PDS-off-take (a negative correlation of 58.89% justifies the relationship). High broad money growth can also cause inflation. If the prices of commodities rise a lot due to inflation, even individuals who used to not rely on the PDS might use it for their needs, leading to an increase in PDS-off-take (34.17% suggests such relationship). Broad money growth may be related to more government spending. If the government chooses to invest more in the PDS (e.g., by increasing the amount or variety of goods offered), this could lead to an increase in PDS-off-take (23.00%).

There are several factors that influence long queues at fair price shops in India [34]: fair price shops, also known as public distribution shops, are part of India's public distribution system. Their main function is to distribute food with high calorific potential at lower-than-market subsidized prices. In the late 1990s, the focus of the PDS shifted to finding, assessing, and targeting the poorest populations. This resulted in state-level reports on households living below the poverty line. Each family below the poverty line is eligible for 35 kg of rice or wheat every month, while a household above the poverty line is entitled to 15 kg of food grain on a monthly basis3. This can lead to high demand and long queues at these shops.

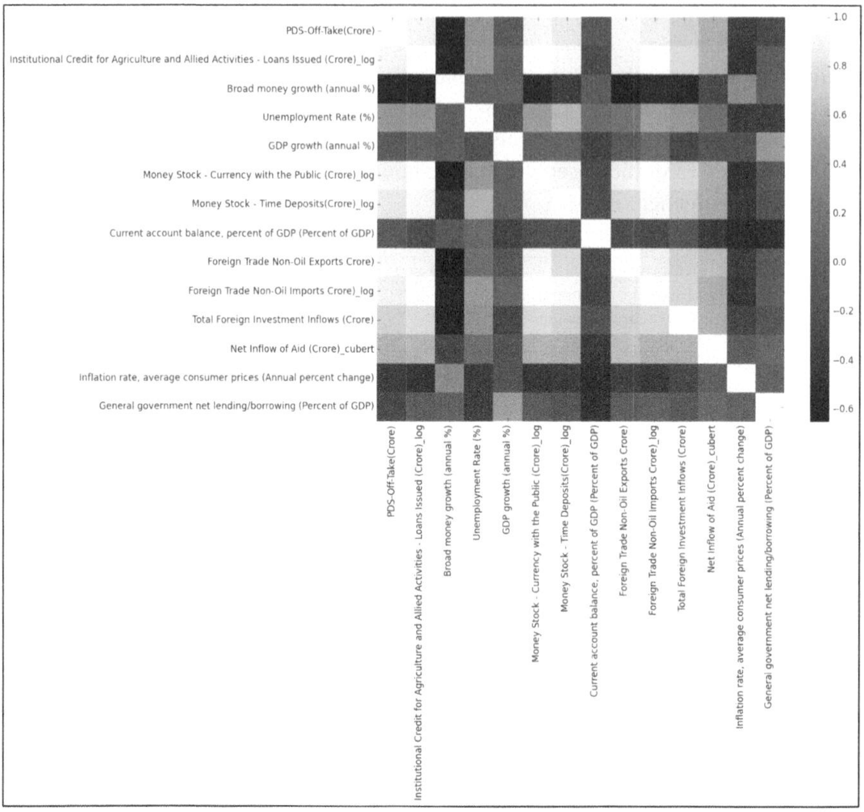

Step 5: Plot Dataset

Plot dataset

```
# Plot
fig, axes = plt.subplots(nrows=4, ncols=4, dpi=120, figsize=(10,16))
for i, ax in enumerate(axes.flatten()):
 data = dataset[dataset.columns[i]]
 ax.plot(data, color='red', linewidth=1)
 ax.set_title(dataset.columns[i])
 ax.xaxis.set_ticks_position('none')
 ax.yaxis.set_ticks_position('none')
 ax.spines["top"].set_alpha(0)
 ax.tick_params(labelsize=6)
plt.tight_layout();
```

Output

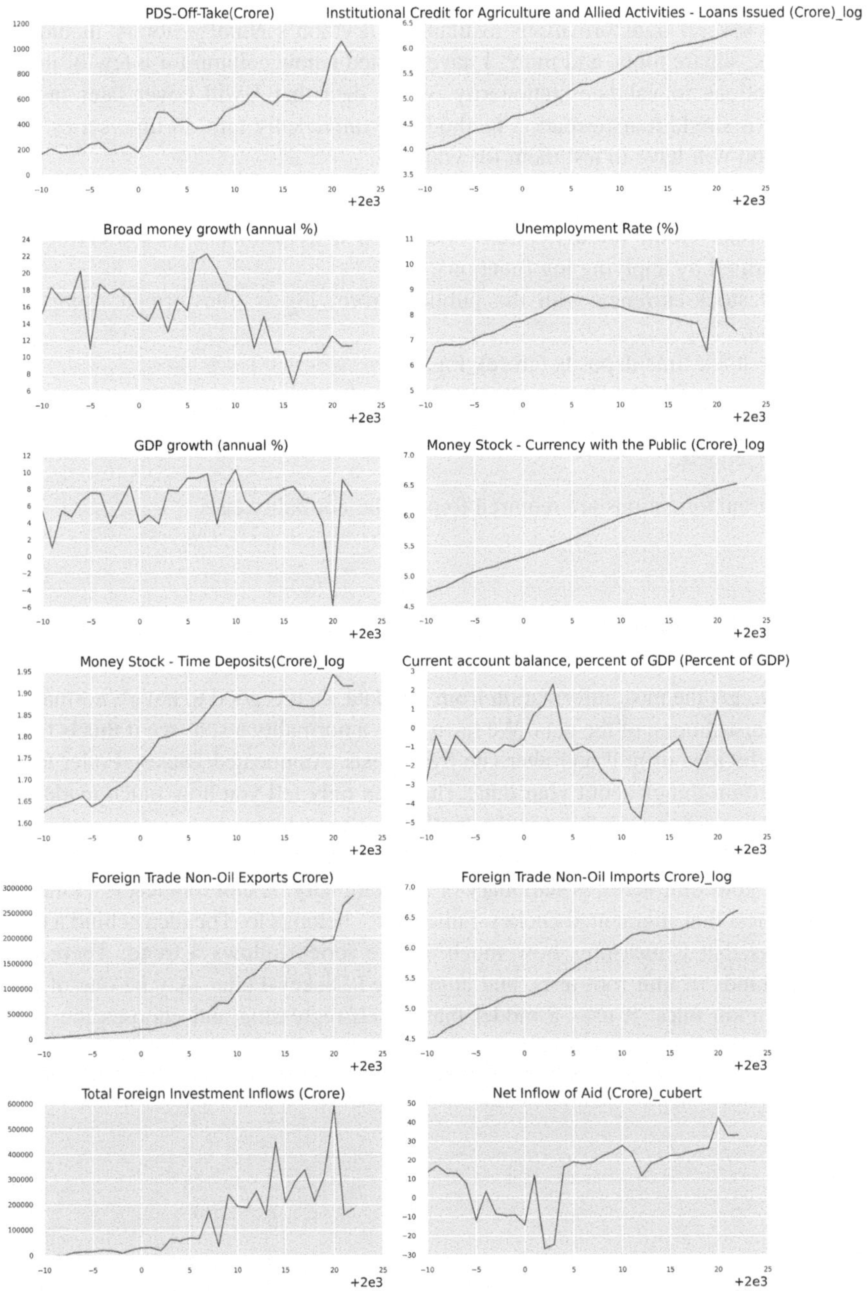
PDS-Off-Take(Crore)
Institutional Credit for Agriculture and Allied Activities - Loans Issued (Crore)_log
Broad money growth (annual %)
Unemployment Rate (%)
GDP growth (annual %)
Money Stock - Currency with the Public (Crore)_log
Money Stock - Time Deposits(Crore)_log
Current account balance, percent of GDP (Percent of GDP)
Foreign Trade Non-Oil Exports Crore)
Foreign Trade Non-Oil Imports Crore)_log
Total Foreign Investment Inflows (Crore)
Net Inflow of Aid (Crore)_cubert
+2e3

Step 6: Transformation

There are several transformations available in Python's NumPy library including logarithms, square roots, and more. I have created a new column for a few of these transformations to validate stationarity of the data that I will cover later in the chapter. No single transformation method will universally turn all-time series stationary; you will have to test them for yourself.

- Institutional credit for agriculture and allied activities-loans issued (crore)_log: institutional credit for agriculture and allied activities-loans issued (crore) is transformed by applying log functions.
- Money stock-currency with the public (Crore)_log is transformed using log function.
- Money stock-time deposits (crore)_log.
- Foreign trade non-oil imports (crore)_log.
- Net inflow of aid (crore) is transformed via cube root to form net inflow of aid (Crore)_cubert.

These transformations are required to pass the normality tests.

Step 7: Normality Test

We want to get the most information from our data, so it is good to have a normal or Gaussian distribution of the data. We have done a normality test to see if this is true, based on the idea of null and alternate hypothesis. Augmented Dickey-Fuller tests assume certain things about your data. They can only tell you how much evidence you have to reject or not reject a null hypothesis. The result has to be interpreted for a specific problem to make sense. However, they can give you a quick check and confirm if your time series is stationary or nonstationary. A unit root test is a kind of statistical test that augmented Dickey-Fuller test[35] belongs to. The idea behind a unit root test is that it measures how much a time series follows a trend. There are different kinds of unit root tests, and augmented Dickey-Fuller may be one of the more common ones. It uses a model that is self-regressing and chooses the best information criteria for different lag values.

[35] How to Check if Time Series Data is Stationary with Python (https://machinelearningmastery.com/time-series-data-stationary-python/)

```
from statsmodels.tsa.stattools import adfuller

def stationary_test(df,col):
    print("Observations of Dickey-fuller test")
    dftest = adfuller(df[col],autolag='AIC')
    dfoutput=pd.Series(dftest[0:4],index=['Test Statistic','p-value','#lags used','number of observations used'])
    for key,value in dftest[4].items():
        dfoutput['critical value (%s)'%key]= value
    print(dfoutput)
    if dfoutput[0] < dfoutput[4] or dfoutput[0] < dfoutput[5] or dfoutput[0] < dfoutput[6]:

        print(f"{col}: The series is stationary ")
    else :
        print(f"{col}: The series is not stationary ")
```

```
for i in dataset.columns:
    stationary_test(dataset,i )
```

- p-value >0.05: fail to reject the null hypothesis (H0); the data has a unit root and is nonstationary.
- p-value <= 0.05: reject the null hypothesis (H0); the data does not have a unit root and is stationary.

Output

Observations of Dickey-fuller test Test Statistic 1.375379 p-value 0.996999 #lags used 9.000000 number of observations used 23.000000 critical value (1%) -3.752928 critical value (5%) -2.998500 critical value (10%) -2.638967 dtype: float64 PDS-Off-Take(Crore): The series is **not stationary**	Observations of Dickey-fuller test Test Statistic -0.695069 p-value 0.848023 #lags used 0.000000 number of observations used 32.000000 critical value (1%) -3.653520 critical value (5%) -2.957219 critical value (10%) -2.617588 dtype: float64 Institutional Credit for Agriculture and Allied Activities - Loans Issued (Crore)_log: The series is **not stationary**

In summary, the following is the stationarity of the variables:

- Institutional credit for agriculture and allied activities-loans issued (crore): the series is not stationary.
- Bank credit to commercial sector (crore): the series is not stationary.
- Broad money growth (annual percentage): the series is not stationary.
- Unemployment rate (%): the series is stationary.
- GDP growth (annual percentage): the series is stationary.
- Money stock-currency with the public (Crore)_log: the series is not stationary.
- Money stock-time deposits (crore)_log: the series is not stationary.

- Current account balance, percent of GDP (percent of GDP): the series is stationary.
- Foreign trade non-oil exports (crore): the series is not stationary.
- Foreign trade non-oil imports (crore)_log: the series is not stationary.
- Total foreign investment inflows (crore): the series is not stationary.
- Net inflow of aid (crore)_cubert: the series is not stationary.
- Inflation rate, average consumer prices (annual percent change): the series is stationary.
- General government net lending/borrowing (percent of GDP): the series is stationary.

Perform kurtosis analysis:

```
import matplotlib.pyplot as plt
import scipy.stats as stats
from scipy.stats import kurtosis
dataset['PDS-Off-Take(Crore)'].plot(kind =
'density')

print('PDS-Off-Take(Crore): Kurtosis of
normal distribution:
{}'.format(stats.kurtosis(dataset['PDS-Off-
Take(Crore)'])))
print('PDS-Off-Take(Crore):Skewness of
normal distribution:
{}'.format(stats.skew(dataset['PDS-Off-
Take(Crore)'])))
```

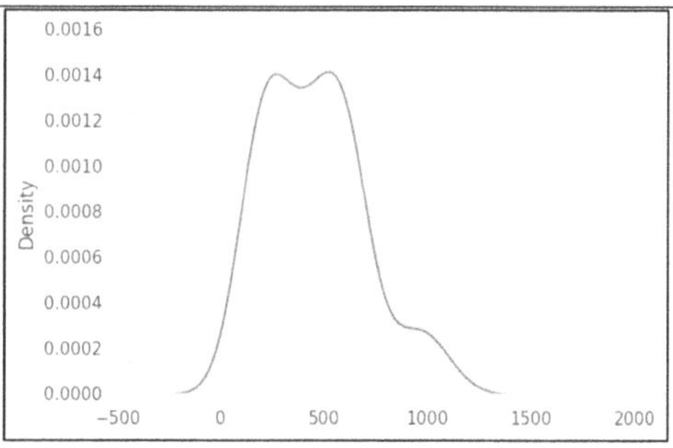

Output

PDS-off-take (crore)—kurtosis of normal distribution: −0.05624453480427949

PDS-off-take (crore)—skewness of normal distribution: 0.7045136806900931

The PDS-off-take (crore) distributions give us some intuition about the distribution of our data. The kurtosis of this dataset is less than 0; it is a light-tailed dataset. It has as much data in each tail as it does in the peak. Moderate skewness refers to the value between −1 and −0.5 or 0.5 and 1.

In terms of a probability density function (PDF), the kurtosis test would be used to determine how the tails of the given distribution deviate from the tails of a normal distribution. High kurtosis might indicate a large number of outliers, while low kurtosis might indicate a lack of outliers as can be seen in PDS-off-take.

Net inflow of aid (crore) kurtosis of normal distribution is 6.826 which shows that outliers exist in the data.

Step 8: Normality Probability Plot

Plot normal probability to see how data looks

```
plt.figure(figsize=(14,6))
plt.subplot(1,2,1)
dataset['PDS-Off-Take(Crore)'].hist(bins=30)
plt.title('PDS-Off-Take(Crore)')
plt.subplot(1,2,2)
stats.probplot(dataset['PDS-Off-Take(Crore)'], plot=plt);
dataset['PDS-Off-Take(Crore)'].describe().T
```

Output

```
count     33.000000
mean     454.348485
std      238.000105
min      164.900000
25%      230.500000
50%      422.500000
75%      602.800000
max     1058.000000
Name: PDS-Off-Take(Crore), dtype: float64
```

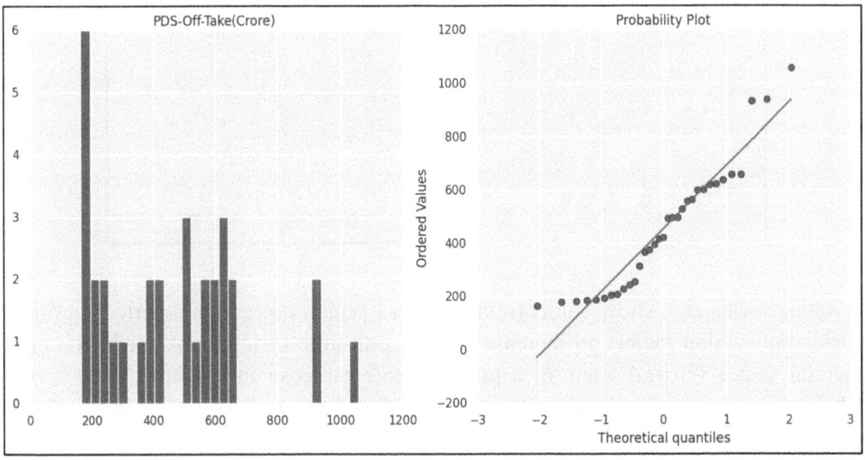

Normal probability plot also shows the data is far from normally distributed.

Step 9: Check for Autocorrelation

It is essential to check for serial correlation within the historical data.

```
import pandas as pd
import matplotlib.pyplot as plt
import statsmodels.api as sm
MAX_LAGS=1
# plots the autocorrelation plots for each stock's price at 150 lags
for i in dataset:
   sm.graphics.tsa.plot_acf(dataset[i])
   plt.title('ACF for %s' % i)
plt.show()plt.show()
```

Output

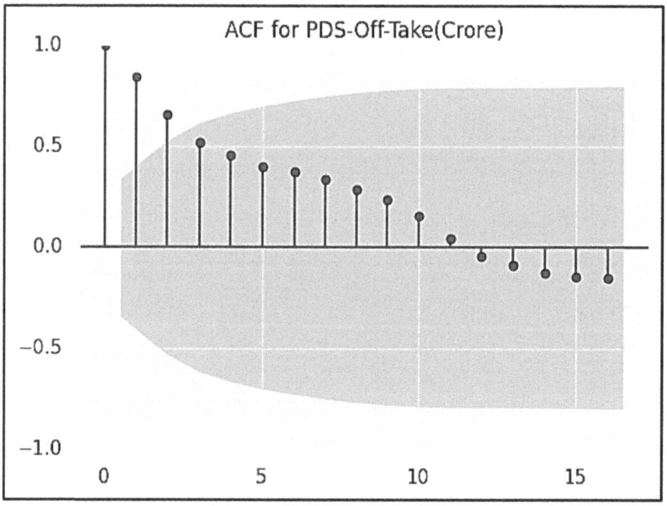

Other series also show autocorrelation of +1 which represents a perfect positive correlation which means an increase seen in one time series (bank credit to commercial sector (crore)) leads to a proportionate increase in the other time series (PDS-off-take (crore)). We need to apply transformation and neutralize this to make the series stationary. It measures linear relationships; even if the autocorrelation is minuscule, there may still be a nonlinear relationship between a time series and a lagged version of itself.

Step 10: Train and Test Data

The VAR model will be fitted on X_train and then used to forecast the next N observations. These forecasts will be compared against the actual present in test data. Observation could vary based on the frequency of the data.

```
n_obs=2
X_train, X_test = dataset[0:-n_obs], dataset[-n_obs:]
print(X_train.shape, X_test.shape)
```

Output

(31,14) (2,14)

Step 11: Transformation

The training set is subjected to the first difference to achieve stationarity for all series. However, if the series remains non-stationary even after the first difference, we may need to apply a second difference or log transformation to normalize the data. This iterative process may be necessary to ensure stationarity.

```
transform_data = X_train.diff().dropna()
transform_data.head()
```

Output

	PDS-Off-Take(Crore)	Institutional Credit for Agriculture and Allied Activities - Loans Issued (Crore)_log	Broad money growth (annual %)	Unemployment Rate (%)	GDP growth (annual %)	Money Stock - Currency with the Public (Crore)_log	Money Stock - Time Deposits(Crore)_log
Year							
1991	42.5	0.056670	3.250339	0.837	-4.4	0.061358	0.012523
1992	-27.9	0.035075	-1.454457	0.078	4.4	0.048222	0.007208
1993	6.5	0.081168	0.145168	-0.017	-0.7	0.081156	0.007210
1994	8.4	0.098592	3.270066	0.032	1.9	0.087542	0.010874
1995	49.1	0.100311	-9.266607	0.184	0.9	0.069883	-0.024019

Step 12: Stationarity Check

Rerun augmented Dick-Fuller test.

```
Iimport statsmodels.tsa.stattools as sm
def augmented_dickey_fuller_statistics(time_series):
  result = sm.adfuller(time_series.values, autolag='AIC')
  print('ADF Statistic: %f' % result[0])
  print('p-value: %f' % result[1])
  print('Critical Values:')
  for key, value in result[4].items():
    print('\t%s: %.3f' % (key, value))
```

```
for i in dataset.columns:
  stationary_test(transform_data,i )
```

Output

```
Observations of Dickey-fuller test
Test Statistic                 -3.147554
p-value                         0.023228
#lags used                      8.000000
number of observations used    21.000000
critical value (1%)            -3.788386
critical value (5%)            -3.013098
critical value (10%)           -2.646397
dtype: float64
PDS-Off-Take(Crore): The series is stationary
```

```
++++++++++++++++++++++++++++++++
Observations of Dickey-fuller test
Test Statistic                -9.463425e+00
p-value                        4.292807e-16
#lags used                     0.000000e+00
number of observations used    2.900000e+01
critical value (1%)           -3.679060e+00
critical value (5%)           -2.967882e+00
critical value (10%)          -2.623158e+00
dtype: float64
Broad money growth (annual %): The series is stationary
```

In summary, the following is the stationarity of the variables:

- Institutional credit for agriculture and allied activities-loans issued (crore)_log: the series is stationary.
- Broad money growth (annual percentage): the series is stationary.
- Unemployment rate (%): the series is stationary.
- GDP growth (annual percentage): the series is stationary.
- Money stock-currency with the public (Crore)_log: the series is stationary.
- Money stock-time deposits (crore)_log: the series is stationary.
- Current account balance, percent of GDP (percent of GDP): the series is stationary.
- Foreign trade non-oil exports (crore): the series is stationary.
- Foreign trade non-oil imports (crore)_log: the series is stationary.
- Total foreign investment inflows (crore): the series is stationary.
- Net inflow of aid (crore)_cubert: the series is stationary.
- Inflation rate, average consumer prices (annual percent change): the series is stationary.
- General government net lending/borrowing (percent of GDP): the series is stationary.

Step 12: Plot Visualization to See Variables Stationarity

```
fig, axes = plt.subplots(nrows=4, ncols=4, dpi=120, figsize=(10,6))
for i, ax in enumerate(axes.flatten()):
    d = transform_data[transform_data.columns[i]]
    ax.plot(d, color='red', linewidth=1)
    # Decorations
    ax.set_title(dataset.columns[i])
    ax.xaxis.set_ticks_position('none')
    ax.yaxis.set_ticks_position('none')
    ax.spines['top'].set_alpha(0)
    ax.tick_params(labelsize=6)
plt.tight_layout();
```

Output

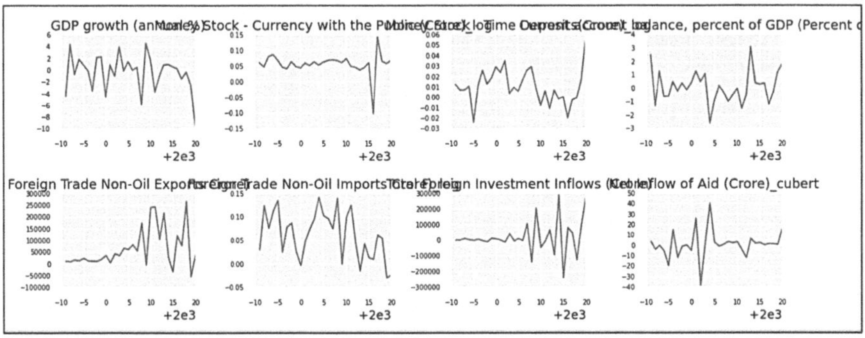

Step 13: Granger's Causality Test

Prof. Clive W.J. Granger, who won the 2003 Nobel Prize in Economics, created the concept of causality to enhance the accuracy of forecasting. It is essentially an econometric speculative test for checking the validity of using one variable to predict another in multivariate time series data with a specific delay.

Before doing the Granger causality test,[36] the data must be stationary, i.e., it should have a fixed mean, fixed variance, and no seasonal component. Make the nonstationary data stationary by subtracting it, either once or twice. Do not continue with the Granger causality test if the data is still not stationary after subtracting it twice.

The formal definition of Granger causality can be explained as whether past values of x aid in the prediction of y(t), conditional on having already accounted for the effects on y(t) of past values of y (and perhaps of past values of other variables).

---

[36] Granger Causality in Time Series—Explained using Chicken and Egg problem (https://www.analyticsvidhya.com/blog/2021/08/granger-causality-in-time-series-explained-using-chicken-and-egg-problem/)

If they do, the x is said to "Granger cause" y. So, the basis behind VAR is that each of the time series in the system influences each other.

Granger's causality test evaluates the null hypothesis that the coefficient of past values in the regression equation is zero. So, if the p-value obtained from the test is lesser than the significance level of 0.05, then, you can safely reject the null hypothesis. This has been performed on the original dataset.

```
import statsmodels.tsa.stattools as sm
maxlag=MAX_LAGS
test = 'ssr-chi2test'
def grangers_causality_matrix(X_train, variables, test = 'ssr_chi2test', verbose=False):
  dataset    =    pd.DataFrame(np.zeros((len(variables),    len(variables))),    columns=variables,
index=variables)
  for c in dataset.columns:
    for r in dataset.index:
      test_result = sm.grangercausalitytests(X_train[[r,c]], maxlag=maxlag, verbose=False)
      p_values = [round(test_result[i+1][0][test][1],4) for i in range(maxlag)]
      if verbose: print(f'Y = {r}, X = {c}, P Values = {p_values}')
      min_p_value = np.min(p_values)
      dataset.loc[r,c] = min_p_value
  dataset.columns = [var + '_x' for var in variables]
  dataset.index = [var + '_y' for var in variables]
  return dataset
grangers_causality_matrix(dataset, variables = dataset.columns)
```

Output

	1	2	3	4	5
		PDS-Off-Take(Crore)_x	**Foreign Trade Non-Oil Exports Crore)_x**	**Foreign Trade Non-Oil Imports Crore)_log_x**	**Total Foreign Investment Inflows (Crore)_x**
1	PDS-Off-Take(Crore)_y	1	**0.0497**	**0.0593**	**0.0014**
2	Broad money growth (annual %)_y	**0.0225**	0.002	0.0277	0.0557
3	**Foreign Trade Non-Oil Exports Crore)_y**	**0.0002**	1	0.4287	0
4	**Foreign Trade Non-Oil Imports Crore)_log_y**	**0.0174**	0.5477	1	0.139
5	**Total Foreign Investment Inflows (Crore)_y**	**0.0301**	0	0	1

I have trimmed the table to emphasize the variables that form bidirectional causation with PDS-off-take. The rows are the response (y), and the columns are the predictor series (x).

- If we take the value 0.0497 in row 1, column 3, it refers to the p-value of the Granger's causality test for foreign trade non-oil exports (crore)_x causing PDS-off-take (crore)_y. The 0.0002 in row 3, column 2, refers to the p-value of foreign trade non-oil exports (crore)_y causing PDS-off-take (crore)_x and so on.

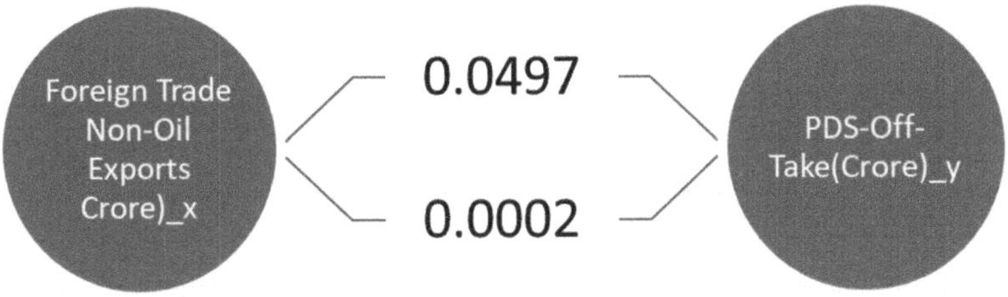

- Our variable of interest is causation of PDS-off-take (crore) on macro, finance, and economic variables—foreign trade non-oil exports, foreign trade non-oil imports (crore), and total foreign investment inflows (crore).

Step 14: VAR Model

VAR requires stationarity of the series which means the meaning of the series does not change over time (we can find this out from the plot drawn next to augmented Dickey-Fuller test). So, I will fit the VAR model on training set and then use the fitted model to forecast the next 14 observations. These forecasts will be compared against the actual present in test data. I have taken the maximum lag (1) to identify the required lags for VAR model [35].

```
import numpy as np
import pandas
import statsmodels.api as sm
from statsmodels.tsa.api import VAR

mod = VAR(transform_data)
results = mod.fit(maxlags=1)
print(results.summary())
```

Output

	Unemployment Rate (%)
PDS-Off-Take(Crore)	0.408267
Institutional Credit for Agriculture and Allied Activities - Loans Issued (Crore)_log	0.294022
Broad money growth (annual %)	-0.069629
Unemployment Rate (%)	1.000000
GDP growth (annual %)	-0.361392
Money Stock - Currency with the Public (Crore)_log	0.163294
Money Stock - Time Deposits(Crore)_log	0.222416
Current account balance, percent of GDP (Percent of GDP)	-0.018800
Foreign Trade Non-Oil Exports Crore)	0.017571
Foreign Trade Non-Oil Imports Crore)_log	0.025346
Total Foreign Investment Inflows (Crore)	0.599059
Net Inflow of Aid (Crore)_cubert	0.186940
Inflation rate, average consumer prices (Annual percent change)	-0.421405
General government net lending/borrowing (Percent of GDP)	-0.280540

Correlation matrix of residuals: with the highest for unemployment rate (%) (40,826%), total foreign investment inflows (27.23%), and negative correlations of −49.00% for consumer prices and −42.966% for net inflow of aid.

	Net Inflow of Aid (Crore)_cubert	Inflation rate average consumer prices (Annual percent change)
PDS-Off-Take(Crore)	-0.429669	-0.490061
Institutional Credit for Agriculture and /	-0.049972	-0.123609
Broad money growth (annual %)	0.257808	0.320646
Unemployment Rate (%)	0.186940	-0.421405
GDP growth (annual %)	0.057549	0.137334
Money Stock - Currency with the Public (Ci	-0.043055	0.213986
Money Stock - Time Deposits(Crore)_log	0.120593	-0.141254
Current account balance, percent of GDP (I	-0.304198	-0.256653
Foreign Trade Non-Oil Exports Crore)	0.064426	0.144793
Foreign Trade Non-Oil Imports Crore)_log	0.109211	0.551575
Total Foreign Investment Inflows (Crore)	0.123563	-0.182735
Net Inflow of Aid (Crore)_cubert	1.000000	0.001761
Inflation rate, average consumer prices (/	0.001761	1.000000
General government net lending/borrowing	0.022013	0.115335

The policy implications are very clear. The uptick in off-take takes place when consumer prices increase. Similarly, when net flow aid or foreign investment takes place, the off-take would decrease given in the increased economic opportunity. To create sustainable social and economic societies for poor and marginalized communities, focusing on economic improvements is the first order of the business.

The policy implications are very clear. The uptick in off-take takes place when consumer prices increase. Similarly, when net flow aid or foreign investment takes place, the off-take would decrease given in the increased economic opportunity. To create sustainable social and economic societies for poor and marginalized communities, focusing on economic improvements is the first order of the business.

Step 15: Residual Plot

Plot the residuals.

```
y_fitted = results.fittedvalues
y_fitted

y_fitted = results.fittedvalues
residuals = results.resid
plt.figure(figsize = (15,5))
plt.plot(residuals, label='resid')
plt.plot(y_fitted, label='VAR prediction')
plt.xlabel('Date')
plt.xticks(rotation=45)
plt.ylabel('Residuals')
plt.grid(True)
```

Output

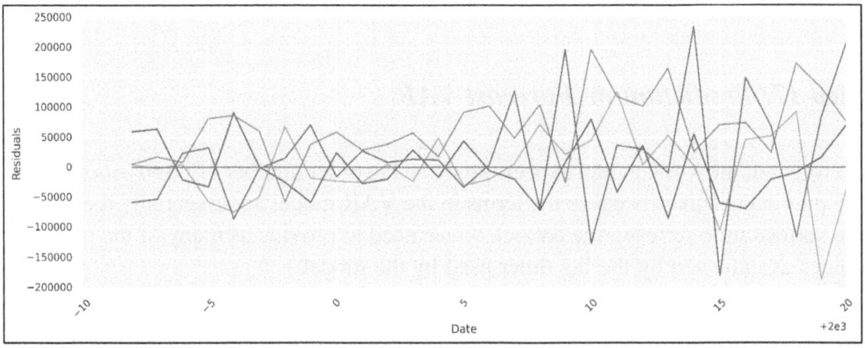

The spikes of prices during certain economic events can be seen.

Step 16: Durbin-Watson Statistics

The Durbin-Watson statistics can only take values from 0 to 4. A value of 2.0 implies that there is no autocorrelation in the sample. Values closer to 0 mean positive autocorrelation, and values closer to 4 mean negative autocorrelations. A rough guideline is that test statistic values between 1.5 and 2.5 are fairly normal. Any value beyond this range might be a problem.

```
from statsmodels.stats.stattools import durbin_watson
out = durbin_watson(results.resid)
for col,val in zip(transform_data.columns, out):
  print((col), ":", round(val,2))
```

Output

```
PDS-Off-Take(Crore) : 1.81
Institutional Credit for Agriculture and Allied Activities - Loans Issued (Crore)_log : 1.88
Broad money growth (annual %) : 1.74
Unemployment Rate (%) : 1.34
GDP growth (annual %) : 1.61
Money Stock - Currency with the Public (Crore)_log : 1.17
Money Stock - Time Deposits(Crore)_log : 1.46
Current account balance, percent of GDP (Percent of GDP) : 1.75
Foreign Trade Non-Oil Exports Crore) : 1.94
Foreign Trade Non-Oil Imports Crore)_log : 1.91
Total Foreign Investment Inflows (Crore) : 2.16
Net Inflow of Aid (Crore)_cubert : 1.63
Inflation rate, average consumer prices (Annual percent change) : 2.25
General government net lending/borrowing (Percent of GDP) : 1.81
```

All the values of DW are between 1.5 and 2.5.

The Durbin-Watson statistics will always have a value between 0 and 4. A value of 2.0 means that there is no autocorrelation.

Step 17: Prediction or Forecast VAR

To forecast, the VAR model expects up to the lag order number of *observations* from the past data. This is because the terms in the VAR model are essentially the lags of the various time series in the dataset, so we need to provide as many of the previous values as indicated by the lag order used by the model.

```
# Get the lag order
lag_order = results.k_ar
print(lag_order)
# Input data for forecasting
input_data = transform_data.values[-lag_order:]
print(input_data)
# forecasting
pred = results.forecast(y=input_data, steps=n_obs)
pred = (pd.DataFrame(pred, index=X_test.index, columns=X_test.columns + '_pred'))
print(pred)
```

Output

```
12
Year      PDS-Off-Take(Crore)_pred      Net Inflow of Aid (Crore)_cubert_pred
2021      -340.665413                   -50.571037
2022       518.462166                    90.236880
```

Step 18: Invert the Transformation

The forecasts are generated, but it is on the scale of the training data used by the model. So, to bring it back up to its original scale, we need to de-difference it. The way to convert the differencing is to add these differences consecutively to the base number. An easy way to do it is to first determine the cumulative sum at index and then add it to the base number. This process can be reversed by adding the observation at the prior time step to the difference value: inverted (ts) = differenced(ts) + observation(ts − 1).

```
def invert_transformation(X_train, pred):
    forecast12 = pred.copy()
    columns = X_train.columns

    for col in columns:
        print('+++++++++++++++')
        print(col)
        forecast12[str(col)+'_pred'] = X_train[col].iloc[-1] + forecast12[str(col)+'_pred'].cumsum()
        print('+++++++++++++++')
    return forecast12

output = invert_transformation(X_train, pred)
output
```

Output

Year	PDS-Off-Take(Crore)_pred	Institutional Credit for Agriculture and Allied Activities - Loans Issued (Crore)_log_pred	Broad money growth (annual %)_pred	Unemployment Rate (%)_pred	GDP growth (annual %)_pred	Money Stock - Currency with the Public (Crore)_log_pred	Money Stock - Time Deposits(Crore)_log_pred	Current account balance, percent of GDP (Percent of GDP)_pred
2021	596.334587	6.151547	7.036524	4.147447	24.138631	6.537153	1.901588	-1.314843
2022	1114.796753	6.463154	8.371193	13.160074	-29.457645	6.392998	1.989379	4.369987

```
#combining predicted and real data set
combine = pd.concat([output['PDS-Off-Take(Crore)_pred'], X_test['PDS-Off-Take(Crore)']], axis=1)
combine['accuracy'] = round(combine.apply(lambda row: row['PDS-Off-Take(Crore)_pred'] /row['PDS-Off-Take(Crore)'] * 100, axis = 1),2)
combine['accuracy'] = pd.Series(["{0:.2f}%".format(val) for val in combine['accuracy']],index = combine.index)
combine = combine.round(decimals=2)
```

Output

Step 19: Evaluation

To evaluate the forecasts, a comprehensive set of metrics, such as MAPE, ME, MAE, MPE, and RMSE can be computed. We have computed some of these below.

```
from sklearn.metrics import mean_absolute_error
from sklearn.metrics import mean_squared_error
import math
#Forecast bias
forecast_errors = [combine['PDS-Off-Take(Crore)'][i]- combine['PDS-Off-Take(Crore)_pred'][i] for i in range(len(combine['PDS-Off-Take(Crore)']))]
bias = sum(forecast_errors) * 1.0/len(combine['PDS-Off-Take(Crore)'])

print('Bias: %f' % bias)
print('Mean absolute error:', mean_absolute_error(combine['PDS-Off-Take(Crore)'].values, combine['PDS-Off-Take(Crore)_pred'].values))
print('Mean squared error:', mean_squared_error(combine['PDS-Off-Take(Crore)'].values, combine['PDS-Off-Take(Crore)_pred'].values))
print('Root mean squared error:', math.sqrt(mean_squared_error(combine['PDS-Off-Take(Crore)'].values, combine['PDS-Off-Take(Crore)_pred'].values)))
```

Output

	Year	PDS-Off-Take(Crore)_pred	PDS-Off-Take(Crore)	accuracy
1	2022	1114.80	931.5	119.68%
0	2021	596.33	1058.0	56.36%

Policy Implication

The main goal of this ML model is to analyze the factors that affect public distribution (food security) of essential food commodities with financial development and economic growth in India using time series analysis so as to sustain the economic and social well-being of poor and marginalized communities. Some findings from the model are, firstly, the time series analysis shows bidirectional causality of macroeconomic variables and money and banking variables with PDS-off-take. In particular, *foreign trade non-oil exports and total foreign investment inflows* macroeconomic variables and *institutional credit for agriculture and allied activities-loans issued, bank credit to commercial sector, and money stock-time deposits* money and banking variables influence public distribution system off-take.

On going one more layer deeper, a time deposit is a type of bank account that earns interest and has a set date of expiry. The money in a time deposit has to remain in the account of the agreed period to get the specified interest rate [36]. Time deposits usually pay more than regular savings accounts, but they have a low return rate and a charge for early withdrawal. The public distribution system (PDS) is a food security system in India that provides food items (rice, wheat, sugar, pulses, edible oils) and nonfood items (kerosene) to India's poor[37] [30]. The term "off-take" in the context of PDS means the amount of food grains taken by the beneficiaries from the fair price shops. The impact of time deposits on PDS-off-take is not directly related as they are two different concepts—one related to banking and the other to food security. However, if we look at a broader economic view, a rise in time deposits could indicate a rise in savings and financial stability among the people. This could possibly lead to a decrease in the reliance on PDS as people might be able to buy food grains from the open market. But this is a very general view, and the actual impact can vary based on many factors such as income distribution, inflation, changes in subsidy policies, etc. [37].

Loans and credit facilities are also very important for agriculture and related activities, and they affect food security in India a lot. Agricultural credit helps to get the capital needed to increase farm output, income, and well-being of farmers, especially small and marginal farmers who do not have the capital to buy essential inputs on time for farming operations. Only 33% of farmers can access credit facilities. Credit access has positive effects on farm investments, such as land improvement, livestock, and machinery. It also affects the farm revenue expenditure positively, including the spending on seeds, machinery, labor, irrigation, plant protection chemicals, and livestock inputs. The financial reforms and credit policy measures in the agricultural sector have helped with the agricultural development as well as the long-standing problems of poverty, food insecurity, and undernourishment in the country. This has allowed the agriculture and related activities sectors to contribute a lot to the country's overall growth, development, and food security [38, 39].

The ML time series results lead to the following inferences. First, the model indicates that the growth of credit from banks and money markets to the commercial sector is the main way that financial development boosts economic growth in the Indian economy. Second, the inflows of total foreign investment and the non-oil exports of foreign trade improve financial development by raising the demand for products or the supply of broad money in the economy. The general implication of this is that financial development can be a policy tool to increase economic growth and economic growth can also be a policy tool to advance financial development in India because a rise in one will directly benefit the other. Lastly, more than 60% of farmers do not have access to credit. There is a need for inclusive and holistic policy interventions to include all farmers in the credit system, specifically offering term loans to small and marginal farmers. It is also suggested that simplified crop loan and Kisan credit card facilities be made available to tenant farmers.

[37] Public Distribution System (PDS) (https://nfsa.gov.in/portal/PDS_page)

In conclusion, while loans and credit facilities have a positive impact on agriculture and food security in India, there is a need for more inclusive policies to ensure that all farmers, especially small and marginal ones, can access these facilities. This will further enhance the positive impact of loans on food security in the country. Overall, the food security that lasts can be attained in India through both economic and financial policy actions. As a policy implication, this model shows that enhancing institutional credit for agriculture and related activities through loans, bank credit to the commercial sector, and attractive time deposits to households and firms will have a positive impact on India's long term economic growth. Moreover, higher foreign investment has an important role in creating jobs and economic activities that help the poor and marginalized communities escape from poverty and ensure overall sustainability.

Policy recommendation: small farmers build the world!

While loans and credit facilities have a positive impact on agriculture and food security in India, there is a need for more inclusive policies to ensure that all farmers, especially small and marginal ones, can access these facilities. This will further enhance the positive impact of loans on food security in the country. Overall, the food security that lasts can be attained in India through both economic and financial policy actions. As a policy implication, this model shows that enhancing institutional credit for agriculture and related activities through loans, bank credit to the commercial sector, and attractive time deposits to households and firms will have a positive impact on India's long term economic growth. Moreover, higher foreign investment has an important role in creating jobs and economic activities that help the poor and marginalized communities escape from poverty and ensure overall sustainability.

Author

Data Science Infused Policy Implications: Large Language Models and Public Services Forecasting Patterns

Large language models and time series forecasting can help us understand the trend and make flexible public service policy based on that. Rather than reacting to emergencies, the public service agencies can prepare for future challenges.

Step 1: Prepare Time Series Prophet Model

Load Python libraries, specifically Prophet time series model (bolded).

```
import numpy as np
import pandas as pd
import functools
from datetime import date
import plotly.graph_objects as go
import seaborn as sns; sns.set()
from sklearn.impute import SimpleImputer
from scipy import stats
from scipy.stats import pearsonr
import matplotlib.pyplot as plt
from sklearn.preprocessing import MinMaxScaler
from sklearn.model_selection import train_test_split
from sklearn.linear_model import LinearRegression
from sklearn.metrics import r2_score
from sklearn import metrics
from sklearn.impute import SimpleImputer
from fbprophet import Prophet
from  fbprophet.diagnostics import cross_validation, performance_metrics
```

Step 2: Prepare Prophet Time Series Variables

Prepare Prophet DataFrame to load key time series attributes that are validated in earlier regressive model.

```
dfProphetDataset = df.copy()
dfProphetdf = dfProphetDataset.copy()
dfProphetdf.rename({'Year': 'ds'}, axis=1, inplace=True)
dfProphetdf.rename({'PDS-Off-Take(Crore)': 'y'}, axis=1, inplace=True)
dfProphetdf.columns
```

Prophet requires two key columns: date time "ds" and Y variable "Yield _Value."
Output

```
Index(['ds', 'y',
    'Institutional Credit for Agriculture and Allied Activities - Loans Issued (Crore)_log',
    'Broad money growth (annual %)', ' Unemployment Rate (%)',
    'GDP growth (annual %)',
    'Money Stock - Currency with the Public (Crore)_log',
    'Money Stock - Time Deposits(Crore)_log',
    'Current account balance, percent of GDP (Percent of GDP)',
    'Foreign Trade Non-Oil Exports Crore)',
    'Foreign Trade Non-Oil Imports Crore)_log',
    'Total Foreign Investment Inflows (Crore)',
    'Net Inflow of Aid (Crore)_cubert',
    'Inflation rate, average consumer prices (Annual percent change)',
    'General government net lending/borrowing (Percent of GDP)'],
   dtype='object')
```

Step 3: Check Any Nulls

```
dfProphetdf.isna().sum()
```

Output

```
ds                                                                                  0
y                                                                                   0
Institutional Credit for Agriculture and Allied Activities - Loans Issued (Crore)_log    0
Broad money growth (annual %)                                                       0
 Unemployment Rate (%)                                                              0
GDP growth (annual %)                                                               0
Money Stock - Currency with the Public (Crore)_log                                  0
Money Stock - Time Deposits(Crore)_log                                              0
Current account balance, percent of GDP (Percent of GDP)                            0
Foreign Trade Non-Oil Exports Crore)                                                0
Foreign Trade Non-Oil Imports Crore)_log                                            0
Total Foreign Investment Inflows (Crore)                                            0
Net Inflow of Aid (Crore)_cubert                                                    0
Inflation rate, average consumer prices (Annual percent change)                     0
General government net lending/borrowing (Percent of GDP)                           0
dtype: int64
```

Step 4: Split Data to Model Time Series

Based on number of observations, split data into train and test data frames.

```
from prophet import Prophet

## test and train data split

n_obs=6
fbProphet_train, fbProphet_test = dfProphetdf[0:-n_obs],dfProphetdf[-n_obs:]
print(fbProphet_train.shape,fbProphet_test.shape)
```

Output

```
(27, 15) (6, 15)
```

Step 5: Prepare Prophet Time Series Model

Prophet provides uncertainty intervals for the forecast yhat by default. These uncertainty intervals are based on several important assumptions. Uncertainty in the trend[38]: the main source of uncertainty in the forecast is the possibility of future changes in the trend. Prophet can find and fit these, but how should we anticipate the trend changes in the future? We don't know for sure, so we do the best thing we can, and we assume that the future will have similar trend changes to the past. We assume that the average frequency and size of trend changes in the future will be the same as what we see in the past. We extend these trend changes forward, and by calculating their distribution, we get uncertainty intervals.

Uncertainty in the trend: the biggest source of uncertainty in the forecast is the potential for future trend changes. The Prophet is able to detect and fit these, but what trend changes should we expect moving forward? It's impossible to know for sure, so we do the most reasonable thing we can, and we assume that the future will see similar trend changes as history. We assume that the average frequency and magnitude of trend changes in the future will be the same as that which we observe in history. We project these trend changes forward, and by computing their distribution, we obtain uncertainty intervals. The width of the uncertainty intervals (by default 80%) can be set using the parameter interval_width:

```
multi_model = Prophet(interval_width = 0.95)
```

We're trending Prophet model that the uncertainty in the data is 95%. Again, these intervals assume that the future heat wave magnitude will see the same frequency and magnitude of rate changes as the past.

Uncertainty in seasonality: by default, Prophet will only return uncertainty in the trend and observation noise. To get uncertainty in seasonality, you must do full Bayesian sampling. This is done using the parameter mcmc.samples (which defaults to 0).

[38] Uncertainty Intervals (https://facebook.github.io/prophet/docs/uncertainty_intervals.html)

```
Prophet_cols = [
    'Institutional Credit for Agriculture and Allied Activities - Loans Issued (Crore)_log',
    'Broad money growth (annual %)', ' Unemployment Rate (%)',
    'GDP growth (annual %)',
    'Money Stock - Currency with the Public (Crore)_log',
    'Money Stock - Time Deposits(Crore)_log',
    'Current account balance, percent of GDP (Percent of GDP)',
    'Foreign Trade Non-Oil Exports Crore)',
    'Foreign Trade Non-Oil Imports Crore)_log',
    'Total Foreign Investment Inflows (Crore)',
    'Net Inflow of Aid (Crore)_cubert',
    'Inflation rate, average consumer prices (Annual percent change)',
    'General government net lending/borrowing (Percent of GDP)']
```

```
# adding all columns in add regressor
for col in Prophet_cols:
  multi_model.add_regressor(col)
```

```
multi_model.fit(fbProphet_train)
```

Next, fit the model.

We make a DataFrame for future predictions. Here, we keep capacity constant at the same value as in history and forecast 5 years into the future:

```
future = multi_model.make_future_dataframe(periods = fbProphet_test.shape[0],freq='Y',
include_history=True)
future.tail(5)
```

```
fbProphet_test['ds']
```

```
# filling all columns data in future frame except for y
for col in Prophet_cols:
  future[col] = dfProphet[col]
```

In the above code, we created data frame and filled all column data in future frame except for y.

Step: Forecast Model

```
forecastProphet = multi_model.predict(future)
forecastProphet
```

```
forecastProphet[['ds','yhat', 'yhat_lower','yhat_upper']].tail()
```

```
fig1 = multi_model.plot(forecastProphet)
```

Output

	ds	yhat	yhat_lower	yhat_upper
28	2017-12-31	592.431621	510.339471	666.308020
29	2018-12-31	509.235173	426.885359	588.055149
30	2019-12-31	754.435191	674.441896	834.107303
31	2020-12-31	855.344512	773.073071	938.359872
32	2021-12-31	833.194010	752.986384	914.173075

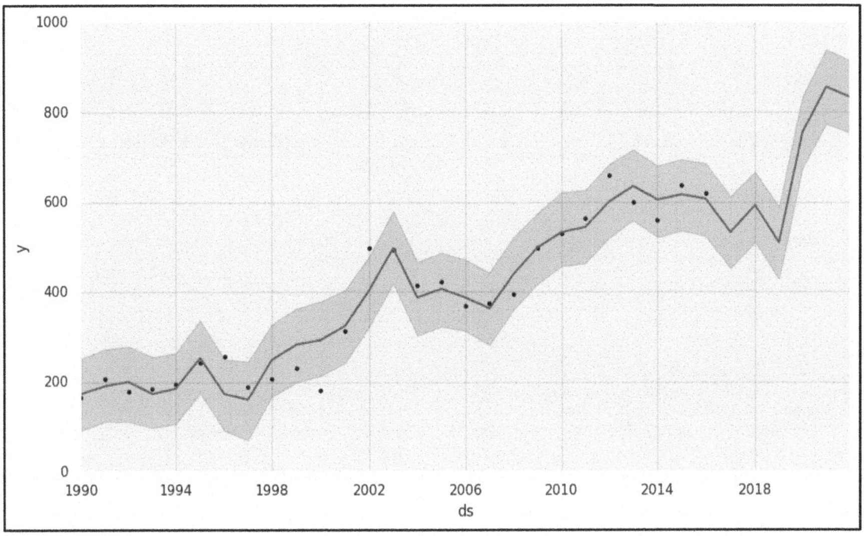

The above graph closely predicts the values from actuals with high degree of accuracy.

```
multi_model.plot_components(forecastProphet)
```

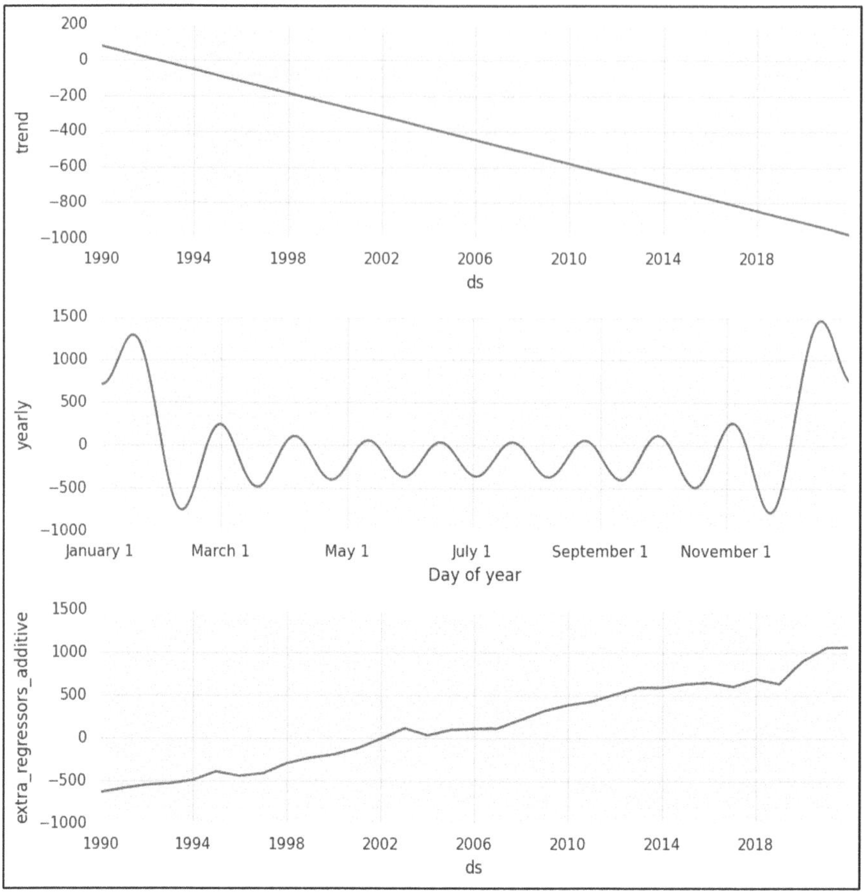

Step: Evaluate the Model

```
#combining predicted and real data set
combineProphetUpper = pd.concat([forecastProphet['yhat'], dfProphet['y']], axis=1)

combineProphetUpper['accuracy'] = round(combineProphetUpper.apply(lambda row: row.yhat /row.y * 100, axis = 1),2)
combineProphetUpper['accuracy'] = pd.Series(["{0:.2f}%".format(val) for val in combineProphetUpper['accuracy']],index = combineProphetUpper.index)
combineProphetUpper = combineProphetUpper.round(decimals=2)
combineProphetUpper
```

Output

	yhat	yhat_upper	accuracy
0	174.03	253.00	68.79%
1	191.63	273.47	70.07%
2	200.88	279.18	71.95%
3	173.88	255.52	68.05%
4	185.69	264.82	70.12%
5	253.70	336.04	75.50%
6	173.53	250.31	69.33%
7	160.09	245.16	65.30%
8	249.08	326.56	76.28%
9	283.33	362.70	78.12%

```
from sklearn.metrics import mean_absolute_error
from matplotlib import pyplot

plt.figure(figsize=(18, 9))
mae = mean_absolute_error(y_true, y_pred)
print('MAE: %.3f' % mae)
# plot expected vs actual
pyplot.plot(y_true, label='Actual')
pyplot.plot(y_pred, label='Predicted')
pyplot.legend()
pyplot.show()
```

Output

MAE: 47.100 crore

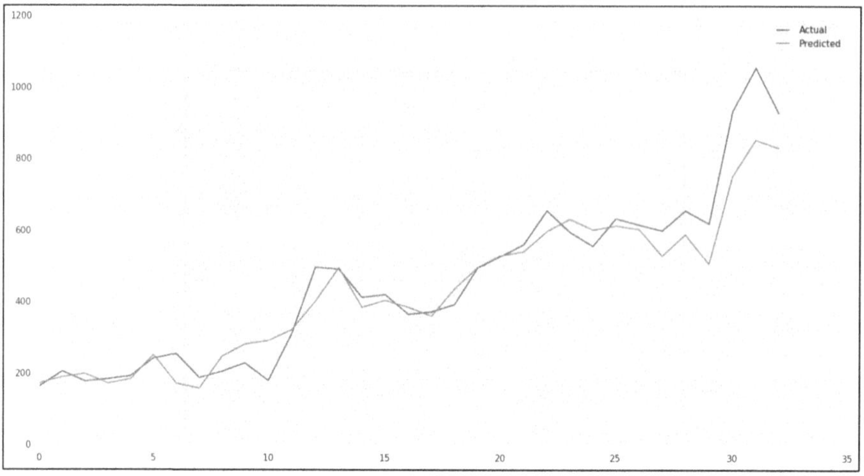


```
fig = go.Figure()

# Add traces
fig.add_trace(go.Scatter(x=dfProphetdf['ds'], y=dfProphetdf['y'],
                mode='lines',
                name='Actual)'))
fig.add_trace(go.Scatter(x=forecastProphet['ds'], y=forecastProphet['yhat'],
                mode='lines',
                name='Forecast'))
fig.update_layout(
    title="actual - forecast")

fig.show()
```

Output

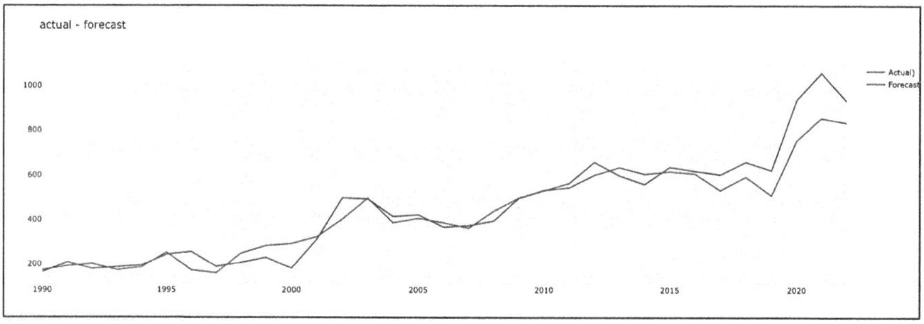

Economic sustainability policy

If India can continue to build on its economic strength, it can be a source of stability for the world economy and provide a safe destination for restless global capital.

Shri Pranab Mukherjee
Minister of Finance, 2012–2013[39]

Agricultural Productivity and Prosperity for Farmers

To ensure prosperity for farmers, availability of free markets and capital/credit is one of the important criteria. The Indian government has focused on liberalizing farm markets, removing distortions, and investing in logistics as noted in 2020 budget. Another important aspect of improving well-being of farmers includes improvement in agricultural productivity. These include sustainable cropping patterns, technology adoption, and cooperation with states that can be promoted to enhance agricultural productivity and income. These are part of the Aspirational India as noted in the budget speech 2020.

Machine learning
Model_PDSProcurementModel_ForPoorerSectionsOftheSociety_India.ipynb

Agricultural Credit

Investment in agriculture is the backbone of the many countries in the world. Given more than half of the youth are employed in agriculture and allied industries, it's pragmatic for the governmental policies to support and increase investments in agriculture. Allocation of a significant portion of investible resources for agriculture and rural development would create an environment of sustained agricultural growth. Additionally, increasing budgetary support for the rural sector to upskill in latest agricultural productivity ensures sustained growth [2]. Lowering the statutory liquidity ratio (SLR) to enhance agricultural lending (please see Fig. 5.11). The government [1] in 1993–1994 has decided to begin a phase reduction in the SLR, which would release funds for banks to expand credit to agriculture and industry. This reduction was in line with the proposed reduction in the fiscal deficit. Is SLR policy effective?

[39] Budget Speech Budget 2012–2013 (https://www.indiabudget.gov.in/doc/bspeech/bs201213.pdf)

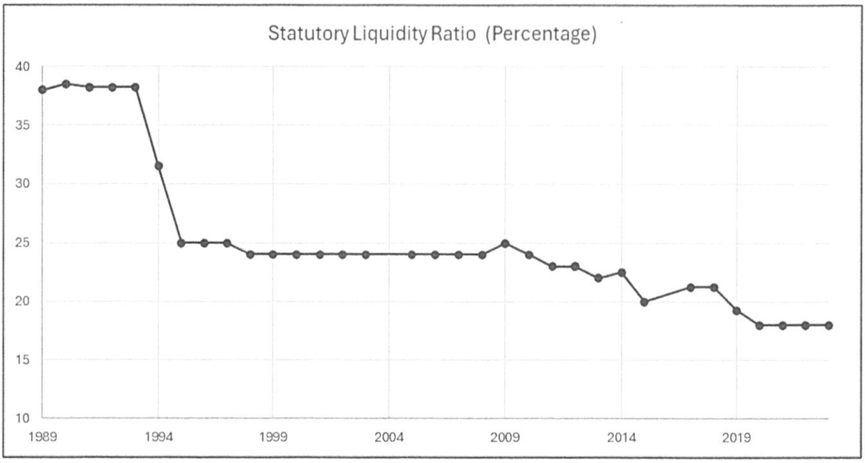

Fig. 5.11 SLR

The statutory liquidity ratio data is updated daily, averaging 19.500% from March 1949 to November 26, 2023, with 27,284 observations. The data reached an all-time high of 38.5% in 1989 and a record low of 18.000% in 2023. India statutory liquidity ratio data[40] is reported by the Reserve Bank of India. Please see SLR Data.[41]

Policy: reduction in the statutory liquidity ratio (SLR) [1]

The government decided to begin a phase reduction in the SLR, which would release funds for banks to expand credit to agriculture and industry. This reduction was in line with the proposed reduction in the fiscal deficit.

Shri Manmohan Singh
Minister of Finance, 1993–1994[42]

Some of the investments in future for agriculture include providing debt relief and fertilizer subsidies to benefit rural areas. Increase investment in irrigation, land development, and soil and moisture conservation to improve agricultural productivity, especially in rainfed and semiarid areas, alleviate social stress and poverty. Promote diversification of agriculture and the development of agro-based processing industries [2]. Agriculture productivity is very necessary and essential as the population increases and urbanization is taking place at a rapid pace. It's now a mandate to produce more with less land to meet the future food security.

[40] SLR (https://www.rbi.org.in/commonman/English/History/Scripts/BankrateCRRandSLRChanges.aspx)

[41] Table 44: Major Monetary Policy Rates and Reserve Requirements - Bank Rate, LAF (Repo, Reverse Repo and MSF) Rates, CRR & SLR (https://rbi.org.in/Scripts/PublicationsView.aspx?id=21850)

[42] India Budget Speech, 1993–1994 (https://www.indiabudget.gov.in/doc/bspeech/bs199394.pdf)

Debt relief for farmers is essential to overcome cycle of indebtedness and low incomes. Agriculture is prone to risk and challenges of climate and extreme weather bore by farmers first in terms of health and second fiscal prudence. The goal of debt relief is to keep farmers encouraged for farming, removing obstacles that currently prevent many farmers from returning to their land, and improving the way that banks and governments approach borrowing and loan servicing in the long term. For many farmers, including those who have been hard hit by pandemic-induced market disruptions exacerbated by more frequent, more intense,[43] climate-driven natural disasters, this debt relief assistance is vital if they are to continue producing the food, fiber, and fuel that are essential to the well-being of not only our rural communities but our nation as a whole. Providing relief to farmers burdened with debt, especially those with loans up to Rs.10,000 (USD $250), would provide immediate relief. Additionally, debt relief should be provided to counter increasing inflationary trends across the world. For instance, the US Department of Agriculture announced the farm loan relief program funded from $3.1 billion set aside in the Inflation Reduction Act allocated toward assisting distressed borrowers of direct or guaranteed loans administered by the USDA[44] [40]. The law was passed by Congress and signed by President Joe Biden in August 2022.

In India, the farm loan waivers, whereby outstanding loans are forgiven if the debt meets criteria set out by the government, are generally succeeds any major natural calamity, market disruption, crop failure, natural disasters, and elections:

- The first national loan waivers date back to 1990, when the central government waived Rs100 billion ($5.6 billion) in farm loans [41].
- In 2008, the government announced a major round of loan waivers totaling Rs 650 billion, equivalent to 1.3% of India's GDP. In 2008, against the backdrop of the 2008–2009 global financial crisis, the Indian government enacted one of the largest borrower bailout programs in history. The program known as the Agricultural Debt Waiver and Debt Relief Scheme (ADWDRS) unconditionally cancelled fully or partially the debts of up to 60 million rural households across the country, amounting to a total volume of US$ 16–17 billion [41, 42].
- Over the 9 years until March 2017, the central and state governments waived nearly Rs890 billion (approximately US$14 billion) in loans to 48 million farmers [41].
- Since April 2017, four states have announced farmer loan waivers, prompting farmers in other states to demand similar measures. The estimated cost of loan waivers announced by Uttar Pradesh, Maharashtra, Punjab, and Karnataka totaled US$13.6 billion. Analysts predict that the loan waivers, if implemented nationally, will ultimately cost between 2 and 2.6% of GDP (approximately US $40–50 billion) [41].

[43] Inflation Reduction Act Assistance for Distressed Borrowers (https://www.farmers.gov/loans/inflation-reduction-investments/assistance)

[44] USDA announces $1 billion in debt relief for 36,000 farmers (https://www.pbs.org/newshour/politics/usda-announces-1-billion-in-debt-relief-for-36000-farmers)

> The USDA announces $1 billion in debt relief for 36,000 farmers, October 16, 2022
>
> Through no fault of their own, our nation's farmers and ranchers have faced incredibly tough circumstances over the last few years. The funding included in today's announcement helps keep our farmers farming and provides a fresh start for producers in challenging positions [40].
>
> Tom Vilsack
> Agriculture Secretary

To enhance agricultural production, investment in irrigation, land development, and water and soil conservation is being prioritized. Expanding the range of crops and developing industries that process agricultural products in areas of high productivity. Creating an Agricultural Policy Resolution to establish the basis of agricultural development [2]. Raising the proportion of investible resources used for the development of agriculture and rural development. Additionally, sustainable agriculture practices would help productivity and farmer incomes.

Sustainable Farming Practices

We are encouraging the balanced use of all kinds of fertilizers, including traditional organic and other innovative fertilizers, to reduce the excessive use of chemical fertilizers and promote sustainable farming practices. We are expanding integrated farming systems in rainfed areas, promoting multitier cropping, beekeeping, solar pumps, solar energy production in non-cropping seasons, and zero-budget natural farming. Implement measures[45] to combat land degradation and water logging, such as promoting sustainable farming practices, watershed management, and efficient irrigation systems [43].

We are implementing the village storage scheme, run by self-help groups, to provide farmers with good holding capacity and reduce their logistics costs, promoting sustainable agriculture practices. We are promoting the development of a seamless national cold supply chain for perishables, including milk, meat, and fish, through the setting up of "Kisan Rail" and the inclusion of refrigerated coaches in express and freight trains.

We are taking steps to eliminate diseases in livestock, such as foot and mouth disease, brucellosis, and pest des petits ruminants (PPR), by 2025, ensuring the health and well-being of animals and improving the livelihoods of farmers.

Credit availability was a perineal issue before independence. In 1947 budget, the Finance Minister clearly stated the state of the economy. "There seems little doubt now that the severity of the last budget, 1946, was defeating its own purpose and was hindering the formation of capital for productive purposes." Post democratic Indian

[45] Budget Speech 2003–2004 (https://www.indiabudget.gov.in/doc/bspeech/bs200405.pdf)

government has taken credit availability on war-front beginning 1947 as failure to achieve sufficient credit dangers the very foundations of society and economic life of the country [44] being undermined by deepening penury and despair.[46]

To provide adequate credit for our farmers, the government has enhanced the agricultural credit target to Rs. 16.5 lakh crore in FY22. Similarly, the allocation to the Rural Infrastructure Development Fund increased from Rs. 30,000 crores (USD $ 3.59 billion) to Rs. 40,000 crores (USD $4.797 billion). The Micro Irrigation Fund, with a corpus of Rs.5000 crore (USD $599 million), which has been created under the National Bank for Agriculture and Rural Development (NABARD) will be doubled [45].

Machine Learning Model: Sustained Agricultural Production and Productivity Enhancement Needs in the Era of Climate Change

Eliminating poverty is essential, as it is a matter of both ethics and economy in our era. India has made great strides from food crisis in the post 1960s to self-reliance achievement in the late 2000s. It's practical and important to maintain a renewed focus and measurable progress on the front of agriculture, as it is the backbone of the nation. Diversification and value chain development in agriculture to address the second-generation issues after the Green Revolution are needed, and integration of technological innovation to improve agriculture is the key to overcome current shortcomings and to achieve sustainable growth for the future.

Prime Minister Lal Bahadur Shastri asked Indians to skip at least one meal a week in 1965 when India was fighting Pakistan. This was one of the many challenges the country faced in supplying food for its growing population. It is a different story that today we produce excess food grains [46] (please see Fig. 5.12).

India's challenge to meet the growing population's need started in 1937 when Burma (now Myanmar) became separate from British India. The problem got worse when India lost West Punjab and East Pakistan during the partition.[47] Burma was an important region for producing pulses, East Pakistan was a rice-rich area, and West Punjab, which had a well-connected canal system, was a wheat hub. A key event happened in 1943 that prolonged India's food shortage until the early 1950s [47]. In 1943, Prime Minister Winston Churchill instructed food grains destined for eastern parts of the country to be redirected to the British troops involved in World War II. It caused a human-made famine in Bengal and the deaths of millions of people [46].

Indian agriculture faced challenges when food output dropped to a low of 62 million tons (mt) in 1957–1958, the lowest in 5 years. A team of US agriculture experts who came to India in 1959 advised Prime Minister Jawaharlal Nehru to give

[46]India Budget Speech, 1947–1948 (https://www.indiabudget.gov.in/doc/bspeech/bs194748.pdf)

[47]Budget Speech 1947–1948 (https://www.indiabudget.gov.in/doc/bspeech/bs194748.pdf)

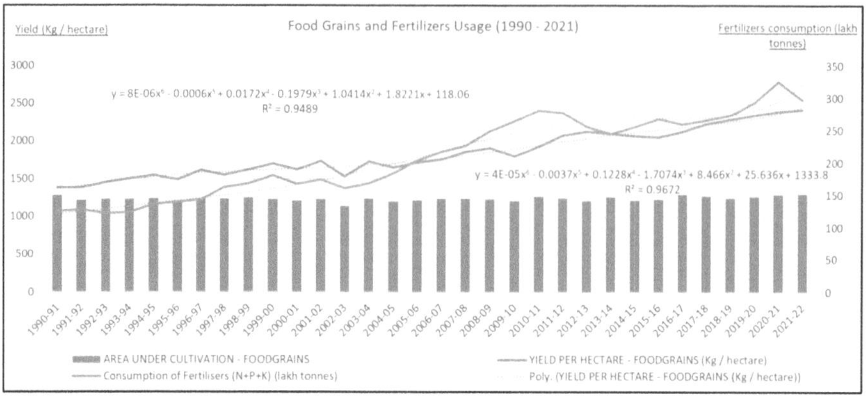

Fig. 5.12 Food grains and fertilizer usage

more priority to agriculture than industry in the second 5-year plan. Earlier, in 1954, India had entered into a long-term agreement [48] with the USA under Public Law (PL) 480 to receive food aid as part of the US government's agricultural trade development assistance. The agreement was renewed several times until the USA stopped it in the late 1960s because Lal Bahadur Shastri and then Indira Gandhi did not agree to make policy changes, especially to allow private sector involvement in the industrial sector, in exchange for food aid [46].

The agriculture sector in India faced difficulties after independence because it was ignored in favor of industries [48]. Farming suffered from the imports of food grains, which discouraged farmers from increasing their output. In his book [49] *Green Revolution*[48], Chidambaram Subramaniam [50], who became the Food and Agriculture Minister under Shastri, wrote that he supported bringing science and technology to farming. He also invited agricultural scientists for an honest and open discussion on the state of agricultural science. "I had to hear a sad story," he wrote. Subramaniam [49] claimed that self-reliance in food production was more important than importing food grains. India required a large amount of fertilizers, which would cost $250 million at that time, 1965. It needed foreign financial help, and India got it from then US President Lydon B Johnson. That's how history unfolded.

Green Revolution

India started the Green Revolution [51] in 1960, which led to a large growth in food grain output, based on the creation of high-yielding types of wheat and pulses, water supply systems, and other infrastructural work.

[48]The Green Revolution in India (https://artsandculture.google.com/story/the-green-revolution-in-india/vwUhSHBIOFTDiA?hl=en)

Agriculture is facing more difficulties that make it less productive and unable to meet future needs. On one hand, population growth requires more land for urbanization and housing. As a result, the land with irrigation facilities for agriculture is decreasing. On the other hand, the food security needs are rising to ensure the nutritional needs per capita, which requires more agricultural output. This means higher yield per hectare is needed to maintain national food security. Moreover, there are multiple challenges facing agriculture, such as extreme weather, climate change, water levels, and human resources availability. To avoid food crises and shortages that India faced in 1964, I think it is necessary to adopt technological revolution as part of agriculture. Continuing with the usual practices will lead to serious problems in the future. India needs second agricultural revolution, the same scientific path as inspired by Chidambaram Subramaniam, with data science that can understand signals from nature and nurture our paths into a sustainable future to meet demands of food of our future generations [46].

Sustainable development and environmental protection [52]

Promoting sustainable development practices and protecting the environment can help alleviate poverty by ensuring long-term access to natural resources and improving resilience to climate change impacts.

In Fig. 5.12, India's agricultural production of food grains[49] 1990:2021 is shown. The area under cultivation (bar graph) has not changed much since 1990 (i.e., an increase of 1.88% since 1990). However, consumption of fertilizers went up by 137.45% and yield went up by 81.15%. In essence, the usage of fertilizers has increased drastically.

To expand the area of irrigation, new lands have to be cleared and farmed, which need investment from farmers. Government, supporting agriculture by providing reliable irrigation, credit, and diversification and creating a market for agricultural products, encourages growth in areas under cultivation that slightly increased in the last four decades. Raising the budget for irrigation projects and enhancing the Command Area Development Programme to facilitate participatory irrigation management. Carrying out schemes and initiatives to foster the development of horticulture, fisheries, and jute industries [53].

Government should focus on improving agriculture productivity for the following reasons:

- Prioritize subsidies: evaluate and prioritize subsidies based on their necessity and deserving recipients. Allocate subsidies only to those who truly need them, ensuring efficient use of limited resources.
- Focus on agriculture: continue supporting the agriculture sector by promoting increased food grain production, exports, and agricultural credit. Implement policies to ensure sustainable growth and improve the livelihoods of farmers.

[49] Table17Agricultural Production—Food grain (https://rbi.org.in/Scripts/PublicationsView.aspx?id=21823)

- Water.
- Land.
- Climate change.

In the 2014–2015 budget, it was highlighted the importance of disseminating real-time information to farmers regarding[50] new farming techniques, water conservation, and organic farming to achieve economic and social sustainability [54].

Rural economic sustainability mantra: increasing farmers' income and promoting agriculture and farmers' welfare; enhancing rural employment and infrastructure development; providing healthcare services and implementing health protection schemes; and Improving education, skills, and job creation to create a productive society.

During the formation and early years of independence in 1947, the first budget speech by the Finance Minister underscored the necessity to increase domestic food production.[51] To achieve the self-sufficiency needed in food grains, the government prioritized and invested in agricultural development. The government achieved this objective by providing subsidies, training, and modern farming techniques to farmers, as well as improving irrigation systems and infrastructure. Second is the necessity to increase vital food grains through partnerships from international agriculture producers. To meet food demand, the newly formed Indian government, 1947, adopted import policy of secure *surplus wheat* from countries like Australia.

Exploring partnerships with other countries for food imports or agricultural technology transfer was a priority that provides economic and social benefits. Third, infrastructure development through the active investment in infrastructure, particularly in areas like river development and dam construction, can have long-term benefits for irrigation, power generation, and flood protection. The government should continue to prioritize and fund such projects.

The imports of policies have paid off. It is being pointed out that the average annual growth rate of gross domestic product (GDP) hit the 5.6% mark in the 1980s, well before the launch of the July 1991 reforms.[52] Alternatively, the growth rate in the 1990s was not much higher. Therefore, liberalization cannot be credited with having made a significant difference to growth in India [55].

The difference between the reforms in the 1980s and those in the 1990s is that the former were limited in scope without a clear road map whereas the latter were systematic and systemic [55]. This said the reforms in the 1980s must be viewed as precursor to those in the 1990s rather than a part of the isolated and sporadic liberalizing actions during the 1960s and 1970s, which were often reversed within a short period. The 1980s reforms proved particularly crucial to building the confidence of politicians regarding the ability of policy changes such as devaluation, trade liberalization, and delicensing of investment to spur growth without disruption. It is

[50] Budget Speech Budget 2014–2015 (https://www.indiabudget.gov.in/doc/bspeech/bs201415.pdf)

[51] India Budget Speech, 1947–1948 (https://www.indiabudget.gov.in/doc/bspeech/bs194748.pdf)

[52] India in the 1980s and 1990s: A Triumph of Reforms (https://www.imf.org/external/pubs/ft/wp/2004/wp0443.pdf)

questionable, for example, whether the July 1991 package would have been politically acceptable in the absence of experience and confidence in liberal policies acquired during the 1980s.

The agricultural credit policy of India has been to increase the access and availability of institutional farm credit. This has led to in general increase in the flow of formal farm credit in the agricultural sector. However, the impact of institutional credit and crop productivity especially for small and marginal farmers across social groups is not well recognized in India using field data. Accordingly, this field data study contributes to the existing research by providing fresh evidence from field across social groups for both kharif and rabi crops using recent survey data from small and marginal farmers which has important policy implications.

India and Wheat Production

Wheat is the principal grain crop in India.[53] The crop covered about 29.8 million hectares of land in 2018 and expanded to 31.6 million hectares in 2021.[54] Wheat production in the country has risen considerably from 75.81 million MT in 2006–2007 and 94.88 million MT in 2011–2012 to an all-time high of 107.85 million MT in 2019–2020. The yield of wheat which was 2602 kg/hectare in 2004–2005 and 3140 kg/hectare in 2011–2012 has grown to 3440 kg/hectare in 2019–20.[55] A significant increase in the yield of wheat has been seen in the states of Haryana, Punjab, and Uttar Pradesh. Higher area under the crop has been reported from Madhya Pradesh (MP) in recent years (please see Table 5.1).

Let's look at consumption of fertilizers (please see Fig. 5.13):

Consumption of fertilizers growing exponentially (although growing at polyphonic at the order).

Indian wheat is mostly a soft/medium hard, medium protein, white bread wheat, similar to US hard white wheat. Wheat from central and western India is usually hard, with high protein and high gluten content. India also grows around 1.0–1.2 million tons of durum wheat, mainly in the state of Madhya Pradesh (MP). Most Indian durum is not sold separately due to mixing problems in the market yards. However, some amounts are bought by the private trade at a higher price, mainly for making higher value/branded products. The production and productivity of wheat crop were very low when India gained independence in 1947. The wheat output was only 6.46 million tons, and yields were barely 663 kg per hectare during 1950–1951,

[53] Farmer Portal (https://farmer.gov.in/M_cropstaticswheat.aspx)

[54] What Area under cultivation (https://www.statista.com/statistics/765704/india-area-of-cultivation-for-wheat/)

[55] Hand Book of Statistics on Indian States (https://www.rbi.org.in/Scripts/AnnualPublications.aspx?head=Handbook+of+Statistics+on+Indian+States)

Table 5.1 Wheat production

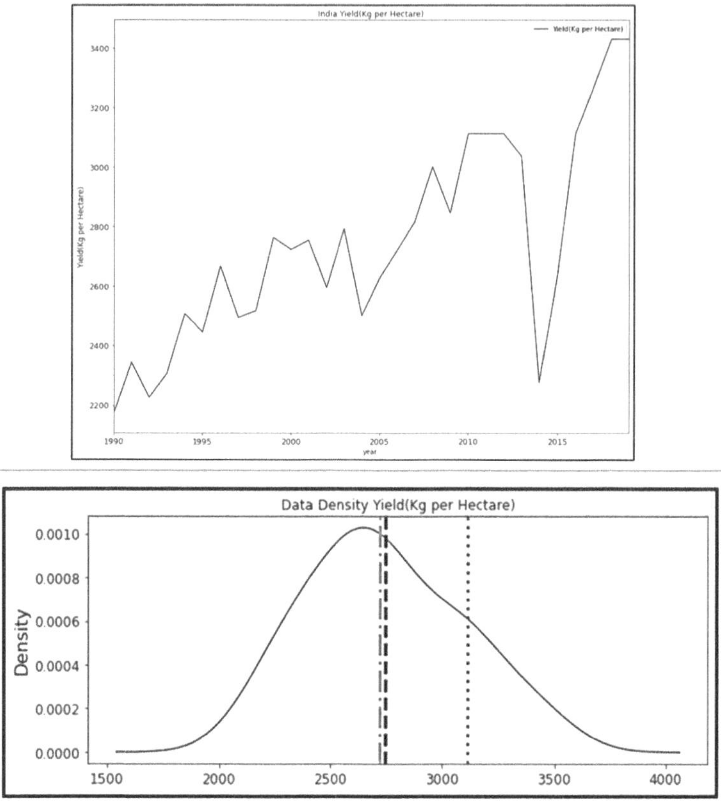

The data distribution of yield (kg per hectare) left skewed with minimum of 2171.00 (kg per hectare), mean of 2745.50 (kg per hectare), median of 2722.50 (kg per hectare), mode of 3113.00, and maximum of 3432.00 (kg per hectare)

which was not enough to feed the Indian population. The country used to import wheat in large quantities to meet the needs of our people from many countries like the USA under PL-480.[56]

The factors that caused wheat to have low production and productivity at that time were the tall plant height that led to lodging, when grown in rich soils, the weak tillering and low grain filling ability of the varieties used, higher vulnerability to diseases, higher response to changes in temperature and light, etc., leading to poor adaptability, and longer crop cycle leading to a long exposure of plants to weather changes and insect pest / disease damage.

[56] USAID and PL–480, 1961–1969 - https://history.state.gov/milestones/1961-1968/pl-480

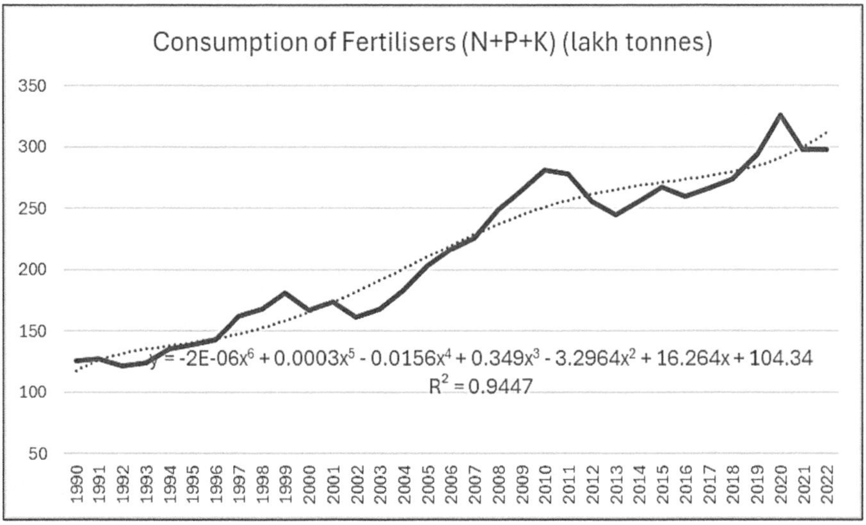

Fig. 5.13 Fertilizer consumed per

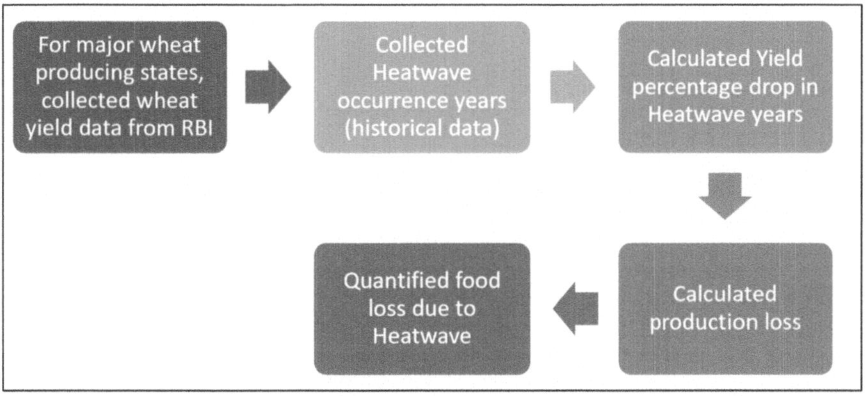

Fig. 5.14 Heatwave modeling

In the following heat wave-yield loss basic bending process (please see Fig. 5.14), progression of food security due to heat wave is outlined. Although the intermediary statistical steps could be performed in a nonsequential order, nonetheless, the inputs and outputs of the process are the start and end of process, respectively. The data is from the Reserve Bank of India (RBI) *Handbook of Statistics on India States.*[57]

[57] Handbook of statistics on India states (https://www.rbi.org.in/Scripts/AnnualPublications.aspx?head=Handbook+of+Statistics+on+Indian+States)

Indian Wheat Growing Zones

The entire wheat growing areas of the country have been categorized into six major zones as follows (please see Table 5.2)[58]:

The text focuses on three main states: Uttar Pradesh, Madya Pradesh, and Punjab. Both Utta Pradesh (UP) and Punjab are known as wheat breadbasket of India, and Punjab is India's "grain bowl" where the government has promoted wheat and rice farming since the 1960s. It usually gives the most to India's national reserves, and the government expected to buy about one-third of this year's crop from the region. But government estimates forecast lower harvests this year, 2022 (see Fig. 5.15), of about 25% less. The situation is similar in other big wheat-producing states like Uttar Pradesh and Madhya Pradesh.

Beyond India, other countries are also grappling with poor harvests that hinder their ability to help offset the potential shortfall of supplies from Russia and Ukraine, normally the world's largest and fifth largest exporters of wheat. According to China's Agriculture Minister, 2022 the winter wheat harvest was likely to be poor, hindered by flooding and by delays in planting (please see Fig. 5.16).

Map: courtesy of Farmer-India.gov website: https://farmer.gov.in/M_cropstaticswheat.aspx (access date: June 11, 2022).

Table 5.2 India—wheat growing zones

Zones	States/regions covered	Approx. area (million ha)
Northern Hill Zone (NHZ)	Hilly areas of J&K (except Jammu, Kathua, and Samba districts), Himachal Pradesh (except Una and Paonta valleys), Uttarakhand (excluding Tarai region), and Sikkim	0.8
North Western Plains Zone (NWPZ)	Punjab, Haryana, Western UP (except Jhansi Div), Rajasthan (excluding Kota and Udaipur div), Delhi, Tarai region of Uttarakhand, Una and Paonta valley of HP, Jammu, Samba and Kathua districts of J&K and Chandigarh	11.55
North Eastern Plains Zone (NEPZ)	Eastern UP (28 dist), Bihar, Jharkhand, West Bengal, Assam, Odisha, and other NE states (except Sikkim)	10.5
Central Zone	MP, Gujarat, Chhattisgarh, Kota and Udaipur div of Rajasthan, and Jhansi div of UP	5.2
Peninsular Zone	Maharashtra, Tamil Nadu (except Nilgiris and Palani hills), Karnataka and Andhra Pradesh	1.6
Southern Hill Zone (SHZ)	Nilgiris and Palani hills of Tamil Nadu	0.1

[58] India Wheat (https://farmer.gov.in/M_cropstaticswheat.aspx)

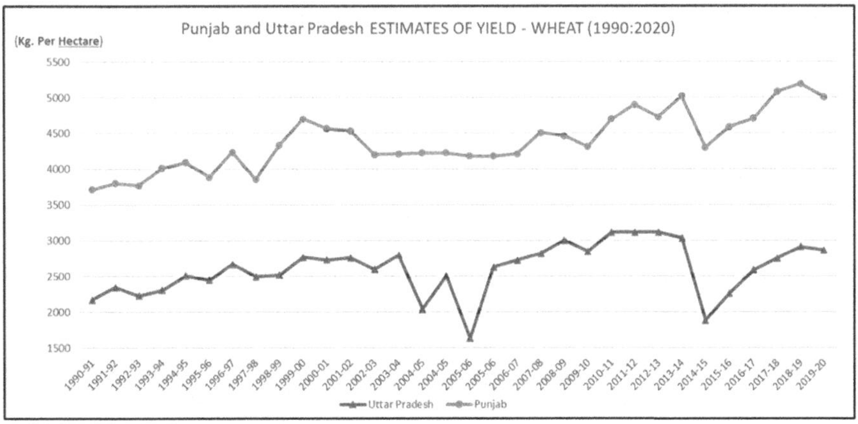

Fig. 5.15 Punjab and Uttar Pradesh wheat yield (1990–2020)

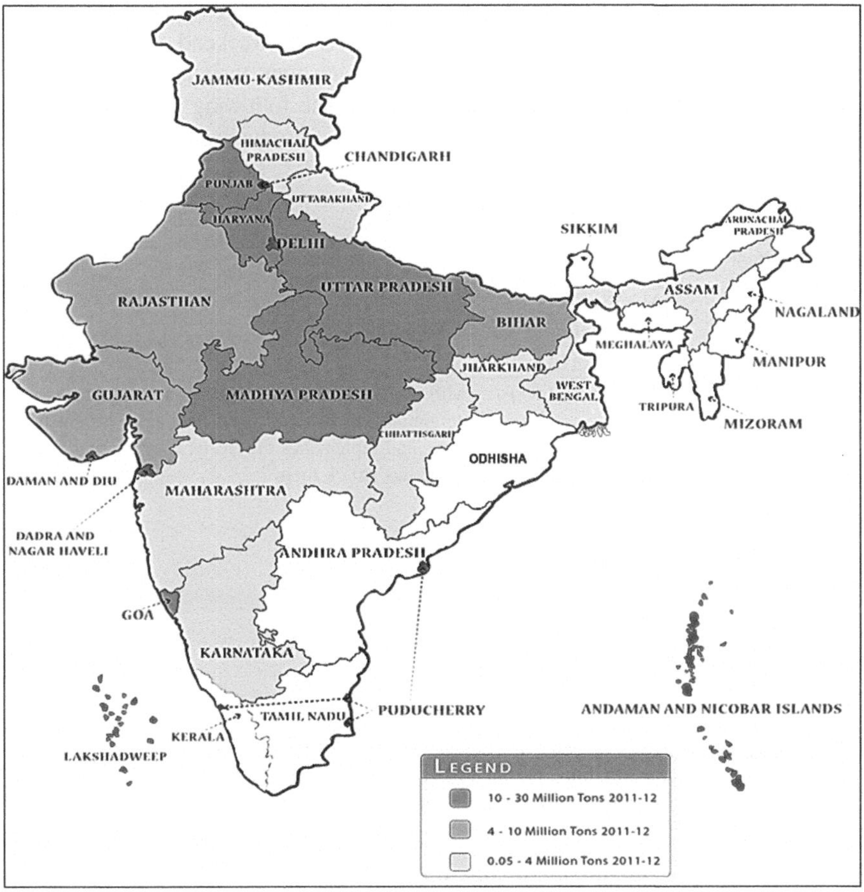

Fig. 5.16 Wheat producing major states map

Climate change and heat waves are attributed to recent increase in loss of productivity and yield drop in major wheat-producing areas of India. To assess the impact, the following statistical process is used:

India and Heat Waves: Wheat Yield Drop

According to a 2021 [56] report, India's risk of extreme heat went up 15% from 1990 to 2019. It is one of the top 5 countries where vulnerable people, like the elderly and the poor, face the most heat exposure. India and Brazil have the highest deaths from heat in the world. An unprecedented heat wave with daytime temperatures soaring up to 46.5 degrees Celsius (114.08 Fahrenheit). Temperatures that reached above 45 °C (113 °F) in nine other cities in India have damaged wheat crops, raising issues about how the country will meet its domestic needs with plans to increase exports and compensate for gaps due to Russia's war in Ukraine [57]. It was the heat in March 2022—the warmest in India since records began in 1901—that hurt crops. Wheat is very sensitive to heat, especially *during the last stage when its grains grow and mature*. Moreover, maximum temperatures are expected to hit 50 ° C (122 °F) in some places later in the week or over the weekend, with very high overnight temperatures staying [58]. Indian farmers plant their crops so that this stage happens with India's usually cooler spring. The following Table 5.3 shows heat wave occurrences in India, especially major heat producing states.

Table 5.3 India heat wave historical table

Year	Duration	Maximum temperature	Agricultural impact
1956		50.6 °C (123.1 °F)[a]	
1995	June	45.5 °C (113.9 °F)	Millions of Indians are desperate for the monsoon, an annual torrent that lowers the heat, extinguishes fires, and replenishes the rivers, after a lethal heat wave has struck them[b]
1998	May–June	49.5 °C (121.1 °F)	
2002	April–May	49 °C (120 °F)	
2015	May–June	49.4 °C (120.9 °F)	
2016	April–May	51.0 °C (123.8 °F)	
2019	May–June	50.8 °C (123.4 °F)	
2022	March–May	50.8 °C (123.4 °F)	

[a]Heat wave in India: 1956 (https://news.google.com/newspapers?nid=1241&dat=19980612&id=FI5TAAAAIBAJ&sjid=L4YDAAAAIBAJ&pg=5859,2125235)
[b]Heat wave in India—1995 (https://news.google.com/newspapers?nid=1980&dat=19950617&id=yDtAAAAAIBAJ&sjid=_K0FAAAAIBAJ&pg=5707,3954685)

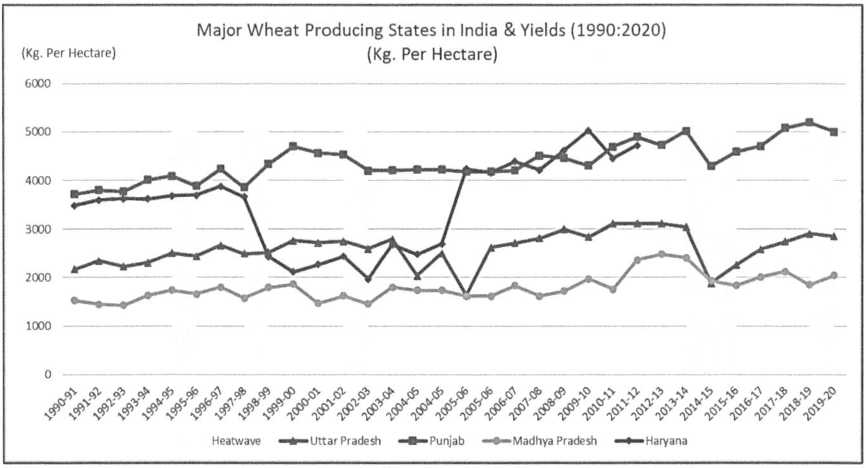

Fig. 5.17 Wheat yield (1990–2020)

Wheat yield declines sharply as heat waves become more intense, both in maximum temperature and duration. The years of heat wave show a lower yield. The decline is related to the timing of the heat wave (May–June), the number of days it lasts, and the winds that help spread the heat wave over larger areas (see Fig. 5.17). The gray areas mark the years of heat wave. Allahabad, in Uttar Pradesh, faced the peak of 2015 (June 30) and it affected the harvest season.

2015 Heat Wave in India and Pakistan

What yields have seen drastic reduction in 2015, especially in Uttar Pradesh (UP)? For much of May, parts of India were gripped in an intense heat wave that has seen the mercury rising above 110 °F (43 °C). Temperatures have been hot enough (over 111 °F, or 44 °C) to melt pavement in the capital city, New Delhi. The searing heat across the country—which has seen the capital observe five consecutive days with high temperatures over 110 °F (43 °C)— is the worst heat event in a decade according to the India Meteorological Department.[59] The impact of heat event can be see in the production of wheat (please see Fig. 5.18). In the below figure, as intensity of heat wave increased, the drop in yield can be seen.

The production of wheat has been greatly affected by a severe drop in yields in 2015, especially in Uttar Pradesh (UP). A lot of India suffered (please see Fig. 27) from an extreme heat wave for much of May, with temperatures reaching over 110 ° F (43 °C) in some areas. The heat was so intense (above 111 °F, or 44 °C) that it

[59] India heat wave kills thousands (https://www.climate.gov/news-features/event-tracker/india-heat-wave-kills-thousands)

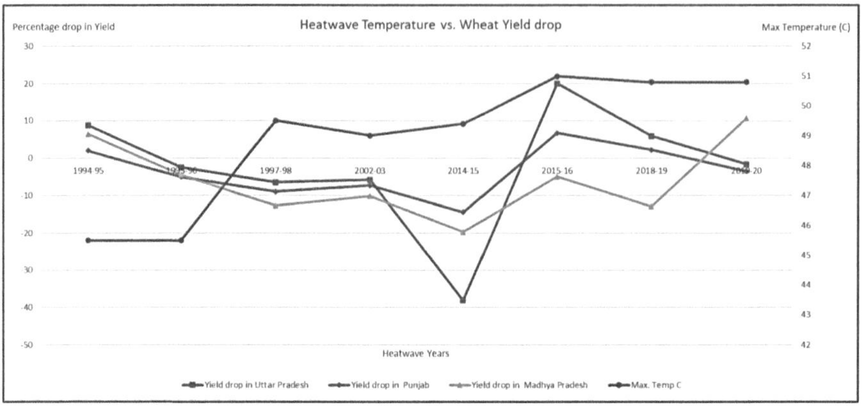

Fig. 5.18 Heatwave & Yield drop

melted the road surface in the capital city, New Delhi. The scorching heat across the nation—which had the capital experience 5 days in a row with highs above 110 °F (43 °C)—is the worst heat event in 10 years according to the India Meteorological Department. This is evident in the wheat output. In the below figure, as intensity of heat wave increased, the drop in yield can be seen.

The southeast Indian states of Odisha, Andhra Pradesh, and Telangana were severely affected by the heat, with temperatures rising above 113 °F (45 °C). The town of Titlagarh in Odisha experienced a scorching 117.7 °F (47.6 °C)! Near the coast, the city of Ongole in Andhra Pradesh had an average high temperature of 110 °F (43.47 °C) between May 24 and 30. Further inland, in Jharkhand, the city of Daltonganj had an average high temperature of 115.6 °F for the whole week of May 24–30, reaching 117 °F (47.2 °C) on May 27. Coastal cities experienced an increased level of discomfort and danger due to the humidity accompanying the extreme heat, which is considered highly unusual even for India (please see Fig. 5.19).

The Intergovernmental Panel on Climate Change (IPCC) warns that more frequent and severe heat waves in Asia (including India) will harm vulnerable communities and raise death rates in the future. In fact, they say that heat waves probably happen more often in Asia now than they did in 1950. If carbon emissions are high in the future, the hottest temperature that happens once every 20 years will likely happen twice as often (a 1-in-10-year event) by the end of this century. Studies that focus only on India also find that heat waves will be longer, stronger, and more frequent in the future. Because heat wave affected major producing states, 2015 wheat yield went down a lot. By supporting sustainable[60] development practices and

[60]Budget Speech 2005–2006 https://www.indiabudget.gov.in/doc/bspeech/bs200506.pdf

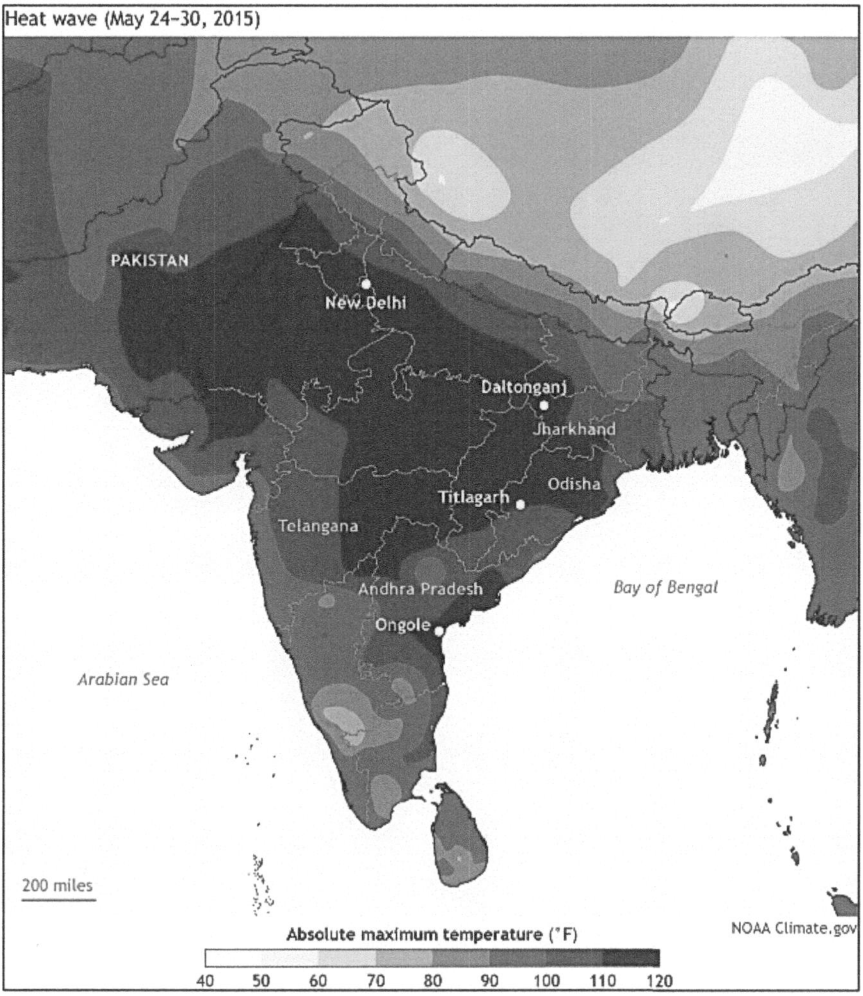

Fig. 5.19 Heat wave, May 24–30, 2015

safeguarding the environment, poverty can be reduced by securing lasting availability of natural resources and enhancing adaptability to climate change effects.

Sustainable development and environmental protection: promoting sustainable development practices and protecting the environment can help alleviate poverty by ensuring long-term access to natural resources and improving resilience to climate change impacts

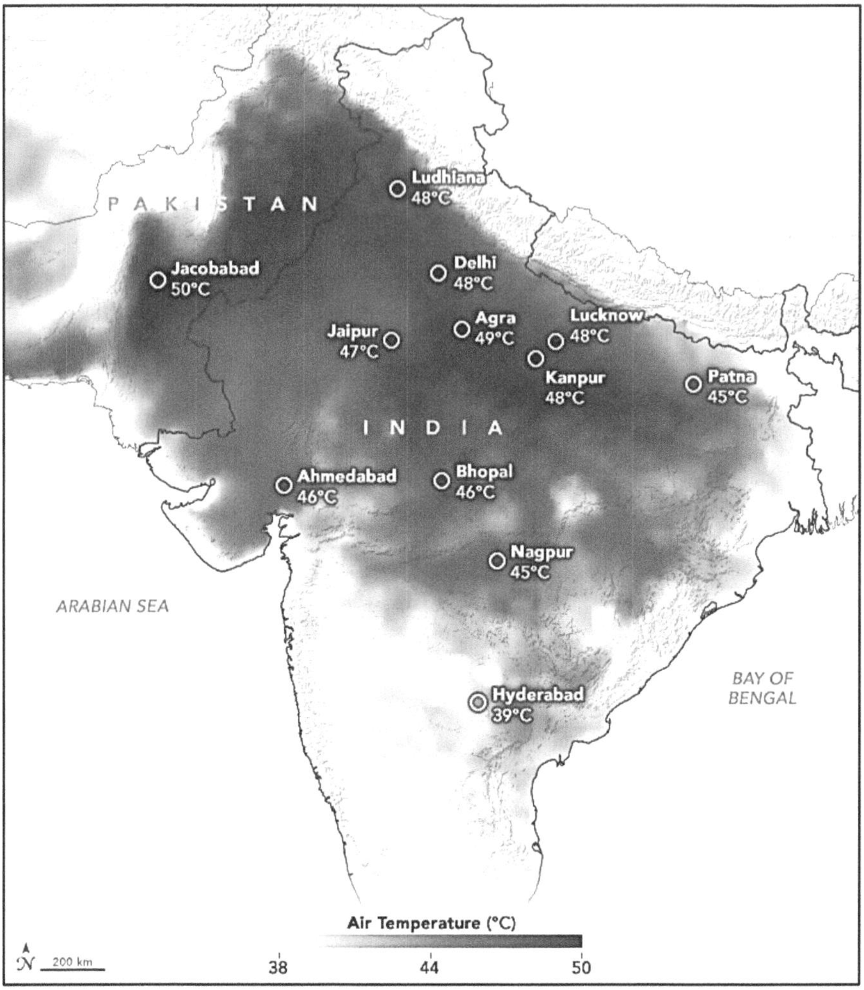

Fig. 5.20 June 10, 2019 – Heatwave India

2019 Heat Wave in India and Pakistan

A severe heat wave hit northern India in early June[61] 2019. Some areas had temperatures above 45 °C (113 °F) for most of 3 weeks. On June 10, Delhi (please see Fig. 5.20) had its hottest day ever recorded for that month, reaching 48 °C (118 °F). The map above shows how hot it was on June 10 in India and Pakistan, which

[61] NASA Earth Observatory (EO) June 10th, 2019 - https://earthobservatory.nasa.gov/images/145167/heatwave-in-india

also faced hot and dry weather for the last 2 months. The map was made from the Goddard Earth Observing System (GEOS) model and shows air temperatures at 2 meters (about 6.5 feet) above the ground. The GEOS-5 model, like other weather and climate models, uses math equations that represent physical processes (like rain and clouds) to figure out what the atmosphere will do. Actual measurements of things like temperature, moisture, and wind are regularly added to the model to keep the simulation as close to reality as possible. Heat wave mainly affected Punjab, Haryana, and can be seen fading away.

Scientists say the climate crisis has increased the probability of extreme heat waves in northwest India and Pakistan by a factor of 100. The study means that sweltering weather that used to occur every 300 years is now likely to happen every 3 years [58]. The models developed using the factors of yield (Hg/Ha), area harvested (Ha), production (Mt) [56], and climate parameters have to be run more often, ironically, to estimate the impact of heat wave related yield loss on food security assessment.

In the following Table 5.4, the wheat data[62] for state-wise production, area harvested, and yield details: the heat wave years and production yield drop are from Reserve Bank of India—Handbook of statistics on India states.[63] The percentage decrease is calculated based on the yield data, while the India Meteorological Department (IMD) records the maximum temperature during the heatwave episode .[64]

Table 5.4 Wheat yield and percentage drop

Yield (kg per hectare)					Percentage of yield drop			Heat wave
Heat wave year	Uttar Pradesh	Punjab	Madhya Pradesh	Haryana	Uttar Pradesh	Punjab	Madhya Pradesh	Max. temp °C
1995–1996	2445	3884	1658	3697	−2.51196	−5.03667	−4.49309	45.5
1997–1998	2495	3853	1573	3660	−6.48426	−8.99858	−12.6596	49.5
2002–2003	2596	4200	1456	1966	−5.77132	−7.32568	−10.1235	49
2014–2015	1881	4294	1931		−38.0843	−14.411	−19.7089	49.4
2015–2016	2258	4583	1835		20.04253	6.730321	−4.97152	51
2018–2019	2910	5188	1847		5.856675	2.186331	−12.9184	50.8
2019–2020	2861	5003	2046		−1.68385	−3.56592	10.77423	50.8

[62] State-wise Estimates of Yield—Wheat (https://www.rbi.org.in/Scripts/AnnualPublications.aspx?head=Handbook+of+Statistics+on+Indian+States)

[63] Handbook of Statistics—India states - (https://www.rbi.org.in/Scripts/AnnualPublications.aspx?head=Handbook+of+Statistics+on+Indian+States)

[64] IMD (https://mausam.imd.gov.in/imd_latest/contents/pdf/pubbrochures/Heat%20Wave%20Warning%20Services.pdf)

We can calculate the average yield loss due to heat wave temperature by using a 0.8 °C increase in temperature that has caused 6.59% decrease in yield. That means if we use the 2020 yields of Uttar Pradesh (2230 kg per hectare), Madya Pradesh (2046 kg per hectare), and Punjab (5003 kg per hectare), the heat wave in 2022 could lead to 203.82 kg per hectare less. The total number of hectares harvested in Uttar Pradesh (9853 hectares), Madya Pradesh (6551 hectares), and Punjab (3521 hectares) adds up to 19,925 hectares or 19.9 million hectares, so the heat wave could cause 4.061 billion kg or 8952972467.328 pounds (8.95 billion pounds) less.

Year	Average drop	Temperature increase
1995–1996	−4.01391	4 °C
1997–1998	−9.38082	−0.5 °C
2002–2003	−7.74016	0.4 °C
2014–2015	−24.0681	1.6 °C
2015–2016	7.267112	−0.2 °C
2018–2019	−1.62514	0 °C

To sum up, heat waves are costly, not to mention the unimaginable human toll they cause in terms of deaths and property damage from wildfires. Food security issues are huge with yield loss resulting in wheat loss that can feed 49.7 million people in the world for a year, assuming 180 pounds[65] (81.6 kg) of wheat per person. With more frequent and hotter heat waves, the decline in yield will lead to serious food insecurity.

> "Climate change has made India's heat wave hotter. Before human activities increased global temperatures, heat waves like year 2022 would have struck India once in about half a century. But now it is a much more common event—we can expect such high temperatures about once every four years."
>
> Friederike Otto,
> A climate scientist at the Imperial College of London

Climate change has caused more frequent heat waves around the world. At the time of writing this book, December 2023, India has just experienced its worst heat wave in decades, and Spain is facing its worst heat wave—Spain suffers from the earliest heat wave in over 40 years.[66] Spain is dealing with its first heat wave of the year, June 2022, with temperatures in some western and southern areas expected to hit 44 °C (111.2 °F).[67] Climate scientists say that global warming is clearly the cause

[65] Per capita wheat flour consumption declines along with other starches (https://www.ers.usda.gov/data-products/chart-gallery/gallery/chart-detail/?chartId=81227)

[66] Spain heat wave (https://www.reuters.com/world/europe/spain-melts-under-earliest-heat-wave-over-40-years-2022-06-13/)

[67] Spain in grip of heat wave with temperatures forecast to hit 44C (https://www.theguardian.com/world/2022/jun/10/spain-heatwave-temperatures-forecast-hit-44c)

of the long-lasting heat wave. The heat wave has led India to take a firm position[68] in Bonn, Germany, where officials are meeting to prepare for the next UN Climate Change Conference in November in Egypt.

"The main input for growing the crop is rain and You don't have any rain, you don't have any grain basically"[69] [59]

Breadbasket to the World

Wheat is a vital and basic food crop. It provides the world with the nutrients that we need for a healthy and balanced life. The countries that produce the most wheat, such as the USA, Brazil, EU, Argentina, Russia and Ukraine, and others, are known as the breadbasket of the world. Ukraine is also an important country in the production of barley—winter barley—a kind of wheat that can grow better in colder climates. The greenhouse gas emissions that result from the war between Ukraine and Russia are causing the temperature to rise according to the weather patterns. This means that the winter barley plant cannot be harvested because of the terrible conditions in the area.[70] Also, the crops such as oats, wheat, corn, and legumes need more irrigation to grow, but they are neglected by the higher temperature and lower rainfall in Ukraine. Ukraine is facing more greenhouse gas emissions and also shortages of food and money because of the war with Russia, which affects the wheat production directly.

Climate change is causing problems for many parts of the world, but Russia [60] is gaining from it. The country's temperature is rising due to climate change, which makes it easier to grow[71] crops in Russia. This gives Russia another source of income and growth. It also leads to more wheat production in the world and food security in the country. However, Russia may be exploiting this situation. Many of their neighbors are suffering from climate change and need to import wheat and other resources from Russia. Russia is enjoying the agricultural boost that climate change is creating for their country.

[68] India takes tough stand at climate talks as Delhi endures brutal heat wave (https://www.theguardian.com/world/2022/jun/14/india-takes-tough-stand-at-climate-talks-as-delhi-endures-brutal-heatwave)

[69] Bumper Wheat Crop Looms Again in Australia on Ample Rains (https://www.bloomberg.com/news/articles/2022-04-05/bumper-wheat-crop-looms-again-in-australia-on-high-prices-rains)

[70] Agriculture and Horticulture Ukraine (https://www.climatechangepost.com/ukraine/agriculture-and-horticulture/)

[71] Climate Change in Russia and the Weaponization of Wheat (https://climateandsecurity.org/2020/08/climate-change-in-russia-and-the-weaponization-of-wheat/)

Production of 2021 Wheat Crop Officially Forecast at Record Level

Harvesting of the 2021, mostly irrigated, main "rabi" wheat crop is about to start, and early official forecasts indicate a record production of 109.2 million tons. The area planted is officially estimated at an all-time high of 34.6 million hectares. Farmers have been encouraged to increase the planted area by the remunerative producer prices guaranteed by the government, coupled with optimal soil moisture conditions at planting time and the timely harvest of previously grown summer crops that made land available. Overall, yields are forecast close to average as weather conditions were generally favorable throughout the cropping season and supplies of agricultural inputs, including irrigation water and fertilizer, were adequate.[72]

Maximum, Minimum, and Average Temperature and Precipitation (1901–2020)

India's temperature from 1901 to 2020 can be downloaded from the Climate Change Knowledge Portal (CCKP).

Maximum temperatures in wheat growing states and India: as can be seen, the temperatures are trending upward year over year. The future temperatures will be much higher ($R^2_{\text{UP}} = 0.2248$ and $R^2_{\text{India}} = 0.4666$) as the prevailing equation predicates (please see Fig. 5.21):

$$Y_{\text{UP}} = 0.0057x + 31.636$$

$$Y_{\text{India}} = 0.007x + 29.965$$

Similar observations can be made for minimum and mean or average temperatures.

Average temperatures play an important role for wheat crops. The average temperatures in wheat growing states and India: as can be seen, the temperatures are trending upward year over year. The future temperatures will be much higher ($R^2_{\text{UP}} = 0.3257$ and $R^2_{\text{India}} = 0.5111$) as the prevailing equation predicates (please see Fig. 5.22):

$$Y_{\text{India}} = 0.0076x + 23.935$$

[72]GIEWS - Global Information and Early Warning System (https://www.fao.org/giews/countrybrief/country.jsp?lang=en&code=IND)

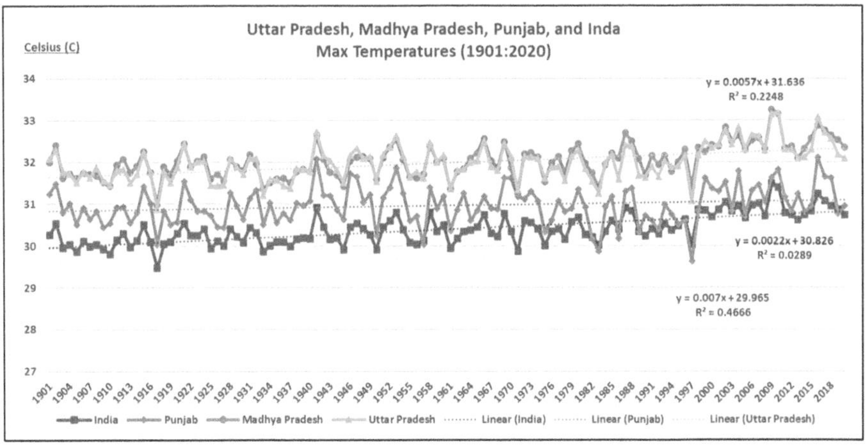

Fig. 5.21 Max temeratures (1901:2020)

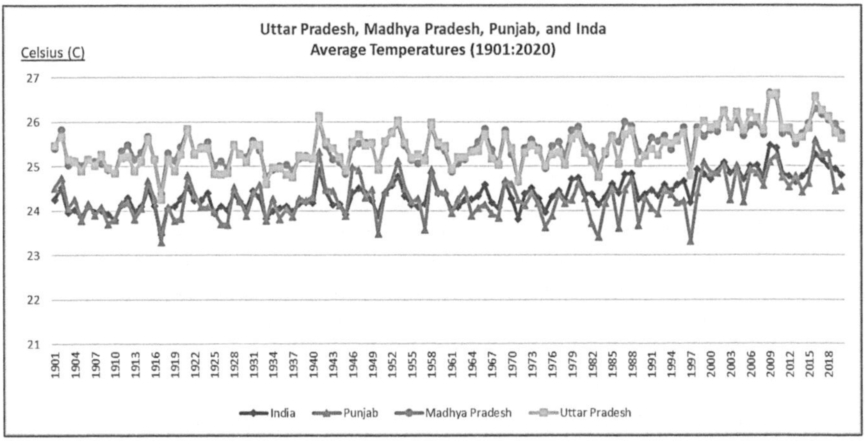

Fig. 5.22 India average temperatures (1901–2020)

$$Y_{\mathrm{UP}} = 0.0072x + 24.981$$

The temperature trends clearly suggest that the elevated temperatures are projected for the upcoming century. Above data from CCKP is downloaded using following options (please see Fig. 5.23):

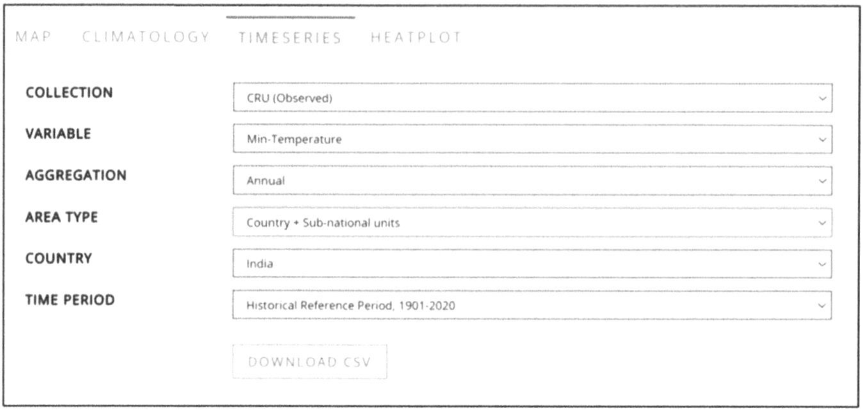

Fig. 5.23 CCKP Portal

Machine Learning Model: Heat Waves and Wheat Yield Production Linkage in India

India has the largest population in the world and one of the world's quickest growing large economies since 2000.[73] India also ranks among the world's biggest producers and consumers of various crop and livestock products. As a lower middle-income country, fast economic growth is increasing and diversifying food demand, putting stress on the farm sector to raise productivity that is currently low, and testing agricultural policies that have traditionally aimed at self-sufficiency [61].

India plays a key role in global markets for commodities that are important to US agriculture and the world at large. India is the top importer of soybean oil in the world, surpassing the European Union and China. India is also the biggest producer and consumer of milk. Milk supply and demand have grown strongly, but forecasts suggest that it will be hard to meet production goals without more improvements in efficiency. Increasing incomes, urbanization, and India's young demographic profile are expected to drive more growth and diversification in consumer food demand. While the consumption per person of staple grains, such as rice and wheat, has not changed much, demand for higher valued foods—fruits, vegetables, edible oils, dairy products, and poultry meat—has been vigorous. However, this growth story also has its challenges, as government estimates show that around 22% of India's population—or about 260 million people—remain poor. According to USDA esti-

[73] India (https://www.ers.usda.gov/topics/international-markets-u-s-trade/countries-regions/india/)

mates, India still has the largest portion of the global population that is food insecure [61]. Moreover, because India has the largest portion of the global population that is food insecure, technologies that increase productivity and sensors that are climate smart are needed to feed future generations.

Improving the efficiency of India's agricultural markets through reforms could increase output and wages across the economy. These reforms could increase the prices that farmers get, lower the prices that consumers pay for food, and boost consumption, especially by poor households.

India should pay attention to two important aspects to improve food security: (a) climate change and (b) data driven approaches to improve farming practices. India has improved its agricultural productivity, but there is still a large gap to fill to ensure food security for the nation. Increasing agricultural productivity is crucial for India and its food security. India needs technologies that can improve productivity, as well as data science and data driven approaches to deal with the challenges of climate change and temperature sensitivity.

Heat waves reduce the amount of crops that can be produced, especially for main grains. India has recently faced difficulties in exporting needed wheat because of heat waves. Wheat is a crop that grows in winter in India and is expected to finish by the end of June. The planted area is officially estimated at 34.3 million hectares, which is similar to the 2021 level. This was supported by the high minimum support price (MSP) from the government and good soil moisture conditions when planting. The weather was good for the main wheat-producing areas from October 2021 to February 2022, and there were hopes for record yields. However, unusually high temperatures and much lower-than-average rainfall from mid-March to April in the northwest of the country damaged crops right before they were harvested. The high temperatures affected wheat crops the most when they were at the stage of filling grains, because they were more sensitive to heat then. As a result, yields are now expected to be lower than the record levels that were predicted earlier. Wheat farming in Punjab, Haryana, and Uttar Pradesh states suffered the most. Based on current official estimates, the 2022 wheat production is expected to be around 106.4 million tons, which is higher than the 5-year average. However, it is possible that production may be lower as the full impact of the heat wave is not yet known. Let's build the model:

Software code for this model
Final_Indian_wheat_model_Punjab.ipynb (Jupyter Notebook Code)

Data Sources

The data is from the Reserve Bank of India (RBI) *Handbook of Statistics on India States*[74]	Table 52: State-wise production of food grains—wheat (1982: 2021) Table 61: State-wise area of food grains—wheat Table 74: State-wise estimates of yield—wheat Table 24: Pattern of land use and select inputs for agricultural production[75] Table 25: Minimum support price for food grains according to crop year (fair average quality)[76] Loans issued and loans outstanding Table 52: Institutional credit for agriculture and allied activities[77] Gold and silver Table 39: Average price of gold and silver in domestic and foreign markets[78]	
NOAA STAR[79]—Global Vegetation Health Products: Province-Averaged VH	Mean data for US province = 17: Kansas, from 1982 to 2022, weekly for cropland[80] area only Year, week, SMN, SMT, VCI, TCI, VHI	

(continued)

[74]Handbook of statistics on India states (https://www.rbi.org.in/Scripts/AnnualPublications.aspx?head=Handbook+of+Statistics+on+Indian+States)

[75]Table 24: Pattern of Land Use and Select Inputs for Agricultural Production (https://rbi.org.in/Scripts/PublicationsView.aspx?id=21830)

[76]Table 25: Minimum Support Price for Food grains According to Crop Year (Fair Average Quality) (https://rbi.org.in/Scripts/PublicationsView.aspx?id=21831)

[77]Table 52: Institutional Credit for Agriculture and Allied Activities (https://rbi.org.in/Scripts/PublicationsView.aspx?id=21858)

[78]Table 39: Average Price of Gold and Silver in Domestic and Foreign Markets (https://rbi.org.in/Scripts/PublicationsView.aspx?id=21845)

[79]NOAA STAR (https://www.star.nesdis.noaa.gov/smcd/emb/vci/VH/vh_adminMean.php?type=Province_Weekly_MeanPlot)

[80]Weekly Crop Land (https://www.star.nesdis.noaa.gov/smcd/emb/vci/VH/get_TS_admin.php?provinceID=17&country=USA&yearlyTag=Weekly&type=Mean&TagCropland=land&year1=1982&year2=2022)

The Climate Change Knowledge Portal (CCKP)[81]	India temperatures[82] Mean temperature Minimum temperature Maximum temperature Precipitation	THE WORLD BANK

Step 1: Load Libraries

Load libraries to process Indian major wheat-producing states data.

```
import pandas as pd
import numpy as np
import plotly.graph_objects as go
import seaborn as sns; sns.set()
import matplotlib.pyplot as plt
from scipy import stats
from scipy.stats import pearsonr
import matplotlib.pyplot as plt
from sklearn.preprocessing import MinMaxScaler
from sklearn.model_selection import train_test_split
from sklearn.linear_model import LinearRegression
from sklearn.ensemble import RandomForestClassifier
from sklearn.metrics import r2_score
from sklearn import metrics

import eli5
import matplotlib.pyplot as plt
import matplotlib as mpl
import numpy as np
import scipy.stats as spstats
import seaborn as sns
import utils

%matplotlib inline
mpl.style.reload_library()
mpl.style.use('classic')
mpl.rcParams['figure.facecolor'] = (1, 1, 1, 0)
mpl.rcParams['figure.figsize'] = [6.0, 4.0]
mpl.rcParams['figure.dpi'] = 100
```

Step 2: Load Wheat Macroeconomic Finance and Monetary Agriculture Data Frame (1961–2020)

Load wheat agricultural dataset with production, yield, and harvested area. In addition, load finance, monetary, and market variables.

[81] The Climate Change Knowledge Portal (https://climateknowledgeportal.worldbank.org/)

[82] The CCKP Time Series (https://climateknowledgeportal.worldbank.org/download-data)

```
wheatMacroFinMonetrayAgLinkageDF= pd.read_csv(Finalwheatproduction.csv')
wheatMacroFinMonetrayAgLinkageDF.tail()
```

Output

Year	Wheat YIELD PER HECTARE - FOODGRAINS (Kg / hectare)	Wheat AREA UNDER CULTIVATION - FOODGRAINS	Consumption of Fertilisers (N+P+K) (lakh tonnes)	MSP - Wheat (Rupee per quintal)	Loans Issued (Crore)	Loans Outstanding (Crore)	Rural population (% of total population)
1990	2281	377	125.5	225	9829	29316	73.809
1991	2394	363	127.3	280	11199	31142	73.601
1992	2327	334	121.6	330	12141	34263	73.393
1993	2380	344	123.7	350	14636	36988	73.183
1994	2559	328	135.6	360	18366	40738	72.972
1995	2483	322	138.8	380	23138	46020	72.760
1996	2679	309	143.1	475	25677	50921	72.547
1997	2485	318	161.9	510	27798	54518	72.333
1998	2590	308	168.0	550	32277	57408	72.082

Step 3: Check Any Column Data Is Null

Check for nulls in data.

```
wheatMacroFinMonetrayAgLinkageDF.isnull().sum()
Output:
Wheat YIELD PER HECTARE - FOODGRAINS (Kg / hectare)  0
Wheat AREA UNDER CULTIVATION - FOODGRAINS             0
Consumption of Fertilisers (N+P+K) (lakh tonnes)    0
MSP - Wheat (Rupee per quintal)                     0
Loans Issued (Crore)                         0
Loans Outstanding  (Crore)                     0
Rural population (% of total population)             0
Gold (Rupees per 10 gms.)                      0
Silver(Rupees per kg.)                      0
Income velocity (GDP/M1)                         0
Income velocity (GDP/C)                        0
HeatwaveYear                              0
```

No nulls in the data.

Step 4: Perform Correlation Analysis

The relationship between wheat yield and various factors such as silver, loans outstanding, wheat area under cultivation, loans issued, MSP wheat, rural population, consumption of fertilizers, gold, and temperature is complex and multifaceted. Let's check correlation among variables.

```
wheatMacroFinMonetrayAgLinkageDF.corr()
plt.figure(figsize=(12, 9))
sns.heatmap(wheatMacroFinMonetrayAgLinkageDF.corr())
result
```

Output

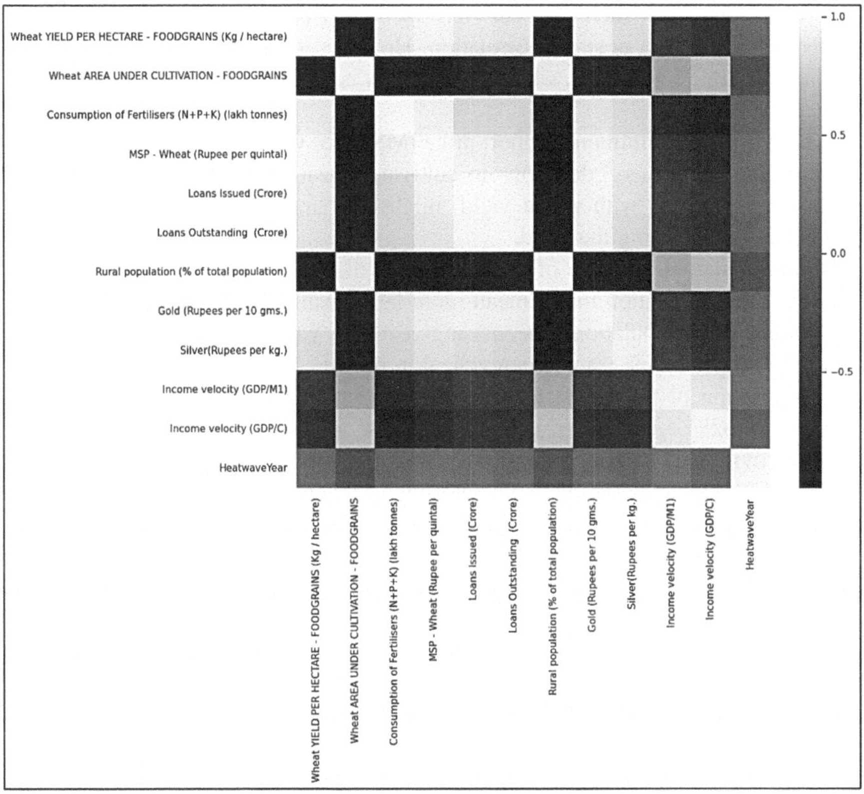

- *Wheat area under cultivation*: the area under wheat cultivation directly impacts the total yield. There exists a negative correlation between the size of the area and potential yield due to adverse weather conditions and inadequate productivity tools. This is a constant challenge to increase productivity in agriculture for India.
  - Implementing land reform policies and improving agricultural productivity can help alleviate rural poverty[83] by providing farmers with secure land tenure, access to credit, and technical support.
  - Enhancing agricultural productivity and rural development: supporting smallholder farmers and investing in agricultural infrastructure, technology, and research can increase agricultural[84] productivity, improve food security, and lift rural communities out of poverty [62].
- Loans issued and outstanding: Farmers can improve their productivity by investing in inputs such as seeds, fertilizers, and equipment through access to credit, leading to a positive correlation. However, outstanding loans can cause financial strain on farmers, which is positively correlated with wheat yield in this scenario.
- MSP wheat: the minimum support price (MSP) for wheat set by the government can influence farmers' decisions to cultivate wheat. A higher MSP (positively correlated 96.71% with wheat yield) might encourage more wheat cultivation, potentially increasing the yield.
- Rural population: the size of the rural population can influence wheat yield. A larger rural population might mean more labor available for farming. However, given the rural population is decreasing year over year, it's negatively correlated with yield (−95.06%), a countercyclical in nature.
- Consumption of fertilizers: the use of fertilizers can significantly boost wheat yield (91.628%). However, overuse can lead to soil degradation and reduced productivity over time.
- Silver and gold: the relationship between wheat yield and the values of silver and gold is direct in small, marginalized farming. These precious metals often resort as farming inputs or economic collateral for marginalized farmers. These metals act as a hedge against crop failures and climate change induced stressors, hence positively correlated.

[83] Budget Speech 2005–2006 (https://www.indiabudget.gov.in/doc/bspeech/bs200506.pdf)

[84] Budget Speech 2008–2008 (https://www.indiabudget.gov.in/budget_archive/ub2008-09/bs/speecha.htm)

Step 5: Train Regressive Model

Prepare regressive model to see the influence of temperature and precipitation coefficients on the wheat yield.

```
X = dataset[[
    'HeatwaveYear',
    'Wheat AREA UNDER CULTIVATION - FOODGRAINS',
    'Consumption of Fertilisers (N+P+K) (lakh tonnes)',
    'MSP - Wheat (Rupee per quintal)', 'Loans Issued (Crore)',
    'Loans Outstanding  (Crore)',
    'Rural population (% of total population)', 'Gold (Rupees per 10 gms.)',
    'Silver(Rupees per kg.)', # 'Income velocity (GDP/M1)', 'Income velocity (GDP/C)'
    ]]
y = dataset['Wheat YIELD PER HECTARE - FOODGRAINS (Kg / hectare)']

from sklearn.model_selection import train_test_split
X_train, X_test, y_train, y_test = train_test_split(X, y, test_size = 0.3, random_state = 0)
```

```
# Train the model
from sklearn.linear_model import LinearRegression

# Fit a linear regression model on the training set
model = LinearRegression().fit(X_train, y_train)
print (model)
```

Output

```
LinearRegression()
```

```
import matplotlib.pyplot as plt

%matplotlib inline

plt.scatter(y_test, predictions)
plt.xlabel('Actual Labels')
plt.ylabel('Predicted Labels')
plt.title('Predictions')
# overlay the regression line
z = np.polyfit(y_test, predictions, 1)
p = np.poly1d(z)
plt.plot(y_test,p(y_test), color='magenta')
plt.show()
```

Output

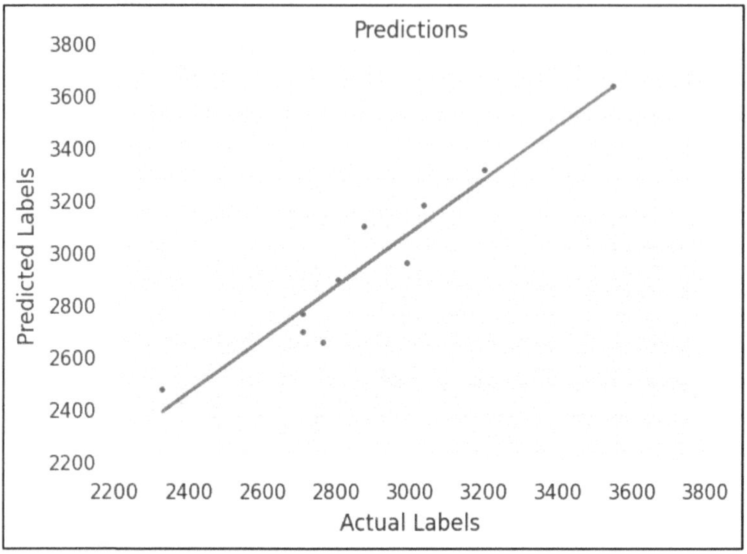

Step 8: Evaluate the Model

Let's calculate model metrics: MSE, RMSE, and MAE.

```
from sklearn.metrics import mean_squared_error, r2_score

mse = mean_squared_error(y_test, predictions)
print("MSE:", mse)

rmse = np.sqrt(mse)
print("RMSE:", rmse)

r2 = r2_score(y_test, predictions)
print("R2:", r2)
```

Output

```
MSE: 15791.643797470451
RMSE: 125.66480731481846
R2: 0.8355592777037588
```

Step 8: Model Explainability

Let's run the model explainability to see the dominant parameter.

```
eli5.explain_weights(model, feature_names = list(X.columns))
```

Output

y top features

Weight?	Feature
+58.162	Rural population (% of total population)
+1.131	MSP - Wheat (Rupee per quintal)
+0.003	Gold (Rupees per 10 gms.)
+0.000	Loans Outstanding (Crore)
-0.000	Loans Issued (Crore)
-0.002	Silver(Rupees per kg.)
-1.372	Consumption of Fertilisers (N+P+K) (lakh tonnes)
-2.008	Wheat AREA UNDER CULTIVATION - FOODGRAINS
-53.052	HeatwaveYear
-1310.286	<BIAS>

Model accuracy is high, and bias is low. MSE is quite large.

Step 9: Model Equation (Y = MX + C)

Let's construct model equations to understand influence of independent variables on the yield.

```
mx=""
for ifeature in range(len(X.columns)):
  if model.coef_[ifeature] <0:
    # format & beautify the equation
    mx += " - " + "{:.2f}".format(abs(model.coef_[ifeature])) + " * " + X.columns[ifeature]
  else:
    if ifeature == 0:
      mx += "{:.2f}".format(model.coef_[ifeature]) + " * " + X.columns[ifeature]
    else:
      mx += " + " + "{:.2f}".format(model.coef_[ifeature]) + " * " + X.columns[ifeature]

print(mx)
```

```
# y=mx+c
if(model.intercept_ <0):
  print("The formula for the " + y.name + " linear regression line (y=mx+c) is = " + " -
{:.2f}".format(abs(model.intercept_)) + " + " + mx )
else:
  print("The formula for the " + y.name + " linear regression line (y=mx+c) is = " +
"{:.2f}".format(model.intercept_) + " + " + mx )
```

Output

The formula for the Wheat YIELD PER HECTARE - FOODGRAINS (Kg / hectare) linear regression line (y=mx+c) is = - 1310.29 + - 53.05 * HeatwaveYear - 2.01 * Wheat AREA UNDER CULTIVATION - FOODGRAINS - 1.37 * Consumption of Fertilisers (N+P+K) (lakh tonnes) + 1.13 * MSP - Wheat (Rupee per quintal) - 0.00 * Loans Issued (Crore) + 0.00 * Loans Outstanding (Crore) + 58.16 * Rural population (% of total population) + 0.00 * Gold (Rupees per 10 gms.) - 0.00 * Silver(Rupees per kg.)

Heat waves have a significant impact on wheat yield in India. Here are some key points:

- Heat waves in India, particularly in the northern and central regions, can damage the country's wheat harvest [63].
- The heat wave conditions experienced in March have led to grain shrinkage as the crop was in the advanced ripening stage [64].
- For every 1 degree increase in temperature, the yields of wheat are expected to decline by 3–7% [65].
- The heat waves and their early onset have resulted in an estimated 10–35% reduction in crop yields in Punjab, Haryana, and Uttar Pradesh [66].
- In 2022, Punjab's wheat yield reduced to 43 quintals per hectare, which is the lowest since the year 2000 [66].
- A farmer from Uttar Pradesh's Etawah district reported that his wheat crop was down 60% compared to normal harvests [63]. As clearly demonstrated by the model equation, yields as much as 53.05% could be affected by heat waves.

These impacts are not only local but also international, as many countries rely on wheat from India. The effects of heat waves on wheat yield are a significant concern given the increasing global temperatures due to climate change.

Step 10: Conclusion

Climate change will have detrimental impact on small farmers and marginalized communities across the world and especially in India. The profound and pervasive effect of climate change on agriculture implies that policy interventions must adopt a multi-pronged approach:

1. Agricultural development: increase investment in irrigation, land development, and soil and moisture conservation to improve agricultural productivity,[85] especially in rainfed and semiarid areas. Promote diversification of agriculture and the development of agro-based processing industries [2].

[85] Budget Speech 1990–1991 (https://www.indiabudget.gov.in/doc/bspeech/bs199091.pdf)

2. Investment in agriculture and rural infrastructure: double agricultural credit in 3 years,[86] accelerate the completion of irrigation projects, and invest in rural infrastructure. This will help improve agricultural productivity, create employment opportunities, and enhance the overall development of rural areas [52].
3. Focus on improving pulses production: address the stagnation[87] in pulses production by expanding the availability and quality of certified seeds and implementing measures to boost production and productivity.
4. To ensure prosperity for farmers,[88] the government can focus on liberalizing farm markets, removing distortions, and investing in logistics, agri-services, and farm-based activities. Additionally, sustainable cropping patterns, technology adoption, and cooperation with states can be promoted to enhance agricultural productivity and income [30].
5. Strengthening the rural economy: focus on improving agriculture and allied activities to boost the rural economy. This can be done through investment in infrastructure and technology and providing support to farmers.
6. Diversification and value chain development in agriculture: encourage farmers to diversify their crops and move up the value chain by adopting modern farming techniques, utilizing technology, and accessing market information to meet the changing demands of consumers.
7. Focus on agriculture: continue supporting the agriculture sector by promoting increased food grain production, exports, and agricultural credit. Implement policies to ensure sustainable growth and improve the livelihoods of farmers.

> Agriculture is the basis of national wealth, and any economic development plan will fail in India if it does not ensure fast growth of output and employment in agriculture. Agriculture has a key role in promoting inclusive and fair development in India, and no economic plan can ignore this. Climate change makes this strategic need even more urgent. Developing models based on climate, using Information Communication Technologies (ICT), bringing data science to the small and marginal farmers, and training and empowering workers on data science who work with small farmers are essential. And government policies should give them high priority.
>
> Author

Finally, improving agriculture productivity is a keyway to raise income, keep fiscal order, balance the budget, lower government debt, increase purchasing power, and enhance living standards for over a billion people in the world. When agriculture productivity improves, it increases farm income that supports the economy at the bottom of the pyramid. It also reduces the dependence on subsidies by applying data science to agriculture and using the synergies to cut the subsidies, which lowers government deficit. Of course, some subsidies, such as food, fertilizer, and petroleum, will continue to be offered to protect the poor, but with less pressure on the government through consistent improvement in farm productivity. Improving agriculture productivity achieves fiscal discipline directly:

[86] Budget Speech 2004–2004 (https://www.indiabudget.gov.in/doc/bspeech/bs200405.pdf)

[87] Budget Speech 2007–2008 (https://www.indiabudget.gov.in/doc/bspeech/bs200708.pdf)

[88] Budget Speech 2020–2021 (https://www.indiabudget.gov.in/doc/bspeech/bs202021.pdf)

Chapter Machine Learning Experiments

ML Experiment 1 Given population growth, agricultural productivity, frequencies of climate change, and public distribution system, models clearly show that Kashmir is the new food basket of India. What would be resulting consequences if India is not enabling agriculture shift?

Whis policy implications need to be addressed in such scenario?

A. The government will use the public distribution system to counter inflation and protect the poorer sections of the population from high prices and shortages. They aimed to ensure that food grains and essential commodities reach the poor and underprivileged in adequate quantities and at affordable prices.
B. Any reform on social protection should come in tandem with carefully crafted reforms on the agricultural and food price risk management sides. This is not a call for inaction, but for concerted cross-sectoral policy. Put it simply, economic sustainability pillar depends on three things: (1) increasing farmers' income and promoting agriculture and farmers' welfare; (2) enhancing rural employment and infrastructure development, providing healthcare services, and implementing health protection schemes; and (3) improving education, skills, and job creation to create a productive society.
C. The policy implications are very clear. The uptick in off-take takes place when consumer prices increase. Similarly, when net flow aid or foreign investment takes place, the off-take would decrease given in the increased economic opportunity. To create sustainable social and economic societies for poor and marginalized communities, focusing on economic improvements is the first order of the business.
D. All the above.

After reading this chapter, you should be able to answer the role of public distribution system off-take on inflation control. Next, the role of agricultural productivity on PDS and climate change was discussed.

References

1. Shri Manmohan Singh, Budget Speech 1993–1994, 27 February 1993, https://www.indiabudget.gov.in/doc/bspeech/bs199394.pdf, Access Date: 11/22/2023
2. PROF. MADHU DANDAVATE, INTRODUCING THE BUDGET FOR THE YEAR 1990–91, PUBLISHED TUE, 1990, https://www.indiabudget.gov.in/doc/bspeech/bs199091.pdf, Access Date: 11/23/2023.
3. Sarma, N. A. "Economic Development in India: The First and the Second Five Year Plans". IMF Staff Papers 1958.001 (1958), A002. < https://doi.org/10.5089/9781451949636.024.A002>. Web. 4 Nov. 2023.

4. Shri Yashwant Sinha, Budget 2000–2001, February 2000, https://www.indiabudget.gov.in/doc/bspeech/bs200001.pdf, Access Date: November 29, 2023.
5. Shri Manmohan Singh, Budget 1992–93, PUBLISHED TUE, 29th February 1992, https://www.indiabudget.gov.in/doc/bspeech/bs199293.pdf, Access Date: 11/23/2023
6. Ayşe Aylin Baya, The impact of financial development on income inequality and poverty, Oct 2023, https://www.ncbi.nlm.nih.gov/pmc/articles/PMC10602251/, Access Date: November 25, 2023.
7. P. Chidambaram, Budget 2007–2008, February 2007, https://www.indiabudget.gov.in/doc/bspeech/bs200708.pdf, Access Date: December 04, 2023,
8. Karl Usuka, An Analysis of the Determinants of Poverty in India and South Africa, Global Majority E-Journal, Vol. 10, No. 2 (December 2019), pp. 100–115, https://www.american.edu/cas/economics/ejournal/upload/usuka_accessible.pdf, Access Date: 11/12/2021
9. R. Radhakrishna (Editor) and Shovan Ray (Editor), Handbook of Poverty in India: Perspectives, Policies, and Programmes 1st Edition, Publisher : Oxford University Press; 1st edition (January 6, 2006), ISBN-13 : 978-0195675139
10. Rob Jenkins (Author), James Manor (Author), Politics and the Right to Work: India's National Rural Employment Guarantee Act, Publisher : Oxford University Press (April 15, 2017), ISBN-13 : 978-0190608309
11. Ministry of Finance, Summary of the Budget 2021–22, 01 FEB 2021 2:10PM by PIB Delhi, https://pib.gov.in/PressReleasePage.aspx?PRID=1693908, Access Date: Novmber 05, 2023
12. Raghav Puri, Harold Alderman, Shrayana Bhattacharya, and Ugo Gentilini, Schemes To Systems | The Public Distribution System: Anatomy of India's Food Subsidy Reforms, February 2019., https://www.worldbank.org/en/news/feature/2019/02/21/schemes-to-systems-public-distribution-system, Access Date: November 26, 2023.
13. Kayode Ajewole and Nathan Childs, India, a major rice and wheat producer, sharply expands subsidized distribution of food grains in response to pandemic, Last updated: Monday, November 30, 2020, https://www.ers.usda.gov/data-products/chart-gallery/gallery/chart-detail/?chartId=99959, Access Date: November 26, 2023.
14. K. R Venugopal (Author), Deliverance from hunger: The public distribution system in India, Publisher: Sage Publications (January 1, 1992), ISBN-13 : 978-8170362623, Access Date: November 25, 2023.
15. N.P. NAWANI, Indian experience on household food and nutrition security, AUGUST 1994, https://www.fao.org/3/x0172e/x0172e00.htm#TopOfPage, Access Date: November 25, 2023.
16. Vasu malai Kannan and Ankita Singh, Wheat Procurement Forecasting for Indian PDS Using Regression Model, August 2022, https://www.technoarete.org/common_abstract/pdf/IJERCSE/v9/i8/Ext_80629.pdf, Access Date: November 25, 2023.
17. Dr.B.Venkatraja, Is Rising Food Subsidy Attributed to Poor Food Management? An Empirical Evidence from India, 2016, https://core.ac.uk/download/pdf/234670954.pdf, Access Date: November 25, 2023.
18. Chetan Ghate, Sargam Gupta, and Debdulal Mallick, The inflationary impact of the grain procurement policy in India, 24 Apr 2017, https://www.theigc.org/blogs/inflationary-impact-grain-procurement-policy-india, Access Date: November 25, 2023.
19. Brian Ames, Ward Brown, Shanta Devarajan, and Alejandro Izquierdo, Macroeconomic Policy and Poverty Reduction, August 2001, https://www.imf.org/external/pubs/ft/exrp/macropol/eng/ , Access Date: 11/20/2023
20. Jerome H. Friedman. "Greedy function approximation: A gradient boosting machine.." Ann. Statist. 29 (5) 1189 - 1232, October 2001. https://doi.org/10.1214/aos/1013203451, Access Date: November 26, 2023.
21. Elif C Arbatli Saxegaard ; Mattia Coppo ; Nasser Khalil ; Shinya Kotera ; Filiz D Unsal, Inequality and Poverty in India: Impact of COVID-19 Pandemic and Policy Response, July 14, 2023, https://www.imf.org/en/Publications/WP/Issues/2023/07/20/Inequality-and-Poverty-in-India-Impact-of-COVID-19-Pandemic-and-Policy-Response-535908, Access Date: November 26, 2023.

22. Ministry of Fisheries, Animal Husbandry & Dairying, Milk Production in India, Posted On: 07 SEP 2022 11:41AM, https://pib.gov.in/FeaturesDeatils.aspx?NoteId=151137, Access Date: December 07, 2023
23. Shri Manmohan Singh, Budget Speech 1991–1992, 24th July, 1991, https://www.indiabudget.gov.in/doc/bspeech/bs199192.pdf, Access Date: November 26, 2023.
24. Rajiv Kumar and Dony Alex, The Great Recession and India's trade collapse, 27 Nov 2009, https://cepr.org/voxeu/columns/great-recession-and-indias-trade-collapse, Access Date: December 07, 2023
25. Mathew Joseph, Global Financial Crisis: How was India Impacted?, September 3–4 2009, https://www.idos-research.de/fileadmin/_migrated/content_uploads/Global_Financial_Crisis_and_Impact_on_India_Berlin030909.pdf, Access Date: December 07, 2023
26. Swaminathan S Anklesaria Aiyar, India weathers 12 months of financial crisis, Sep 13 2009, https://economictimes.indiatimes.com/swaminathan-s-a-aiyar/india-weathers-12-months-of-financial-crisis/articleshow/5005007.cms, Access Date: December 07, 2023
27. Swati Gupta and Hanna Ziady, India suffers first recession in decades, Updated 11:45 AM EST, Fri November 27, 2020, https://edition.cnn.com/2020/11/27/economy/india-recession/index.html, Access Date: December 07, 2023
28. IMF (2020), 'World Economic Outlook', International Monetary Fund, Washington D.C., April 2020, https://rbidocs.rbi.org.in/rdocs/AnnualReport/PDFs/IIECONOMICREVIEW650084A9F3494152939D99449E63D040.PDF, Access Date: December 07, 2023
29. IMF (2021), Fiscal Monitor- A Fair Shot', International Monetary Fund, Washington D.C., April. 2021, https://rbidocs.rbi.org.in/rdocs/AnnualReport/PDFs/IIECONOMICREVIEW650084A9F3494152939D99449E63D040.PDF, Access Date: December 07, 2023
30. Nirmala Sitharaman, Budget Speech 2020–2021, February 2020, https://www.indiabudget.gov.in/doc/bspeech/bs202021.pdf, Access Date: 11/22/2023
31. Pranab Mukherjee, Budget 2012–2013, March 16, 2012, https://www.indiabudget.gov.in/doc/bspeech/bs201213.pdf, Access Date: December 07, 2023
32. Stijn Claessens, M. Ayhan Kose and Marco E. Terrones, What Happens During Recessions, Crunches and Busts?, December 2008, https://www.imf.org/external/pubs/ft/wp/2008/wp08274.pdf, Access Date: December 07, 2023
33. Dua, P. Monetary policy framework in India. Ind. Econ. Rev. 55, 117–154 (2020). https://doi.org/10.1007/s41775-020-00085-3, Access Date: December 08, 2023
34. Gabrielle Harris, History of India's Public Distribution Shops aka Fair Price Shops, Nov 2020, https://conversationforsix.com/conversations/india-fair-price-shops, Access Date: December 08, 2023
35. Sarit Maitra, Vector Autoregressive for Forecasting Time Series, Jun 18, 2020, https://towardsdatascience.com/vector-autoregressive-for-forecasting-time-series-a60e6f168c70, Access Date: December 08, 2023
36. JULIA KAGAN and GORDON SCOTT, Time Deposit (aka Term Deposit) Definition and How Does It Work?, Updated March 11 2020, https://www.investopedia.com/terms/t/timedeposit.asp, Access Date: December 08, 2023
37. Jasudha Kirpalani and Arundathi Abyshekar, Citi Launches Sustainable Deposits Solution for Institutional Clients in India, June 14 2023, https://www.citigroup.com/global/news/press-release/2023/-citi-launches-sustainable-deposits-solution-for-institutional-clients-in-india, Access Date: December 08, 2023
38. Ramasamy, P., Malaiarasan, U. Agricultural credit in India: determinants and effects. Ind. Econ. Rev. 58, 169–195 (2023). https://doi.org/10.1007/s41775-023-00187-8, Access Date: December 06, 2023
39. Ministry of Finance, AGRICULTURE SECTOR REMAINS BUOYANT WITH 4.6 PERCENT ANNUAL GROWTH DURING LAST SIX YEARS, Posted On: 31 JAN 2023 1:21PM by PIB Delhi, https://pib.gov.in/PressReleasePage.aspx?PRID=1894900, Access Date: December 08, 2023

40. David Pitt, Associated Press, USDA announces $1 billion in debt relief for 36,000 farmers, Oct 18, 2022 9:02 PM EST, https://www.pbs.org/newshour/politics/usda-announces-1-billion-in-debt-relief-for-36000-farmers, Access Date: 11/23/2023
41. Susan Desai, India's Farm Loan Waiver Crisis, September 11 2017, https://www.frbsf.org/banking/asia-program/pacific-exchange-blog/indias-farm-loan-waiver-crisis/, Access Date: 11/23/2023
42. XAVIER GINE and MARTIN KANZ, The economic effects of India's farm loan bailout: business as usual?, MARCH 16 2015, https://blogs.worldbank.org/allaboutfinance/economic-effects-india-s-farm-loan-bailout-business-usual, Access Date: 11/23/2023
43. Shri Jaswant Singh, Budget 2003–2004, February 2003, https://www.indiabudget.gov.in/doc/bspeech/bs200304.pdf, Access Date: November 30, 2023.
44. NIRMALA SITHARAMAN, BUDGET 2023–2024, YEAR 2023–2024, https://www.indiabudget.gov.in/doc/budget_speech.pdf, Access Date: 11/5/2021
45. Ministry of Finance, SUMMARY OF UNION BUDGET 2022–23, 01 FEB 2022 1:19PM by PIB Delhi, https://pib.gov.in/PressReleasePage.aspx?PRID=1794167, Access Date: 11/5/2021
46. SUBRAMANI RA MANCOMBU, How India overcame food emergency, attained self-sufficiency, Updated - August 15, 2022 at 11:34 AM, https://www.thehindubusinessline.com/india-at-75/overcoming-food-emergencies-through-imports-from-us-via-pl480/article65753881.ece, Access Date: November 30, 2023.
47. SHRI R.K. SHANMUKHAM CHETTY, THE BUDGET FOR THE YEAR 1947–1948, YEAR 1947–1948, https://www.indiabudget.gov.in/doc/bspeech/bs194748.pdf, Access Date: 11/5/2021
48. Bernard Weinraub, NDIA REQUESTING FOOD AID FROM U.S., Sep 1974, https://www.nytimes.com/1974/09/03/archives/india-requesting-food-aid-from-us-seeks-emergency-help-but-shuns-a.html, Access Date: November 30, 2023.
49. Chidambaram Subramaniam, Hand of Destiny Vol. 2: The Green Revolution, ISBN : 81-8276-049-3
50. Celia W. Dugger, Chidambaram Subramaniam, India's 'Green' Rebel, 90, Dies, Nov. 10, 2000, https://www.nytimes.com/2000/11/10/world/chidambaram-subramaniam-india-s-green-rebel-90-dies.html, Access Date: November 30, 2023.
51. Aaryama Somyaji, Initiation of the Green Revolution, https://artsandculture.google.com/asset/initiation-of-the-green-revolution-aaryama-somyaji/ggFoygZPEbbhvA?hl=en, Access Date: November 30, 2023.
52. P. Chidambaram, Budget 2004–2005, July 8, 2004, https://www.indiabudget.gov.in/doc/bspeech/bs200405.pdf, Access Date: November 30, 2023.
53. P. Chidambaram, Budget 2006–2007, February 2006, https://www.indiabudget.gov.in/doc/bspeech/bs200607.pdf, Access Date: December 02, 2023
54. Shri Arun Jaitley, Budget 2014–2015, July 2014, https://www.indiabudget.gov.in/doc/bspeech/bs201415.pdf, Access Date: December 07, 2023
55. Arvind Panagariya and Raghuram Rajan, IMF Working Paper India in the 1980s and 1990s: A Triumph of Reforms, March 2004, https://www.imf.org/external/pubs/ft/wp/2004/wp0443.pdf, Access Date: 11/11/2021
56. FAO, Trade Reforms and Food Security, 2003, https://www.fao.org/3/y4671e/y4671e.pdf, Access Date: June 10, 2022
57. Aniruddha Ghosal Ap Science Writer, Heat wave scorches India's wheat crop, snags export plans, April 29, 2022, 10:11 AM, https://abcnews.go.com/International/wireStory/heat-wave-scorches-indias-wheat-crop-snags-export-84389631, Access Date: June 11, 2022
58. Damian Carrington, Climate crisis makes extreme Indian heatwaves 100 times more likely – study, Wed 18 May 2022 08.29 EDT, https://www.theguardian.com/environment/2022/may/18/climate-crisis-makes-extreme-indian-heatwaves-100-times-more-likely-study, Access Date: June 11, 2022
59. Sybilla Gross, Bumper Wheat Crop Looms Again in Australia on Ample Rains, April 5, 2022, 2:00 PM PDT Updated on April 6, 2022, 12:40 AM PDT, https://www.bloomberg.com/news/

articles/2022-04-05/bumper-wheat-crop-looms-again-in-australia-on-high-prices-rains, Access Date: April 06 2022.
60. Leah Emanuel, Climate Change in Russia and the Weaponization of Wheat, July 2021, https://climateandsecurity.org/2020/08/goodman-and-bergenas-why-the-us-needs-to-restructure-its-national-security-strategy/, Access Date: June 12, 2022
61. Kayode Ajewole, India, Thursday, Last updated: Thursday, October 08, 2020, https://www.ers.usda.gov/topics/international-markets-u-s-trade/countries-regions/india/, Access Date: June 09, 2022
62. P. Chidambaram, Budget 2008–2009, February 2008, https://www.indiabudget.gov.in/budget_archive/ub2008-09/bs/speecha.htm, Access Date: December 05, 2023
63. Patrick Galey, Heat wave in India threatens residents and crucial wheat harvest, April 26, 2022, 9:56 AM PDT, https://www.nbcnews.com/science/environment/heat-wave-india-threatens-residents-crucial-wheat-harvest-rcna26032, Access Date: December 11, 2023
64. Significance of Wheat in the Indian Economy & Effect of Heatwaves on Wheat Yield, Last updated on Jul 31 2023, https://testbook.com/ias-preparation/wheat-crop, Access Date: December 11, 2023
65. Deepak Birewar,What are heat waves, El Nino and how does it affect agriculture and your pocket?, Mar 2023, https://economictimes.indiatimes.com/news/how-to/what-are-heat-waves-el-nino-and-how-does-it-affect-agriculture-and-your-pocket/articleshow/99069251.cms, Access Date: December 11, 2023
66. Kapil Kajal, Heatwave takes a toll on north India's wheat yield, 8 June 2022, https://india.mongabay.com/2022/06/heatwave-takes-a-toll-on-north-indias-wheat-yield/, Access Date: December 11, 2023

Chapter 6
Social Sustainability

This chapter covers the following:

- Two approaches to measuring poverty and inequality
  - Approach 1—the Consumer Pyramids Household Survey (CPHS)
  - Approach 2—IMF Working Paper and Poverty Estimates
- Pro-poor and marginalized inclusive growth machine learning model: poverty and macroeconomic linkage model.
- Policy implications.
- To promote poverty reduction and establish a sustainable environment for growth, policymakers should develop economic and social policies that lead to increased growth, reduced inflation, responsible fiscal policies, decreased fiscal debt, limited subsidies, higher tax to GDP ratios, strengthened banking systems, and promotion of entrepreneurship and start-up companies. If developing countries can maintain their economic strength, they can become a key stabilizing force in the global economy and a secure destination for international capital.

This chapter introduces approaches to calculate poverty. Second, it goes through approach 1 or the Consumer Pyramids Household Survey (CPHS) and approach 2 or the IMF Working Paper and Poverty Estimates. Model developed Pro-poor and Marginalized Inclusive growth Machine Learning Model: Poverty and Macroeconomic Linkage. Finally, the chapter ends with policy recommendations.

In the first postindependence budget speech in 1947, Shri R.K. Shanmukham Chetty, the first Indian Finance Minister, has committed social sustainability by ensuring that the fruits of economic reforms are shared by all sections of society, especially those living in rural areas and more particularly the Scheduled Castes,

© The Author(s), under exclusive license to Springer Nature Switzerland AG 2024

C. Vuppalapati, *Assessing Policy Effectiveness using AI and Language Models*,
International Series in Operations Research & Management Science 354,
https://doi.org/10.1007/978-3-031-56097-2_6

Scheduled Tribes, and Other Backward Classes [1]. Five elements of social and economic infrastructure are critical to the quality of life, as per Shri R.K. Shanmukham Chetty, especially in rural areas: health, education, drinking water, housing, and roads.

> Preservation of the rural environment will raise the living standards of millions belonging to the weakest sections of our society.
>
> Shri R.K. Shanmukham Chetty
> Minister of Finance, 1947–1948[1]

Rural communities are essential parts of achieving social sustainability. The well-being of rural, poor, and marginalized communities is the standard for evaluating policy effectiveness and social sustainability. Enhancing wealth creation and capital accumulation as part of economic sustainability should be accompanied by the development of new attitude toward wealth, seeing it as a social product and using it for the benefit of society, especially the disadvantaged. Adopting and advancing the philosophy of trusteeship, where wealth owners hold it as a trust and use it in the interest of society; investing in education, especially elementary education, to develop human resources and provide quality education to all, especially children from marginalized groups, which is the long term sustained policy; and lastly, at the policy level, increasing the allocation for the programs of the Ministry of Welfare, Department of Women and Child Development, and Health and Family Welfare are necessary practical measures [2].

> Policy: wealth as a social product
>
> Developing a new attitude toward wealth, considering it as a social product, and using it for the benefit of society, especially the underprivileged. Practicing and promoting the philosophy of trusteeship, where wealth owners hold it as a trust and use it in the interest of society.
>
> Shri Manmohan Singh
> Minister of Finance, 1991–1992[2]

Preservation of the rural environment will raise the living standards of millions belonging to the weakest sections of our society. Environmental conversation is among the important sustainability aspects, given more than half of the population depend on agriculture. Implement measures to address soil erosion, salinity, and loss of tree cover. Promote sustainable practices in agriculture and forestry. Encourage the use of renewable energy sources, and reduce pollution from industries and transportation [3]. The task that the democratically elected leaders of newly independent India embarked on in the early 1950s was not for the faint of heart [4]. It was to lift living standards of people accounting for one-seventh of the world's popula-

[1]India Budget Speech, 1947–1948—https://www.indiabudget.gov.in/doc/bspeech/bs194748.pdf

[2]India Budget Speech, 1991–1992—https://www.indiabudget.gov.in/doc/bspeech/bs199293.pdf

tion. Three-fourths of the Indian people were engaged in agriculture working with primitive tools and techniques, as either destitute landless laborers, highly insecure tenants-at-will, or small-plot holders eking out subsistence living from their meager plots. The literacy rate stood at 14%, and the average life expectancy was 32 years [4].

Implementing land reform policies and improving agricultural productivity can help alleviate rural poverty by providing farmers with secure land tenure, access to credit, and technical support [5]. Ensuring that marginalized and vulnerable groups have equal access to resources, services, and opportunities is essential in poverty reduction. This includes addressing issues of discrimination, social exclusion, and inequality.

Another way to improve social sustainability for small farmers is to encourage horticulture. As a way to increase revenue, they should diversify their agriculture and grow fruits, vegetables, flowers, dairy, poultry, fisheries, pulses, and oilseeds as advised in 2005. Even after two decades later, the demand for specialty crops is persistently higher as India's edible oil import has increased exponentially for the current MY compared to MY 2021/2022. Crude palm oil remains the primarily consumed edible oil with record imports from Indonesia and Malaysia estimated at 10.5 MMT for MY 2022/2023.[3] The agriculture sector, which supports the livelihoods of a large part of the population, needs urgent action. Diversifying with horticulture will raise income [5]. The National Horticulture Mission covers everything from research to production to postharvest management to processing to marketing [6].

In order to return to a path of high sustained growth with reasonable price stability and greater social equity post 1991 economic crisis, the task for the planners of Indian economy has become imperative to be internationally competitive economy. This is when opening of economy to international investment took place. Furthermore, before 1991, there was insufficient focus on reducing costs, upgrading technology, and enhancing living standards. The lack of such competitive measures made the local companies noncompetitive to the world economy, and knowledge training of Indian workforce has become a necessity. The provision of the National Renewal Fund is to ensure that the cost of technical change and modernization does not devolve on the workers. The fund will not merely provide ameliorative measures for the workers affected in the course of technical change but, more importantly, provide retraining to them. The basic challenge of our times is to ensure that wealth creation is not only tempered by equity and justice but is harnessed to the goal of removal of poverty and development for all. Our approach to development has to combine efficiency with austerity, holding our society together in pursuit of the noble goal of banishing poverty, hunger, and disease [2].

[3]India: Oilseeds and Products Update—https://fas.usda.gov/data/india-oilseeds-and-products-update-26

In 2023, 75 years after the independence, the priorities for Indian government have remained the same as outlined by the Indian Finance Minister,[4] under aspirational districts and blocks programs:

- Improve socioeconomic conditions of vulnerable tribal groups.
- Provide basic facilities such as housing, clean water, education, and healthcare.
- Enhance road and telecom connectivity in tribal areas.
- Create sustainable livelihood opportunities for tribal communities.

So, what it tells us is the social sustainability is still a work in progress and policies related to social sustainability remain fully yet to realize the overall growth for the economy and honorable citizens of India.

Poverty elimination has remained a major challenge right from independence and lies at the core of India's national development agenda to create a just and equitable society.[5] Given the limited resources, reliable estimation of poverty is the first step toward eradication of poverty as a basic input for design, implementation, and monitoring of anti-poverty programs [7]. Poverty measurement is also important to serve as a barometer of the extent of the success of strategies for inclusive growth and poverty reduction.

Some strategies to reduce poverty are to reform and modernize the tax system, improve the fiscal balance, strengthen industrial growth, and change development policies and programs to deal with poverty, unemployment, and social exclusion. Address both social and economic inequalities by implementing reforms that promote inclusive growth[6] and ensure a wider distributive spread of national wealth. This may include targeted social welfare programs, skill development initiatives, and measures to enhance income levels for all citizens [8]. Additionally, maintain and enhance the progress in improving the banking system's health, reduce dependence on foreign loans through foreign direct and portfolio investment, pay off high-interest debt and let companies repay their previous foreign loans earlier, and restructure regional rural banks (RRBs) comprehensively to make them more efficient and sustainable. These strategies are some of the important ones implemented during 1993 economic restructuring [9].

Some of the poverty reduction strategies include the following (please see Fig. 6.1) [10]:

- Pursuing macroeconomic policies to sustain a growth rate of at least 9% per annum over an extended period of time
- Strengthening mechanisms for inclusive growth and creating new job opportunities
- Expanding employment generation schemes and strengthening social security nets

---

[4]Budget-Speech, 2023–2024—https://www.indiabudget.gov.in/doc/budget_speech.pdf

[5]POVERTY MEASUREMENT IN INDIA: A STATUS UPDATE https://rural.nic.in/sites/default/files/WorkingPaper_Poverty_DoRD_Sept_2020.pdf

[6]Budget 2003–2004—https://www.indiabudget.gov.in/doc/bspeech/bs200405.pdf

Fig. 6.1 Reinforcement poverty reducation strategies

- Increasing plan allocation for agriculture and implementing schemes to support farmers
- Ensuring food security and maintaining affordable food prices for the poor
- Giving highest priority to rural development and implementing various programs to improve living conditions in rural areas.
- Implementing the National Rural Employment Guarantee Scheme to generate wage employment
- Improving access to quality elementary education through social programs
- Providing midday meals in schools to enhance school participation and reduce hunger
- Universalizing the Integrated Child Development Scheme to reduce malnutrition levels
- Focusing on integrated development of urban infrastructure and improving rural economies
- Ensuring safe drinking water supply and sanitation in rural areas through various programs
- Continuously working on rural sanitation programs to improve hygiene and sanitation in rural areas

Poverty

Poverty can be defined as a condition in which an individual or household lacks the financial resources to afford a basic minimum standard of living. However, the perception regarding what constitutes poverty may vary over time and across countries. The conventional approach to measuring poverty is to specify a minimum expenditure (or income) required to purchase a basket of goods and services necessary to satisfy basic human needs. This expenditure is called the poverty line. The basket of goods and services necessary to satisfy basic human needs is the poverty line basket (PLB). Poverty can be measured in terms of the number of people living below this line (with the incidence of poverty expressed as the head count ratio (HCR) or the poverty ratio—number of poor to the total population expressed as percentage). You will be on BPL list if your annual income is Rs.27,000 [11]. Globally, countries use different measures for measuring poverty, but the underlying principle remains the same—a poverty line is calculated based on consumption required for maintaining some minimum standard of living in the country [7].

The poverty ratio and the head count index are two common measures of poverty, but they are not the same. The poverty ratio is the percentage of the population living below the national poverty line, while the head count index is the number of people living below the international poverty line of $1.90 a day (2011 PPP). The poverty ratio is based on the country-specific poverty line basket, which reflects the minimum expenditure required to satisfy basic human needs in that country. The head count index is based on a common standard of living across countries, which may not capture the multidimensional aspects of poverty.

In India, the National Sample Survey (NSS) conducts large scale surveys on household consumer expenditure *every 5 years*, which are used to estimate the poverty ratio for India and its states and union territories. The National Sample Survey Office (NSSO) conducts Household Consumer Expenditure Surveys (HCESs) to estimate the poverty ratio in India.[7] The poverty ratio is based on the per capita consumption expenditure over a month and the poverty line determined by various expert groups [7] most recently including the Tendulkar Expert Group (2009) and Rangarajan Committee (2014). The latest survey in this series was conducted in 2011–2012 (NSS 68th round), which repeated the survey of 2009–2010 (NSS 66th round) as it was not a normal year due to a severe drought [7]. The following table lists the poverty ratio estimates for different years based on the NSS surveys:

[7] National Sample Survey Office (NSSO)—https://mospi.gov.in/NSSOa

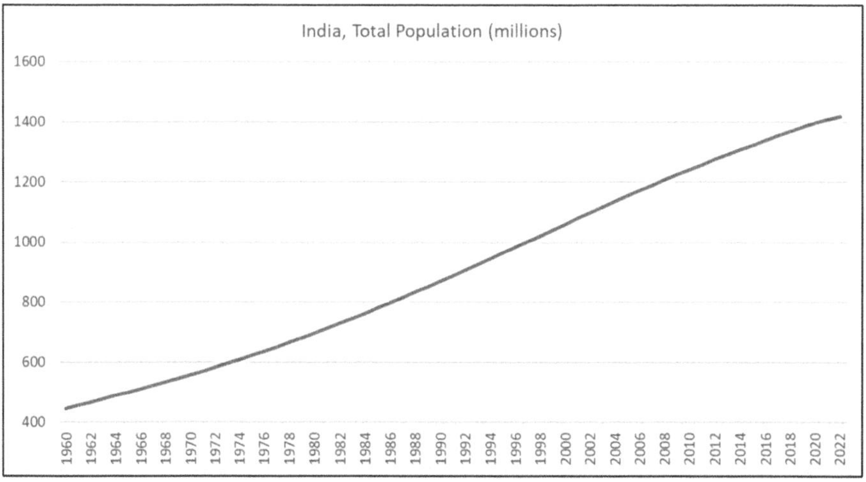

Fig. 6.2 Total Population

	Poverty ratio (%)			Million	
Year	All India	Rural	Urban	Population[8]	Number of poor[9]
1973–1974	54.9	56.4	49.0	596.107	327.262
1977–1978	51.3	53.1	45.2	651.685	334.314
1983	44.5	45.7	40.8	729.169	324.480
1987–1988	38.9	39.1	38.2	815.716	317.313
1993–1994	36.0	37.3	32.4	926.351	333.486
2004–2005	27.5			1136.264	312.472
2011–2012	21.9			1257.621	275.418

The estimates of poverty in the country are made at the national and state level by the Planning Commission at an interval of approximately 5 years from the large sample survey data on consumer expenditure conducted by the National Sample Survey Organization (NSSO). Comparable estimates based on a consistent methodology and data set are available until 1993–1994. These estimates show a secular decline in the poverty ratio, though the number of poor remained stable (please see Fig. 6.2) for a fairly long period of two decades (1973–1993[10]) (above table), mainly due to increase in population [12]. Poverty in India has declined over the last decade but not as much as previously thought [13]. The NSS surveys stopped in 2012.

[8] Population, total—India—https://data.worldbank.org/indicator/SP.POP.TOTL?locations=IN

[9] Table 7: Number of poor in India: 1951–1992—https://documents1.worldbank.org/curated/en/253831468049773494/pdf/900360WP0Pover00Box385299B00PUBLIC0.pdf

[10] Poverty {Page 193}—https://www.indiabudget.gov.in/budget_archive/es2000-01/chap1003.pdf

Although official data on poverty in India for the year 2023 is not available, we can refer to the available data sources. Here is a table displaying poverty measures for India from 1951 to 2023, which have been calculated using various methods and sources:

Year	Poverty line (Rs. per month)	Poverty ratio (%)	Number of poor (millions)	Sources
1951	20.00	47.0	171.0	[12]
1961	22.00	45.0	190.0	
1971	20.00	55.0	320.0	
1981	49.09	44.5	323.9	
1991	81.00	37.3	320.3	
2001	328.80	26.1	260.3	
2011[a]	816.00	21.9	269.3	
2017	N/A	N/A	N/A	
2019	N/A	27.9	372.1	[12]
2023[b]	N/A	24.1	322.6	

[a]The poverty line for the NSS HCES is based on the Tendulkar Committee methodology, which was adopted by the Indian government in 2009

[b]The MPI projection for 2023 assumes that the MPI will decline at the same rate as it did from 2015–2016 to 2019–2021, which was 0.019 points per year. The number of poor for the MPI estimates is calculated by multiplying the poverty incidence (H) by the population of India, which was 1334.4 million in 2019 and is projected to be 1337.8 million in 2023

The poverty estimates for the NSS HCES and the MPI are not directly comparable, as they use different data sources, methods, and reference periods [12]. According to the Planning Commission, the poverty ratio for India was 21.9% in 2011–2012,[11] based on the Tendulkar Committee methodology. This means that about 270 million people were living below the poverty line, which was Rs. 816 per capita per month for rural areas and Rs. 1000 for urban areas. The poverty ratio declined from 37.2% in 2004–2005 and 45.3% in 1993–1994, showing a significant improvement in the level of living of the population. For NSS Survey Poverty Index,[12] please see Appendix C.

A total of 415 million people moved out of poverty in India within just 15 years from 2005/2006 to 2019/2021, remarkable achievement by the world's most populous nation.

The United Nation Development Program (UNDP)[13]

[11]Poverty headcount ratio at national poverty lines (% of population)—https://data.worldbank.org/indicator/SI.POV.NAHC?end=2021&start=2021&type=shaded&view=map

[12]India—Achievements and challenges in reducing poverty (English)—https://www.google.com/books/edition/India/JTvw9hd_omsC?hl=en&gbpv=1&printsec=frontcover

[13]25 Countries Halved Multidimensional Poverty Within 15 Years, but 1.1 Billion Remain Poor—https://www.undp.org/press-releases/25-countries-halved-multidimensional-poverty-within-15-years-11-billion-remain-poor

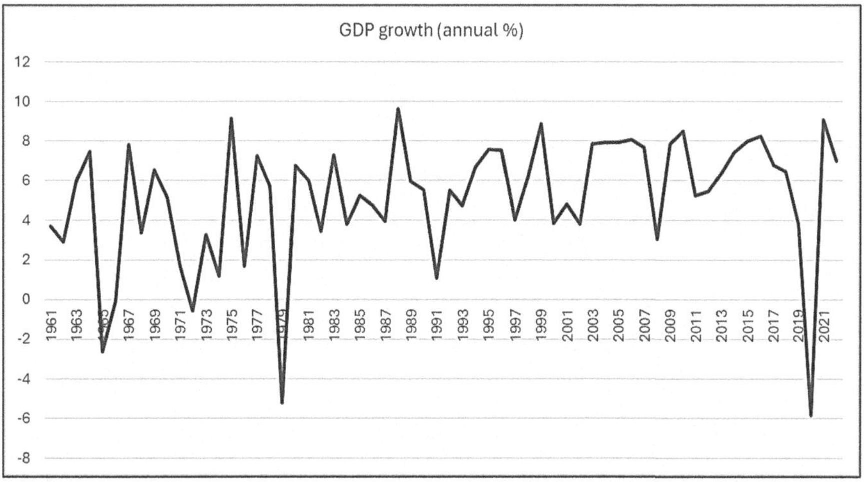

Fig. 6.3 GDP growth

The World Bank also provides data on the head count index for India and other countries, based on the international poverty line of $1.90 a day (2011 PPP). According to the World Bank, the head count index for India was 13.4% in 2011 (please see Fig. 6.3), down from 45.9% in 19,943. This means that about 170 million people were living below the international poverty line in 2011, compared to 450 million in 1994. The head count index for India was lower than the average for South Asia (15.1%) and the world (10.7%) in 2011.

The poverty ratio and the head count index are two common measures of poverty, but they are not the same. The poverty ratio is the percentage of the population living below the national poverty line, while the head count index is the number of people living below the international poverty line of $1.90 a day (2011 PPP). The poverty ratio is based on the country-specific poverty line basket, which reflects the minimum expenditure required to satisfy basic human needs in that country. The head count index is based on a common standard of living across countries, which may not capture the multidimensional aspects of poverty.

The World Bank has poverty rates for all countries using two new international poverty lines: a lower middle-income line, set at $3.20 per day, and an upper middle-income line, set at $5.50 per day. These are in addition to the main poverty line of $1.90 per day. The new lines are supposed to serve two purposes. One, they account for the fact that "achieving the same set of capabilities may need a different set of goods and services in different countries" and, specifically, a costlier set in richer countries. Second, "they allow for cross-country comparisons and benchmarking both within and across developing regions." Using the $1.90 line, the incidence of poverty (please see Fig. 6.4) in lower middle-income countries is 15.5%, as against 45.8% in low-income countries. However, using the $3.20 line, 46.7% of the population of lower middle-income countries is poor. Similarly, for upper middle-

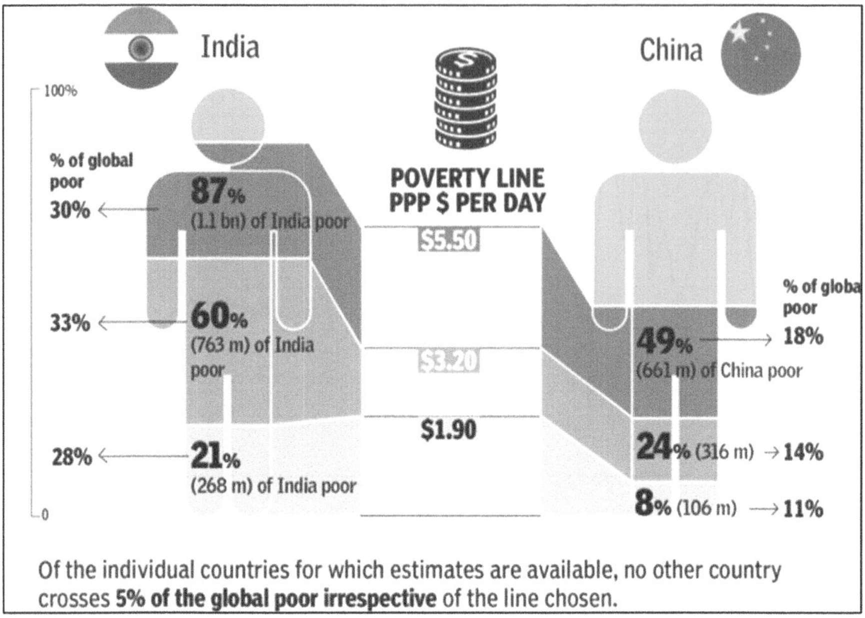

Fig. 6.4 Poor head count (India-China)

income countries, the proportion of the poor at $1.90 is just 2.3%, but at $ 5.50, it is 29.2% [14].

India has the largest chunk of global poor [15].

Please note that the last Consumer Expenditure Survey data was released in 2011, and lack of NSS survey data has left a void in analyzing poverty rates [13, 16–19].

Two Approaches to Measuring Poverty and Inequality

Approach 1: CPHS

CPHS imputes NSS-type household consumption into the Consumer Pyramids Household Survey (CPHS) using predictors of household consumption that are available in both surveys [13, 16, 18].

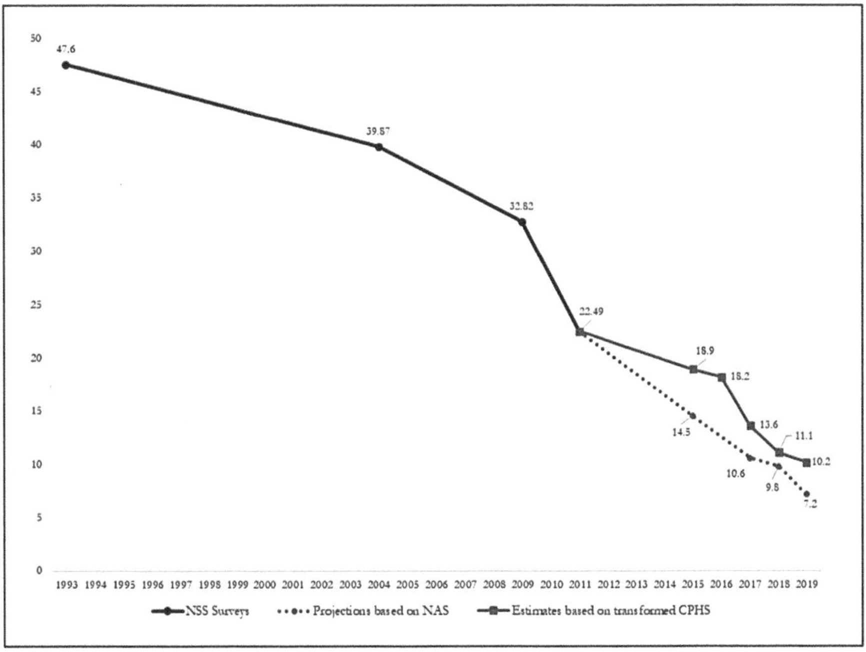

Fig. 6.5 NAS poverty head count

There were good ex ante reasons to think the projection based on the national accounts system (NAS) shown by the dotted line in Fig. 6.5 was too optimistic—more on that below. But consumption growth in the CPHS survey also looks surprisingly high in the latter years, especially 2019–2020. That year, the index of industrial production for consumer goods registered an absolute decline of 3.8% (even more in per capita terms), real rural wages declined, and real urban wages grew less than 2%[14]—meanwhile, the CPHS showed growth in real per capita consumption of 6.2% [16].

Indian poverty (<$1.90 PPP)	
Year	Poverty rate
1993	47.6
1994	46.8
1995	46.01
1996	45.01
1997	44.02

(continued)

[14] Poverty in India Has Declined over the Last Decade But Not As Much As Previously Thought—https://openknowledge.worldbank.org/server/api/core/bitstreams/2938e47d-f0d3-56aa-89d7-54855d7e3f85/content

Indian poverty (<$1.90 PPP)	
Year	Poverty rate
1998	43.80
1999	42.03
2000	40.39
2001	40.01
2002	39.89
2003	39.57
2004	39.87
2005	38.01
2006	37.90
2007	36.88
2008	34.43
2009	32.82
2010	28.77
2011	22.49
2012	21.09
2013	20.00
2014	19.01
2015	18.9
2016	18.2
2017	13.6
2018	11.1
2019	10.2
2020	9.1

Real inequality—Gini coefficient [13]

Year	Gini coefficient
1993	31.7
1994	31.92
1995	32.14
1996	32.36
1997	32.58
1998	32.8
1999	33.02
2000	33.24
2001	33.46
2002	33.68
2003	33.9
2004	34.41
2005	34.70
2006	35.00
2007	35.29

(continued)

Year	Gini coefficient
2008	35.59
2009	35.58
2010	35.67
2011	35.71
2012	35.34
2013	34.97
2014	34.78
2015	34.59
2016	34.48
2017	34.59
2018	33.75
2019	34.75
2020	33.80

Approach 2: IMF Working Paper and Poverty Estimates

This section provides detailed poverty estimates for the World Bank's extreme poverty line and the lower middle-income poverty line of PPP$ 1.9 and PPP$ 3.2, respectively [20]. The estimates are for both the Modified Mixed Recall Period (MMRP) and the Uniform Recall Period (URP), with and without food-subsidy transfers.[15]

Poverty rate (%)—without food transfers				
Year	PPP$ 1.9		PPP$ 3.2	
Recall method				
	MMRP	URP	MMRP	URP
2004	32.7	45.5	73.8	80.3
2005	29.2	41.8	70.9	78.0
2006	25.8	38.0	67.9	75.5
2007	22.1	33.6	64.6	72.9
2008	20.0	31.6	63.0	71.7
2009	21.3	33.0	64.3	72.7
2010	16.6	27.9	59.7	69.0
2011	12.2	21.8	53.6	64.0
2012	11.0	20.1	51.3	62.0
2013	9.2	17.7	47.6	58.9
2014	7.4	14.6	43.3	55.4

(continued)

[15] Pandemic, Poverty and Inequality: Evidence from India., https://www.elibrary.imf.org/downloadpdf/journals/001/2022/069/001.2022.issue-069-en.xml

Poverty rate (%)—without food transfers				
Year	PPP$ 1.9		PPP$ 3.2	
Recall method				
	MMRP	URP	MMRP	URP
2015	5.7	12.0	38.9	51.2
2016	3.9	8.9	32.7	45.8
2017	2.9	7.1	29.0	41.7
2018	2.0	4.9	23.0	35.2
2019	1.4	3.4	18.5	30.4
2020	2.5	6.1	26.5	38.9

Source: Authors [20] Computations using the 2011–12 NSO Consumer Expenditure Survey & National Accounts Statistics

Poverty rate (%)—with food transfers				
Year	PPP$ 1.9		PPP$ 3.2	
Recall method				
	MMRP	URP	MMRP	URP
2004	31.9	44.7	73.5	80.1
2005	28.2	40.7	70.5	77.6
2006	24.9	36.9	67.5	75.2
2007	21.0	32.6	64.0	72.5
2008	18.7	30.1	62.3	71.2
2009	19.6	31.3	63.4	72.0
2010	15.0	25.7	58.4	67.9
2011	10.8	19.9	52.2	62.9
2012	9.2	17.7	49.3	60.4
2013	7.8	15.2	45.8	57.4
2014	5.1	10.9	39.7	52.0
2015	3.6	8.4	34.6	47.6
2016	2.3	5.8	28.8	41.6
2017	1.9	4.6	25.2	37.7
2018	1.1	2.7	18.7	30.8
2019	0.76	1.9	14.8	25.5
2020	0.86	2.1	18.1	29.9

Source: Authors [20] Computations using the 2011–12 NSO Consumer Expenditure Survey & National Accounts Statistics

Real inequality—Gini coefficient [20]

	Gini	
Year	Without transfers	With transfers
2004	31.3	31.1
2011	31.3	30.9
2014	31.6	30.6
2015	31.7	30.6

(continued)

Year	Gini	
	Without transfers	With transfers
2016	31.7	30.7
2017	31.6	30.7
2018	31.6	30.6
2019	31.4	30.4
2020	31.5	29.4

Notes: (1) For the updated consumption distribution, we used the PFCE growth rates for the adjustment

(2) The consumption distribution used here is the type 2 (Modified Mixed Recall Period)

If the World Bank paper paints a "mildly optimistic" picture of poverty levels, the IMF paper contends that extreme poverty has been nearly eliminated. The IMF paper [20] presents a consistent time series of inequality and poverty levels for both the extreme poverty PPP$1.9 and low- middle-income poverty (PPP$ 3.2) levels for each of the years 2004–2005 to 2020–2021 [16]. These estimates include, for the first time, the effect of in-kind food subsidies on poverty and inequality. Extreme poverty was as low as 0.8% in the pre-pandemic year 2019, and food transfers were instrumental in ensuring that it remained at that low level in the pandemic year 2020.

Effectiveness of Public Distribution System (PDS)

The Indian government has a food security program called the public distribution system (PDS). It is a targeted program, which means it is for people who are below the poverty line (BPL). The PDS gives these people food grains at low prices. The PDS started in the 1950s and has changed several times since then. The latest change was in 2013, when the government introduced the National Food Security Act (NFSA). Under the NFSA, up to 75% of the rural people and up to 50% of the urban people can get subsidized food grains through the PDS. Antyodaya Anna Yojana[16] is part of the PDS.[17] Under the scheme, priority households living below the poverty line are entitled to get 25 kg of food grain per month per family. Increasing foodgrain stocks to achieve successful PDS. In 1990 budget speech, the government gave high priority to stepping up procurement efforts and rebuilding stocks of food grains to maintain price stability and economic security [21]. During high inflationary[18] times, for instance, 1991 with 16.7%, the government used the

[16] Antyodaya Anna Yojana—https://dfpd.gov.in/aay_C.htm

[17] Budget 2007–2008—https://www.indiabudget.gov.in/doc/bspeech/bs200708.pdf

[18] Inflation, consumer prices (annual %)—India—https://data.worldbank.org/indicator/FP.CPI.TOTL.ZG?locations=IN

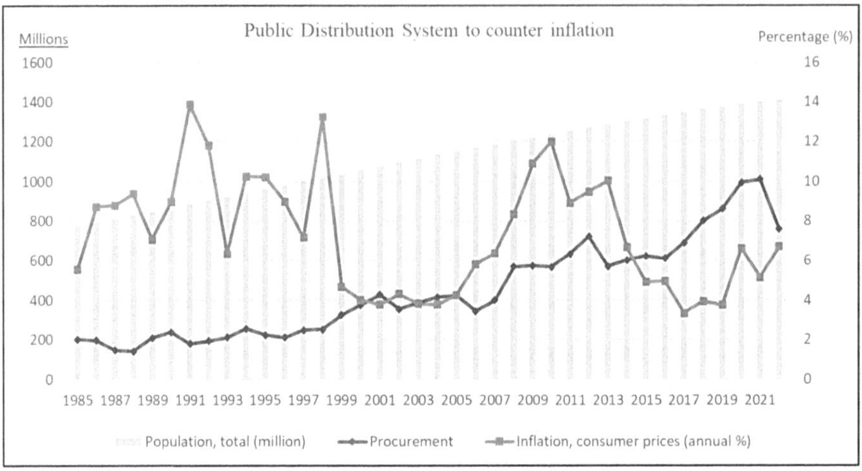

Fig. 6.6 PDS to inflation

public distribution system[19] to counter inflation (please see Fig. 6.6) and protect the poorer sections of the population[20] from high prices and shortages [2].

The procurement is function of inflation plus population. As population grows, the procurement is function of inflation, among others.

Credit to Agriculture

Agricultural credit[21] aims to improve agriculture and rural economy. Supporting small-scale farmers and investing in rural development can help reduce poverty in rural areas by increasing agricultural productivity, improving access to markets, and creating employment opportunities. Expanding access to financial services, such as savings accounts, credit, and insurance, can help individuals and families living in poverty build assets, manage risks, and invest in income-generating activities. Agriculture indebtedness is one of the main reasons to expand agriculture credit as addressing the indebtedness issue of farmer debt and finding solutions to alleviate their financial burden [22].

Credit to agriculture shows the amount of money that banks lend to farmers or rural households, to agricultural cooperatives, or to any businesses related to agriculture. Formal credit is very important for farmers to buy inputs like seeds,

[19] Table 27: Public Distribution System—Procurement, Off-Take and Stocks—https://rbi.org.in/Scripts/PublicationsView.aspx?id=21833

[20] Population—https://data.worldbank.org/indicator/SP.POP.TOTL?locations=IN

[21] Credit to agriculture—https://www.fao.org/documents/card/en/c/cb8790en

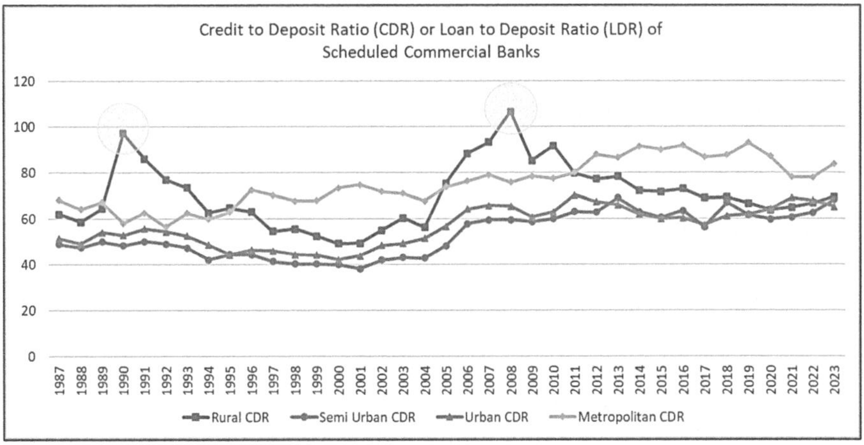

Fig. 6.7 Credit to Deposit Ratio

fertilizers, plant protection materials, or animal feed. If they do not have personal savings, they may have to borrow from informal sources (such as moneylenders, relatives, and friends) that may charge very high interest rates and impose unfair conditions, which may make many agricultural activities unprofitable. The access to credit is especially hard for farmers because there is a delay between the time they spend money on growing crops or raising livestock and the time they earn money from selling the products [23].

Focus on rural development is best policy for ensuring sustainable agriculture and socioeconomic development (please see Fig. 6.7). This includes budgetary support, debt relief, and fertilizer subsidies specifically targeted at benefiting rural communities. Focus on improving agriculture and allied activities to boost the rural economy can be done through investment in infrastructure and technology and providing support to farmers. Microfinance is an effective tool for poverty alleviation [24].

Policy: focus on rural development

Rural development is the key to rural sustainability. The government should pledge to allocate a large part of its resources for the improvement of agriculture and rural areas. This includes budgetary support, debt relief, and fertilizer subsidies that are aimed at helping rural communities.

The government has benefited from the banking and monetary policies and has increased money supply to carry out programs such as a comprehensive policy on agricultural credit, accountability of sponsor banks for regional rural banks (RRBs), and microfinance through self-help groups (SHGs) [5]. The comprehensive policy on agricultural credit aimed to double the credit flow in 3 years, from 2004 to 2007. This policy involved the participation of public and private sector banks, regional rural banks (RRBs), and cooperative banks in executing the policy. The purpose was to enhance the social and economic well-being of rural areas. With the availability of money, the government provided insurance for farming and livestock, ensuring the viability of the agricultural sector.

Macroeconomic Model: Employment and GDP

Between 2000 and 2021, India enjoyed rapid economic growth averaging 6.6% per annum. GDP per capita more than doubled, and extreme poverty (at $2.15 2017 PPP) fell from 39.9% in 2004 to 11.9% in 2021.[22] This strong performance reflected a demographic dividend, steps taken to integrate India into the global economy and improve the business environment, and prudent monetary and fiscal management.

Meanwhile, moderate poverty (at $3.65 2017 PPP[23]—the threshold typically used for L-MICs) fell from 76.6% in 2004 to 46.5% in 2021. Even though growth helped reduce monetary poverty, 16.4% of the population faces deprivations that classify them as multidimensionally poor, and one in three Indians is malnourished. Consumption inequality remained stagnant at 0.35 (using the Gini coefficient). The Gini index, or Gini coefficient, measures income distribution across a population. Developed by Italian statistician Corrado Gini in 1912, it often serves (please see Fig. 6.8) as a gauge of economic inequality, measuring income distribution or, less commonly, wealth distribution among a population [25].

The Gini index[24] is often represented graphically through the Lorenz curve, as depicted below, which shows income (or wealth) distribution by plotting the population percentile by income on the horizontal axis and cumulative income on the vertical axis. The Gini coefficient is equal to the area below the line of perfect equality (0.5 by definition) minus the area below the Lorenz curve, divided by the area below the line of perfect equality. In other words, it is double the area between the Lorenz curve and the line of perfect equality.

Extreme poverty hovers around 11%, while moderate poverty is around 45%. To increase potential growth and reduce poverty, structural reforms should seek to improve human capital, key infrastructure, the business environment, and jobs, particularly for women. Other machine learning features include global growth, oil prices, inflation, reserves, employment, fiscal deficit, services exports, remittance inflows, balance of trade, net foreign direct investment, foreign exchange reserves [26].

---

[22] Poverty headcount ratio at $2.15 a day (2017 PPP) (% of population)—India—https://data.worldbank.org/indicator/SI.POV.DDAY?locations=IN

[23] Poverty headcount ratio at $3.65 a day (2017 PPP) (% of population)—India—https://data.worldbank.org/indicator/SI.POV.LMIC?locations=IN

[24] Gini index—https://databank.worldbank.org/metadataglossary/gender-statistics/series/SI.POV.GINI

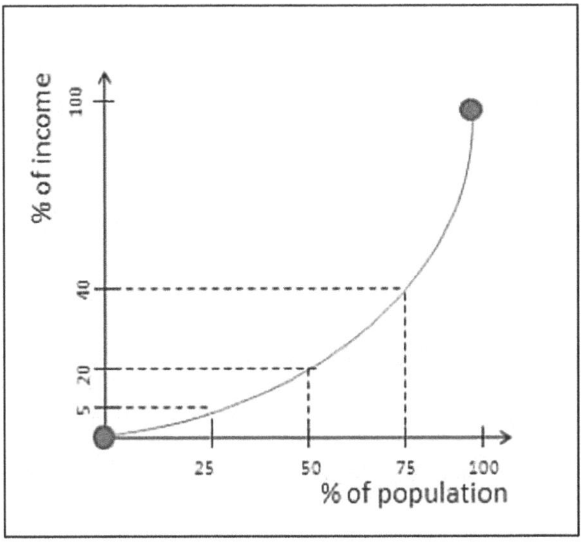

Fig. 6.8 Population vs. income

Most of the previous studies on these issues have used mainly cross-country comparisons, often with single cross sections (such as in the many tests of the Kuznets hypothesis) but sometimes with panel data, though the short time series usually means that the cross-country variation is dominant. (While one can measure economic growth annually for almost all countries, the household surveys needed to monitor living standards are much less frequent.) To address these issues, it is obviously desirable to have a fairly long series of surveys; a short series can mislead for inferring a trend.

For India to achieve the national goal of becoming a high-income country by 2047, real growth would need to rise sustainably to around 8% per annum on average. This requires more ambitious structural reforms to enhance the quality of education, generate more and better jobs (in terms of remuneration, stability, and labor conditions), increase the economic participation of women and youth, improve infrastructure, and strengthen the business environment [26].

India has the most extensive set of national household surveys that can be used to monitor living conditions among developing countries. The surveys are fairly consistent over time because the basic survey tools and methods have not changed much (though we acknowledge, and deal with, some compatibility issues). India therefore offers rich evidence over time—uniquely among developing countries—for examining and measuring the link between living standards of the poor and macroeconomic indicators. Except 2019, highlighting a dramatic fall in the monthly per capita consumption expenditure (MPCE), the first such drop in MPCE levels since 1972–1973, till date drift in the data is minimal[25] [16, 17]. A mathematical model that measures poverty along with a national survey that collects data can help researchers, the public, and policymakers to have more information in the gaps between surveys, as macroeconomics, lifestyle, agriculture, weather, climate change, politics, commerce, and others are constantly changing and affecting the economic results that influence poverty at different levels of the economic system that we live and breathe in.

The general development literature has discussed all the important issues related to growth and poverty, and these have been also relevant for the literature and policy debates on India. The postindependence planners of India had hoped that the country's industrialization process based on urban areas would benefit poor people in the long run, including through absorbing rural labor. But this hope was mostly disappointed by the evidence of the slow rate of poverty reduction from independence until the 1980s. Many observers attributed this to the slow rate of labor transfer from agriculture that resulted from the more inward-focused and capital-intensive development path of this period.

Foreign exchange reserves cover: as a result, foreign exchange reserves fell to US$578 billion, equivalent to 7 months of import cover [26].

Many people thought that India's poverty reduction was slow because of low economic growth. But a closer look at the data shows that the type of growth in different sectors also mattered. Using data until the early 1990s, they found that rural economic growth reduced poverty more, and so did growth in the primary (mainly agriculture) and tertiary (mainly services) sectors compared to the secondary (mainly manufacturing and construction) sector. They also found that positive effects across sectors made rural economic growth more important for reducing national poverty. Urban growth and secondary sector growth had negative effects on the distribution of income that reduced the benefits for the urban poor, and urban growth hardly helped the rural poor at all. The slow improvement in poverty was due to both a lack of overall growth and a type of growth that did not favor poor people (please see Fig. 6.9).

Many people in India hoped that the economic reforms that began in the early 1990s would lead to a quicker reduction of poverty. They achieved higher growth rates as a result of these reforms. But there have also been indications of increased inequality after the reforms, making it unclear how much the poor have benefited from higher growth rates.

[25] What is Consumer Expenditure Survey, and why was its 2017–2018 data withheld?—https://www.thehindu.com/business/Economy/what-is-consumer-expenditure-survey-and-why-was-its-2017-2018-data-withheld/article61611662.ece

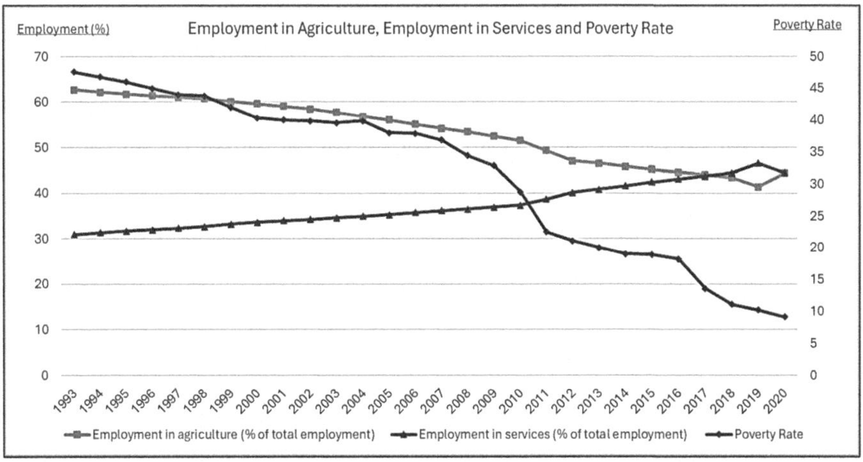

Fig. 6.9 Employment

India's labor markets have undergone significant changes since around 2000, which have important consequences. Rural casual labor markets have become tighter, with higher real wages, and the difference between urban and rural wages has shrunk. Three factors seem to be involved here.

- First, more people have gone to school,[26] thus lowering the availability of unskilled labor, especially in rural areas [27]. For instance, the gross school enrollment[27] increased from 77% in 1971 to 108% in 2022. Improved education has been statistically proven as one of the most significant factors in raising people out of poverty within India [27]. Having diverse educational concentrations in higher grades helped attract many and opened economic opportunities.

Education emphasis

Prioritize and generate additional funds for education, with a special emphasis on providing basic education to all children. Upgrade technical education institutes to meet industry demands and promote public-private partnerships in this sector. Also, ensure access to quality healthcare through health insurance schemes.

P. Chidambaram
Minister of Finance, July 2004

- Second, the rate of women participating in the labor force has also fallen. As household incomes rise, women are dropping out of India's workforce simply because they can afford to. Many no longer have to do backbreaking tasks in

[26] School enrollment, primary (% net)—https://databank.worldbank.org/metadataglossary/millennium-development-goals/series/SE.PRM.NENR

[27] School enrollment, primary (% gross)—https://databank.worldbank.org/metadataglossary/gender-statistics/series/SE.PRM.ENRR

agriculture or other manual labor [28]. One way to promote more women participation would be through gender-sensitive budgeting. The government can continue to prioritize gender budgeting by allocating funds to schemes that benefit women or have a significant allocation targeted toward women. Ministries and departments can also establish gender budgeting cells to ensure a gender perspective in their budgets [29].

- Third, there has been a surge in construction activity across India, especially in (rural and urban) infrastructure, which had been neglected for a long time. In 1993–1994, only 3.2% of rural males worked in the construction sector, but this increased to 13% by 2011–2012. The movement of labor from agriculture to the nonfarm sectors has been faster since the 1990s. The movement has been to construction and services, and also to manufacturing, to a lesser extent, suggesting that higher labor demand from construction has raised the wages of unskilled labor compared to skilled labor within rural areas, as well as rural relative to urban wages (for male workers). Revival of growth in manufacturing and infrastructure is one of the poverty reduction strategies. Manufacturing is the main driver of growth, and this augurs well for the future [22].

Author home construction

In 1993–1994, when I finished my undergraduate studies, I had to help my family with building a house in Secunderabad, India. I have witnessed how construction workers gained from the rising demand for homes in urban areas in India and the income that home construction provided after the economic reforms of the 1990s. Many of the construction workers were farmers, especially men who migrated from rural areas to make money to support themselves economically.

The combination of a lower supply of unskilled labor and rising demand for that labor in construction, transport, and other services is likely to have been a driving force in higher casual wages, in both farm and nonfarm sectors and compressing the urban-rural wage gap. It is unclear how permanent this change will prove to be. It may be conjectured that (like China) India has reached its Lewis turning point. However, that is a conjecture, since there may well be other factors leading to higher wages even though there is still rural underemployment. And temporary reversals might be expected, notably if the current construction boom does not continue.

We use two poverty lines to report our findings. One line is the same as the one that the Planning Commission of 1979 and 1993 used. This poverty line is based on a nutritional standard of 2400 calories per person per day in rural areas and 2100 calories for urban areas, and it means a monthly spending of Rs. 617 and Rs. 922 per person in rural and urban areas, respectively, at 2011–2012 prices. The other poverty line has the same value as the international poverty line of $1.25 per person per day at 2005 PPP dollars, and they are equal to monthly expenditures of Rs. 732 and Rs. 1115 per person in rural and urban areas at 2011–2012 prices. The second set of lines are about 20% more than the first set of lines.

This long account of poverty, urbanization, and growth in India has taught us some things. Using newer data, our analysis has verified some results from previous research and also given us some new perspectives. India's steady progress against absolute poverty is clear from our findings using data covering almost six decades from 1957 to 2012. The national poverty rate for our upper line went down by 0.65% points per year, adding up to a large drop in the poverty rate of more than 35% points. In relative terms, poverty incidence went down by 1.3% per year. Long-run poverty reduction is seen in both urban and rural areas and is higher for the poverty gap and squared poverty gap indices, showing gains to those living far below the poverty line. Rural poverty measures, which were historically higher than for urban areas, have been getting closer to urban measures over time, and the (distribution-sensitive) squared poverty gap index for urban India has actually passed that for rural India in recent years. The share of the poor living in urban areas in India has increased a lot over this period, from about one in eight of the poor living in urban areas in the early 1950s to one in three today.

A trend of decreasing poverty started around the early 1970s, but the year 1991–1992—the year of economic reforms in India—was a turning point. There are many signs of structural change, though the link to the reforms is not clear. There was a significant increase in economic growth, led by growth in the service sector and, to some degree, the industrial sector (please see Fig. 6.9). The economic stability strategy outlined in the 1990–1991 budget seems to play: for greater absorption of labor in agriculture has to go hand-in-hand with faster growth of industry and balanced development of infrastructure [3].

Macroeconomic stability measure [3]

A strategy for greater absorption of labor in agriculture has to go hand-in-hand with faster growth of industry and balanced development of infrastructure.

Professor Madhu Dandavate
Budget Speech 1990–1991, Finance Minister, India

The speed of poverty reduction also increased, with a three to four times higher rate of decline in the post-1991 period. The rural poverty decline was even faster than the urban poverty decline. This happened along with a significant rise in inequality both within and between urban and rural areas, unlike a fall in rural inequality and no trend in urban inequality pre-1991. Despite the rise in inequality, we find that poverty was more sensitive to growth after 1991 in total, no matter if growth is measured based on national accounts or survey-based consumption. So, faster growth also seems to have been more favorable to the poor when the latter is measured by the growth elasticity of poverty reduction.

What is especially remarkable is the structural change in how poverty relates to the type of growth. Both urban-rural and sectoral (output) breakdowns indicate stronger connections between different sectors, so that growth in one sector benefits other sectors as well. After 1991, urban growth has become the main factor for reducing poverty. Urban poverty has become much more sensitive to urban growth, but (more importantly) urban growth has also become much more effective in reducing rural poverty since 1991. This is very different from the pattern before 1991, when urban growth did not affect rural poverty at all. A notable observation is that the growth in services and manufacturing has reduced rural poverty rates. It would be pragmatic to promote manufacturing and service sectors for job creation to reduce the overall poverty rates [29]. Of course, it's not the only way!

Policy recommendation: promote planned urbanization.

As a strategy, one of the pro-growth recommendations is to promote urban areas. The government can actively promote the establishment of new towns, preferably focused on specific industries or themes. This can attract investment, create employment opportunities, and improve the quality of life in these areas [29].

One notable finding is that, after 1991, our data show no difference (by output) in how growth in net domestic products reduces poverty across sectors. Unlike the pre-1991 period, when only primary and tertiary sector growth helped lower poverty, all three sectors have had a significant effect since then. The tertiary sector has the highest (absolute) growth elasticity of poverty reduction, about double those for the primary and secondary sector. This reflects the changing nature of the growth process as well as the large structural transformation of the Indian economy over the last two decades with the secondary and tertiary sectors now having much larger shares of national output and employment. Similarly, the post-1991 sector indifference of marginal poverty impacts may not last as the process of structural change in the economy goes on.

Our poverty-growth accounting exercises for the two periods show the results of these significant differences in how different sources of growth affect poverty. Before 1991, rural growth and favorable changes in distribution were almost the only drivers of poverty reduction; urban growth had very little impact. After 1991, urban growth became the main factor behind the observed (and faster) poverty reduction, even though it worsened distribution. Rural growth was still relevant, but less so. If we look at growth by output sectors, the primary sector's role has quickly decreased from explaining about 40% of the total poverty reduction before 1991 to less than 10% of the total (and bigger) poverty reduction after 1991. The tertiary sector alone has accounted for over 60% of the poverty reduction after 91. The secondary sector growth has explained about a fourth. India's construction boom since 2000 has clearly helped make the growth process from the secondary sector more pro-poor.

Our analysis of poverty and growth also helps us examine the role of urbanization. The classic Kuznets process per se, which involves a movement of people from rural to urban areas without changing the rural or urban distributions, has had very little impact on poverty reduction in either period. This partly reflects the relatively slow increase in the proportion of India's urban population and the relatively small and decreasing differences in rural and urban poverty rates. But the population shift has been linked to significant distributional changes that were positive during the pre-1991 phase of slower growth and negative in the later phase of faster growth. Therefore, the role of urbanization as a population shift seems to have been a mixed one. However, if the urbanization process is understood broadly to include urban economic growth as well as population growth, its contribution to poverty reduction in post-reform India has been clearly significant.

The distribution of employment in India across different industries, typically measured per 1000 individuals, is regularly assessed in various rounds of data collection.[28]

Poverty in India, like its population, is predominantly rural but remarkably responsive to economic growth.[29] As the economy grows, poverty diminishes, faster when growth is strong, slower when it is not. The National Sample Survey Office (NSSO) in 2016, when GDP is above 8%, shows how technology is making inroads into villages, against the backdrop of India notching up its billionth mobile phone subscriber in October [30]. India's progress in reducing poverty has been uneven as well as slow. The range in poverty reduction among states is so wide that Kerala's progress in lowering the head count index of poverty is more than 120 times that of Bihar and more than 4 times that of Rajasthan. Even more noticeable than geographic differences in rates of progress in reducing poverty incidence are the inequalities that persist across gender, castes, and ethnic groups. Social indicators for women are markedly lower than for men, and the level of scheduled castes and tribes in both economic and social achievements is still below the national average. Whether male or female or low caste, landless wage-earners continue to be at the bottom of the economic ladder. Besides slow growth, another important force accounting for the slow progress in poverty reduction is investment in human resources. India's record in this respect has been spotty at best. Healthcare and education for the poor have been neglected. As a result, half of India's labor force is illiterate, and tens of millions of primary-school-age children are working rather than learning while their parents, especially their mothers, languish in illiteracy. A clear antidote is the national program of midday meals that was implemented 62 years, 2009, after independence in schools that has contributed to enhancement of school participation and reduction in classroom hunger [10]. The reforms India started in 1991 hold the promise of considerable improvements in the living standards of the country's 300 million poor. By reexamining its commitment to economic liberalization and redirecting toward infrastructure, health, and education the large resources now absorbed by subsidies, India can give its long battle to reduce poverty a new impetus [31].

I witnessed the benefits of free snacks and milk program myself when I was in second grade in 1976. Many of my classmates who came from low-income and struggling socioeconomic backgrounds showed up for classes to get the milk and snacks, boosting attendance levels.

Author

[28] Table 152—https://www.rbi.org.in/scripts/PublicationsView.aspx?id=21249

[29] India—Achievements and challenges in reducing poverty (English)—https://documents.worldbank.org/en/publication/documents-reports/documentdetail/232941468756917265/india-achievements-and-challenges-in-reducing-poverty

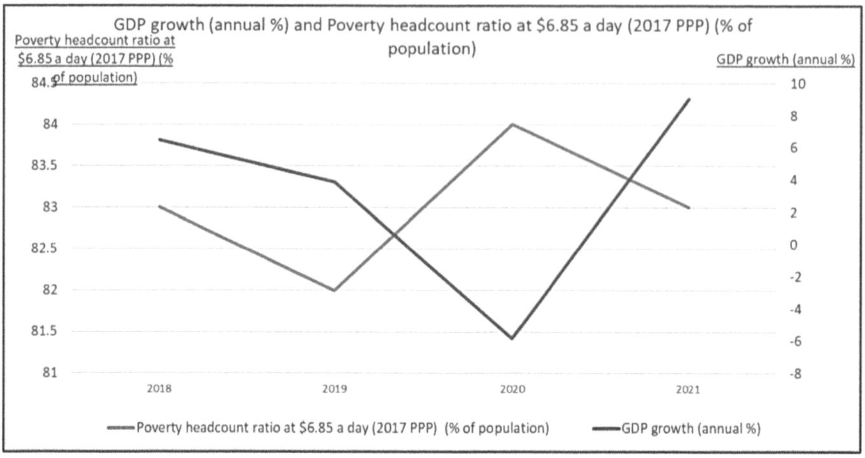

Fig. 6.10 GDP growth

Growth and Poverty

India's poverty rate drops, but few join the middle class.[30] India's poverty rate went down from 40.01% in 2001 to 22.49% in 2011. This means that 131 million Indians left poverty in that decade, the second biggest decline in the world after China. However, the fall in poverty only led to a rise of 269 million in the low-income population, whose share increased from 63% in 2001 to 76% in 2011 (please see Fig. 6.10).

The middle-income population in India barely budged during the decade. Its share increased from 1% in 2001 to 3% in 2011, still small by any measure. The number of middle-income people grew by 17 million, paltry compared with the increase in the low-income population. From these trends, the middle-income threshold appears more like a barrier as only a small share in India stepped across the line from 2001 to 2011 [32]. Creating an ecosystem for the middle class to tap opportunities is an important poverty reduction strategy (please see Figs. 6.11 and 6.12).

Despite poverty's plunge, middle-class status remains out of reach for many [32]. Creating an ecosystem for the middle class to tap opportunities is an important poverty reduction strategy.

[30]Despite Poverty's Plunge, Middle-Class Status Remains Out of Reach for Many—https://www.pewresearch.org/global/2015/07/08/despite-povertys-plunge-middle-class-status-remains-out-of-reach-for-many/

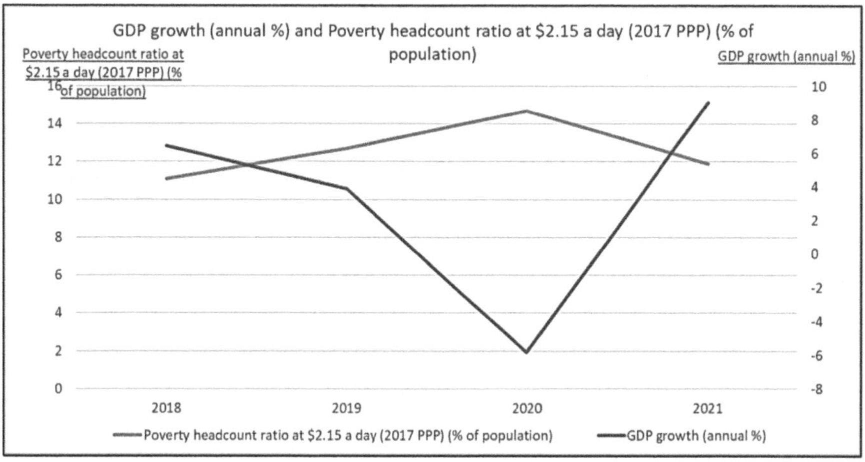

Fig. 6.11 Head count $2.15

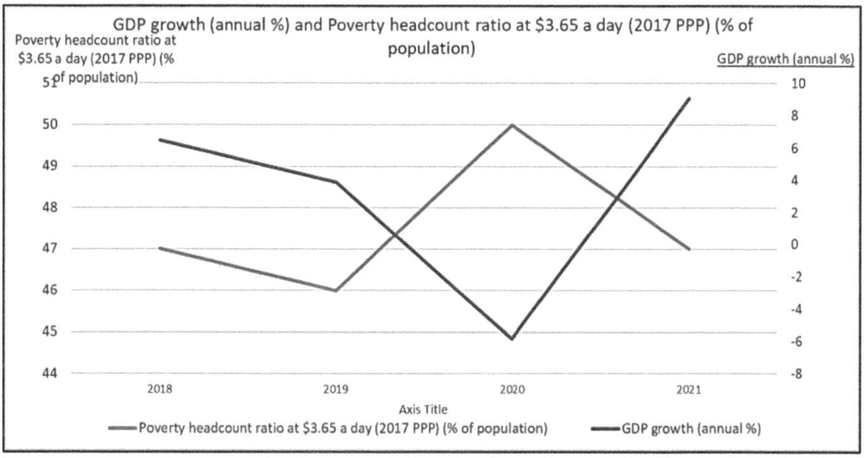

Fig. 6.12 Head count $3.65

Social Sustainability Measures

Education, health, and economic opportunities for able bodies and the youth are key to achieving social sustainability. Human development can be improved by providing better educational opportunities and social security programs. Investing in education and skill development programs can improve the quality of the workforce. For elderly people who are most at risk, increasing social assistance programs can help achieve balanced sustainability. A policy recommendation is that the government can increase the pension amount for poor people above the age of 65 years and create a system where the pension is directly transferred to the beneficiary's account in a post office or bank [29]. The government can also encourage state governments to match the pension amount. Implementing social security programs that offer a safety net for vulnerable populations can increase long term employability. This is a way of designing social security programs for developing countries.

In 2001–2002,[31] the government has launched schemes for women's empowerment, providing microcredit to poor assetless women and assisting women in difficult circumstances. Additionally, the "Janashree Bima Yojana" was launched to extend social security cover to the poorest sections of society [33]. A special scheme for landless agricultural laborers, the Khetihar Mazdoor Bima Yojana, will provide benefits of insurance cover and a pension to beneficiaries. A Shiksha Sahyog Yojana will provide an education allowance to children of parents living below the poverty line.

Agricultural Credit

Agriculture holds key to both economic and social sustainability. Speeding up agricultural sector reforms and better management of the food economy can be achieved by implementing policies that improve irrigation systems [33], provide better access to credit for farmers, and promote the use of advanced farming techniques. Additionally, the use of credit in the agriculture sector, particularly through the implementation of Negotiable[32] Warehousing Receipts (e-NWR) Integrated with e-NAM, has been identified as a significant driver of economic and social sustainability. This credit policy aims to facilitate financing for farmers and improve their access to credit [34]. Sustainable farming is an integral policy of agriculture. We are encouraging the balanced use of all kinds of fertilizers, including traditional organic and other innovative fertilizers, to reduce the excessive use of chemical fertilizers and promote sustainable farming practices [34]. We are expanding integrated farming systems in rainfed areas, promoting multitier cropping, beekeeping, solar pumps, solar energy production in non-cropping seasons, and zero-budget natural farming [34]. We are implementing the village storage scheme, run by self-help groups, to provide farmers with good holding capacity and reduce their logistics costs, promoting sustainable agriculture practices [34].

We are promoting the development of a seamless national cold supply chain for perishables, including milk, meat, and fish, through the setting up of "Kisan Rail" and the inclusion of refrigerated coaches in express and freight trains [34].

We are taking steps to eliminate diseases in livestock, such as foot and mouth disease, brucellosis, and pest des petits ruminants (PPR), by 2025, ensuring the health and well-being of animals and improving the livelihoods of farmers [34].

[31] Budget Speech 2001–2002—https://www.indiabudget.gov.in/doc/bspeech/bs200102.pdf

[32] Budget Speech 2020–2021—https://www.indiabudget.gov.in/doc/bspeech/bs202021.pdf

Machine Learning Model_Poverty_and_Macroeconomic Linkage_Model_for_India.ipynb

Pro-poor and Marginalized Inclusive Growth Machine Learning Model: Poverty and Macroeconomic Linkage Model

Poverty has many aspects, shown by conditions such as hunger, poor housing, unhygienic living environments, inadequate and scarce access to clean water, ineffective solid waste disposal, low educational attainment and lack of quality education, frequent illness, and high crime rates.[33] In 2000, 191 UN member states agreed to reduce poverty by signing the Millennium Development Declaration. However, poverty is hard to define or measure, and tracking poverty in its different forms is a complicated task both theoretically and practically. However, using advanced data science and large language models, we have created machine learning models that use survey data and generate models to estimate poverty [35].

From a policy perspective, besides measuring poverty, we must also perform some essential analyses on how policies and poverty measures are connected. Macroeconomic links to policy outcomes serve two purposes: one, it helps to assess the impact of policies on socioeconomics and, second, helps to provide feedback loops to optimize any future economic and governmental policies. We may want to examine with a regression model what factors[34] affect poverty at the household level in the form of poverty profiles or how sensitive poverty is to economic growth while adjusting for other factors. We may also want to understand how macro variables such as employment in agriculture (% of total employment), Gini coefficient, employment in services, CPI, economic events, general election years, weather and extreme events, wholesale price index, industrial production, public expenditure, gross savings, population growth, demonetization, etc. influence multidimensional poverty levels or changes among groups or regions—and over time. Through regression analysis, we can partly study these connections by looking at what determines multidimensional poverty. In a regression model, we can account for the impact or the "magnitude" of determinants of multidimensional poverty, which would not be possible with a purely descriptive analysis [36].

---

[33] HANDBOOK ON POVERTY STATISTICS: CONCEPTS, METHODS AND POLICY USE—https://unstats.un.org/unsd/methods/poverty/pdf/un_book%20final%2030%20dec%2005.pdf

[34] Multidimensional Poverty Measurement and Analysis: Chap. 10—Some Regression Models for AF Measures—https://www.ophi.org.uk/wp-content/uploads/OPHIWP091_Ch10.pdf

The formula for the poverty rate (<$1.90) linear regression line is = −625.843 − 0.416 * GDP growth (annual %) + 5.180 * employment in agriculture (% of total employment) + 0.099 * Gini coefficient + 9.167 * employment in services (% of total employment) + 0.770 * inflation, consumer prices (annual %) + 1.231 * current account balance (% of GDP) + 0.943 * EconomicEvents −0.668 * GeneralElectionYear +3.491 * DroughtYear −4.063 * Wholesale Prices_Industry Aggregates_Manufactured Products + 0.242 * Production: Industry: Total Industry Excluding Construction −0.242 * Gross national expenditure (% of GDP) + 1.984 * Gross savings (% of GDP) + 52.327 * Population growth (annual %)

There are several factors that contribute to poverty in India:

- Rapid population growth: high population growth is one of the major reasons for poverty in India. This further leads to a high level of illiteracy, poor healthcare facilities, and lack of access to financial resources [37]. India's population has been increasing steadily. This rapid growth outstrips economic development and resources, leading to increased demand for consumption goods.
- Low productivity in agriculture: the agriculture sector in India suffers from low productivity due to fragmented and subdivided landholdings, lack of capital, illiteracy about new technologies in farming, and the use of traditional methods of cultivation.[35]
- Inflation, particularly food price inflation, can significantly influence poverty rates in India. While inflation is not the sole cause of poverty, it can significantly contribute to it by reducing purchasing power, increasing the cost of living, and eroding saving [38–40]. Here's how [41]:
  - Decreased purchasing power: inflation lowers the purchasing power of money. This means that the same amount of money can buy fewer goods and services as prices rise. For the poor, who often live on very tight budgets, this can lead to increased hardship as they struggle to afford the same basic necessities.
  - Increased food prices and higher inflation: in India, the costs of essential food products have soared between 2015 and 2022. This has a significant impact on India's lower middle class and lower class, especially in rural areas. While the prices of many common items have nearly doubled, the "real wage rate" has increased by just 22% since 2015. This disparity between wage growth and price increases [32] can push more people into poverty.[36]
  - High inflation can lead to economic instability, which can exacerbate poverty. For example, the COVID-19 pandemic and the Russia-Ukraine war significantly impacted poverty and inflation in India [32]. The number of people who are poor in India (with incomes of $2 or less a day) is estimated to have increased by 75 million because of the COVID-19 recession [39, 40, 42].

[35] National Agriculture Market (eNAM)—https://www.enam.gov.in/web/

[36] In the pandemic, India's middle class shrinks and poverty spreads while China sees smaller changes—https://www.pewresearch.org/short-reads/2021/03/18/in-the-pandemic-indias-middle-class-shrinks-and-poverty-spreads-while-china-sees-smaller-changes/

  - Impact on savings: inflation erodes the value of money over time, which can have a detrimental effect on the savings of the poor. This can push people who are marginally above the poverty line into poverty [40].

- Gross savings: in general, gross domestic saving and gross capital formation are proxy of how well economy is doing; increase indicates positive economic growth and development [29]. Nevertheless, the relationship between gross savings and poverty rate in India is complex. Here's how an increase in gross savings could potentially lead to an increase in poverty rate.

  - Income inequality: if the increase in gross savings is primarily driven by wealthier segments of society, it could exacerbate income inequality [42]. This is because the rich are able to save and accumulate wealth, while the poor continue to struggle with low incomes and high expenses [42].
  - Reduced consumption: an increase in savings could indicate a decrease in consumption [42, 43]. If households are saving more because they are spending less on essentials and nonessentials, this could be a sign of economic hardship. Reduced consumption can lead to slower economic growth, which can in turn lead to higher poverty rates.
  - Inflation: if the increase in savings is due to inflation, this could lead to an increase in poverty. Inflation erodes the purchasing power of money, which can push people who are marginally above the poverty line into poverty.
  - Economic instability: if the increase in gross savings is a response to economic instability or uncertainty, this could also lead to an increase in poverty. During times of economic instability, people often choose to save more and spend less, which can slow economic growth and lead to higher unemployment and poverty rates.

- Wholesale prices: while wholesale prices are not the sole determinant of poverty, they play a significant role in influencing poverty rates in India.

  - Inflation: wholesale prices are a key indicator of inflation [44]. When wholesale prices rise, it often leads to increased prices for consumers. For the poor, who often live on very tight budgets, this can lead to increased hardship as they struggle to afford the same basic necessities [45].
  - Shortage of capital and able entrepreneurship: capital and able entrepreneurship have important role in accelerating growth. But these are in short supply making it difficult to increase production significantly, when compared to other developing countries [46].

- Economic slowdown: the economy had been slowing for nine quarters prior to the outbreak of the novel coronavirus pandemic. Unemployment reached a 45-year high in 2017–18[2]. This economic slowdown could have potentially slowed the rate of poverty reduction [47].

What can policymakers do: inflation could wreak vengeance on the world's poor?

Subsidies have been used by governments to reduce the shock on households. Subsidies can sometimes be a useful temporary measure to ease the effects of shocks. But they often last longer than they should, leading to negative consequences. Subsidies can take away from spending on infrastructure, health, and education. Energy subsidies tend to benefit richer households more than poorer households and promote wasteful consumption.

Many governments are thinking about using trade limits and export prohibitions to secure local food supplies. This is a bad idea. Policies like these that may seem suitable at the national level often have awful global effects. During the 2010–2011 food price surge, trade limits worsened the rise in world prices and made millions of people poorer, even though they reduced domestic price increases.

The best way to help the poorest cope with rising prices is to use social welfare policies. These policies could include focused safety nets such as cash transfers, food, and in-kind transfers, school feeding programs, and public works programs. The public distribution system (PDS) is a food security measure implemented by the Indian government to alleviate the influence of inflation. Estimating inflation indexes for different income groups gives better information on how inflation affects the poor and should guide the development of social safety nets. International cooperation and communication are important to prevent retaliatory actions.

To study poverty, we can use two types of regressions: micro and macro regressions. The word "micro" means the analysis is done for an individual or a household; the word "macro" means the analysis is done for a group, such as a district, state, province, or country [36]. The model for machine learning development is based on the idea that policies can create economic opportunities at microlevel and reduce poverty through a trickle-down effect. This is the true meaning of market driven democratic pro-poor and marginalized inclusive economic development goals! This section gives a general model for examining the factors that affect poverty measures, at macro levels of analysis. This is the essence for achieving Sustainable Development Goal (SDG) No Poverty.

Inclusive growth, poverty alleviation, and sustainable development are associated with each other, while poverty reduction is considered a sustainable development principle.[37] Poverty in all shapes remains a vulnerable challenge facing humanity. The Sustainable Development Goals (SDGs) are a daring pledge to finish poverty in all forms and dimensions through 2030. Worldwide life-threatening poverty deep-rooted in 2020 due to the interruption of the COVID-19. During the pandemic of COVID-19 and the worldwide economic recessions, the pace of poverty alleviation slowed down as compared to the desired level. Poverty-fighting lost its battle during this COVID-19 pandemic; it is likely that development[38] will be set back for decades [48]. In several countries, poverty is causing millions of individuals to face food shortages and a lack of basic necessities. Financial inclusions has been identified as a potential solution to alleviate poverty and reduce income inequality in developing nations.

[37] Assessing the Relationship between Poverty and Economic Growth: Does Sustainable Development Goal Can be Achieved?—https://www.ncbi.nlm.nih.gov/pmc/articles/PMC8726512/

[38] The Contribution of Financial Inclusion in Reducing Poverty and Income Inequality in Developing Countries—https://archive.aessweb.com/index.php/5002/article/view/1999

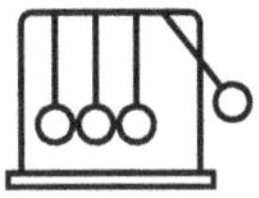

In India, poverty elimination has remained a major challenge and lies at the core of India's national development agenda to create a just and equitable society. Reliable estimation of poverty is the first step toward eradication of poverty as a basic input for design, implementation, and monitoring of anti-poverty programs.

Macro poverty outlook[39] (MPO) indicators [49]

- Real GDP growth, at constant market prices
  - Private consumption
  - Government consumption
  - Gross fixed capital investment
  - Exports, goods and services
  - Imports, goods and services
- Real GDP growth, at constant factor prices
  - Agriculture
  - Industry
  - Services
- GDP per capita [48]
- Government expenses [48]
- Inflation (consumer price index) [48]
- Current account balance (% of GDP)
- Net foreign direct investment inflow (% of GDP)
- Fiscal balance (% of GDP)
- Revenues (% of GDP)
- Debt (% of GDP)
- Primary balance (% of GDP)
- International poverty rate ($2.15 in 2017 PPP) [48]
- Lower middle-income poverty rate ($3.65 in 2017 PPP) [48]
- GHG emissions growth (mtCO2e)
- Energy related GHG emissions (% of total)
- Financial inclusion [48]
  - ATM—automated teller machines per 100,000 adults
  - BRCH—measured as bank branches per 100,000 adults
  - BRW—commercial bank borrowers per 1000 adults
  - DPTR—commercial bank deposit accounts per 1000 adults

Government should pursue poverty reduction policies that're beneficial in encouraging economic growth even if they do not decrease income inequality [50].

[39] MPO—https://www.worldbank.org/en/publication/macro-poverty-outlook/mpo_sar

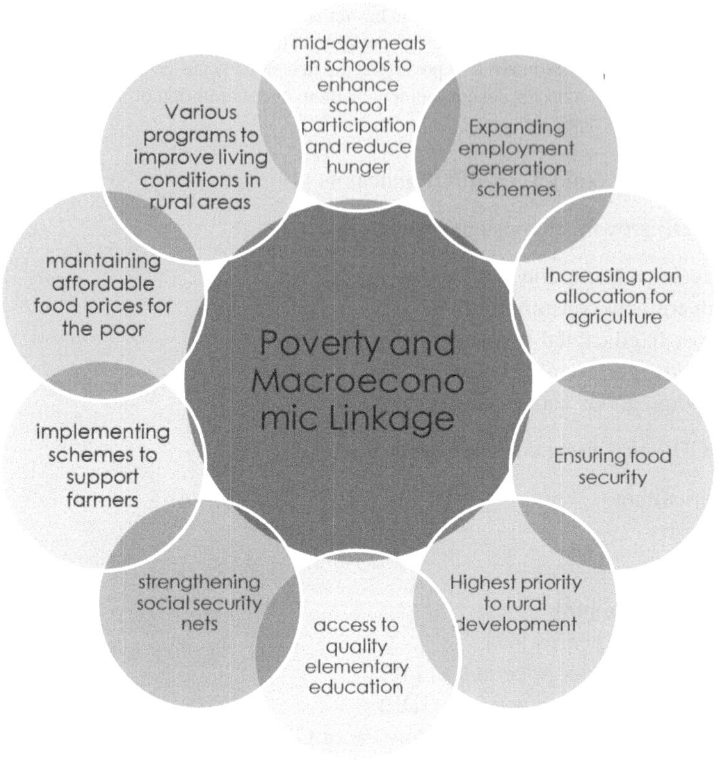

Fig. 6.13 Poverty and macroeconomic linkage

Inflation increases poverty rate. As the model clearly showed, inflation increases poverty rate. The model output strongly correlates with the research that inflation makes the poor worse off [51].

The primary evidence[40] comes from the answers to an international poll of 31,869 respondents in 38 countries. These show that the disadvantages in a number of dimensions, the poor, the uneducated, and the unskilled (blue-collar) workers, are relatively more likely to mention inflation as a top concern than the advantage on these dimensions. Each dimension is significant when controlling for others, suggesting that the different components of being disadvantaged have independent effects on attitudes to inflation.

Our analysis focuses on the effects of inflation on direct poverty measures and their correlation with inflation. We discovered that high inflation rates were associated with a decrease in the share of the bottom quintile and real minimum wage, while also leading to an increase in poverty levels. Moreover, our findings indicate that per capita incomes of the poor are significantly impacted by inflation. Ultimately, the model provides mathematical evidence that the poor are more adversely affected by inflation than the rich [51] (Fig. 6.13).

[40] Inflation and the Poor—https://economics.mit.edu/sites/default/files/2023-05/fischer_inflation_poor.pdf

Step 1: Load Python Libraries to Run the Experiment

```
import numpy as np
import pandas as pd
from functools import reduce
from sklearn.model_selection import train_test_split
from sklearn.linear_model import LinearRegression
from sklearn.metrics import r2_score
from sklearn import metrics
from sklearn.impute import SimpleImputer
import eli5
import warnings
warnings.filterwarnings("ignore")

import numpy as np
import pandas as pd
import functools
from datetime import date
import plotly.graph_objects as go
import seaborn as sns; sns.set()
import statsmodels.api as sm
from statsmodels.tsa.arima_process import ArmaProcess
from scipy import stats
from scipy.stats import pearsonr
import matplotlib.pyplot as plt
import plotly.express as px
from sklearn.preprocessing import MinMaxScaler
from sklearn.model_selection import train_test_split
from sklearn.linear_model import LinearRegression
from sklearn.metrics import r2_score
from sklearn import metrics
from sklearn.impute import SimpleImputer
import eli5
```

Above code sets up experiments with required libraries.

Step 2: GDP Growth (Annual Percentage) and Time Series Plot

Load GDP growth (annual percentage)[41] from the World Bank.

[41] GDP Growth (annual %)—https://data.worldbank.org/indicator/NY.GDP.MKTP.KD.ZG?locations=IN

```
GDPGrowthAnnualPercentage_df= pd.read_csv("GDP growth (annual %).csv")
GDPGrowthAnnualPercentage_df
```

```
GDPGrowthAnnualPercentage_df["Year"] = GDPGrowthAnnualPercentage_df["Year"].astype(int)
```

```
GDPGrowthAnnualPercentage_df.dtypes
```

```
from matplotlib import pyplot as plt
```

```
import seaborn as sns
def _plot_series(series, series_name, series_index=0):
  from matplotlib import pyplot as plt
  import seaborn as sns
  palette = list(sns.palettes.mpl_palette('Dark2'))
  xs = series['Year']
  ys = series['GDP growth (annual %)']

  plt.plot(xs, ys, label=series_name, color=palette[series_index % len(palette)])

fig, ax = plt.subplots(figsize=(10, 5.2), layout='constrained')
df_sorted = GDPGrowthAnnualPercentage_df.sort_values('Year', ascending=True)
_plot_series(df_sorted, '')
sns.despine(fig=fig, ax=ax)
plt.xlabel('Year')
_ = plt.ylabel('GDP growth (annual %)')
```

Plot Output
The following is the output of the above code.

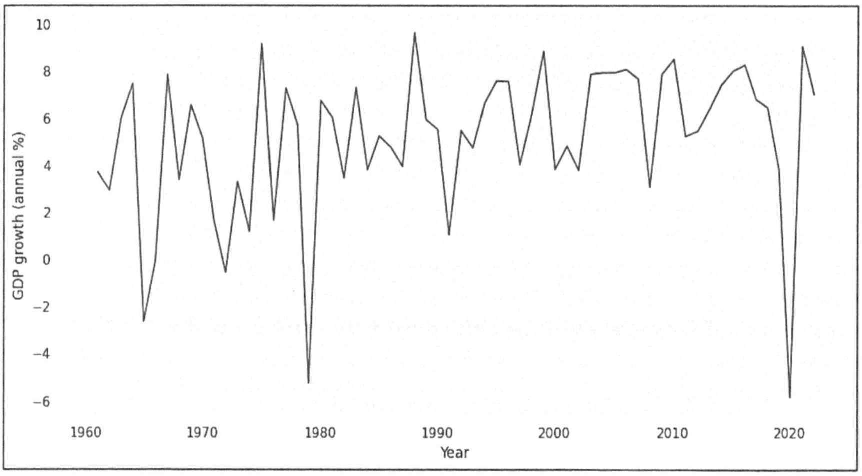

As can be seen, India has experienced drastic economic contractions in 1980, 1990, and 2020.

Step 3: Employment in Agricultures

Load employment in agriculture for India from the World Bank Dataset.[42]

```
EmpInAG_df= pd.read_csv('Employment in agriculture (% of total employment).csv")
EmpInAG_df
```

```
EmpInAG_df.dtypes
```

Output

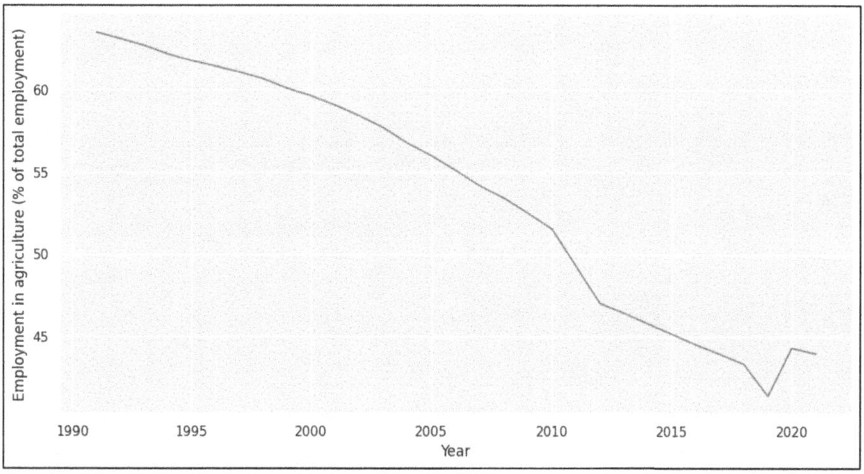

As it can be seen, employment in agriculture as percentage of total employment is trending in downward direction.

The latest PLFS (Periodic Labour Force Survey) report shows a sharp increase in employment in agriculture from 42.5% of the total employment in 2018–2019 to 45.6% in 2019–2020.[43]

[42]Employment in agriculture (% of total employment) (modeled ILO estimate)—India—https://data.worldbank.org/indicator/SL.AGR.EMPL.ZS?locations=IN

[43]Employment in AG % of total employment—https://economictimes.indiatimes.com/news/economy/indicators/share-of-agriculture-sector-in-employment-sees-steady-increase-cmie/articleshow/85266073.cms

Step 4: Employment in Services

Load employment in services for India from the World Bank Dataset.[44]

EmpInServices _df= pd.read_csvEmployment in services (% of total employment).csv") EmpInServices_df
Plot the data frame

Output

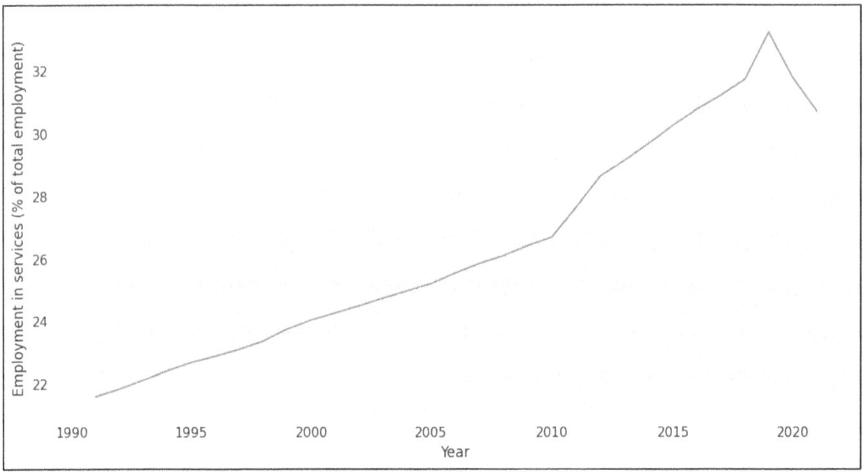

Step 5: Current Account Balance

Load India's current account balance from the World Bank Dataset.[45]

The current account balance as a fraction of GDP shows how competitive a country is internationally. Typically, countries with a large current account surplus rely heavily on exports for their income and have high savings rates but low domestic demand. Conversely, countries with a current account deficit import a lot and have low savings rates and high personal consumption rates as a part of their disposable incomes.

---

[44]Employment in services (% of total employment) (modeled ILO estimate)—India—https://data.worldbank.org/indicator/SL.SRV.EMPL.ZS?locations=IN

[45]Current account balance (% of GDP)—India—https://data.worldbank.org/indicator/BN.CAB.XOKA.GD.ZS?locations=IN

```
BOP_df= pd.read_csv("/GDP-PPP/Current account balance (% of GDP).csv")
BOP_df
```

```
Plot BOP
```

Output

India had a current account shortfall of 2.60% of its gross domestic product in 2022. From 1970 to 2022, the average current account to GDP in India was −1.14% of GDP, with the highest value of 2.30% of GDP in 2003 and the lowest value of −4.80% of GDP in 2012.

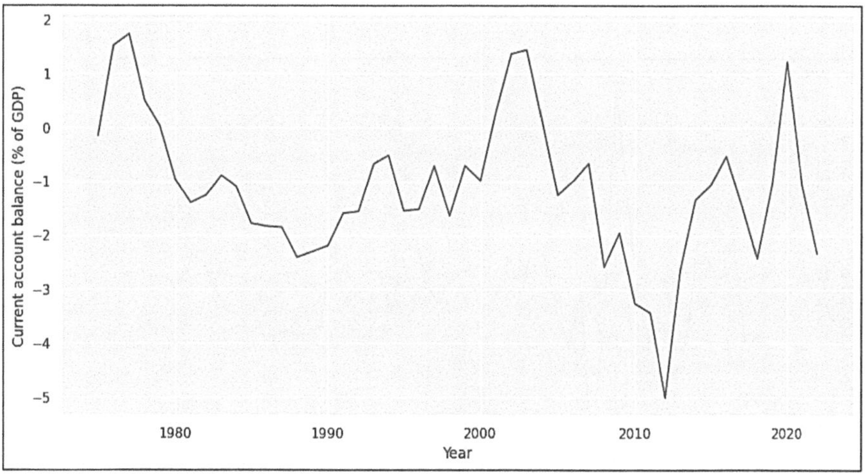

Step 6: Gini Index

Load India's Gini index from the World Bank Dataset.[46]

```
Gini_df = pd.read_excel"PPPIndia.xlsx",sheet_name="GiniCoefficient")
Gini_df
```

```
Gini_df.dtypes
```

Loads of Gini index data

[46] Gini Index—https://data.worldbank.org/indicator/SI.POV.GINI?locations=IN

Step 7: Consumer Prices (Annual Percentage)

Load India's consumer prices (annual percentage) from the World Bank Dataset[47].

```
CPI_df= pd.read_csv("Inflation, consumer prices (annual %).csv")
CPI_df
```

```
CPI_df
```

Outputs: CPI for India

As discussed in earlier chapters, CPI can be influenced by balance of payment and bank rates or cash with the public.

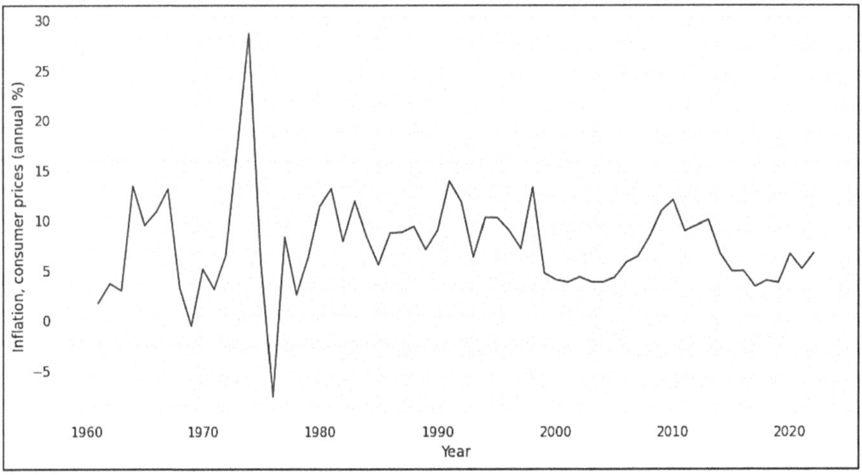

Step 8: Gross Savings (% of GDP)

Load India's gross savings (% of GDP).[48]

[47] Gini Index—https://data.worldbank.org/indicator/SI.POV.GINI?locations=IN

[48] Gini Index—https://data.worldbank.org/indicator/SI.POV.GINI?locations=IN

GROSSSAVINGSPERGDP_df= pd.read_csv("/content/drive/My Drive/GDP-PPP/Gross savings (% of GDP).csv") GROSSSAVINGSPERGDP_df
GROSSSAVINGSPERGDP_df["Year"] = GROSSSAVINGSPERGDP_df["Year"].astype(int)
Plot Gross Savings

Output

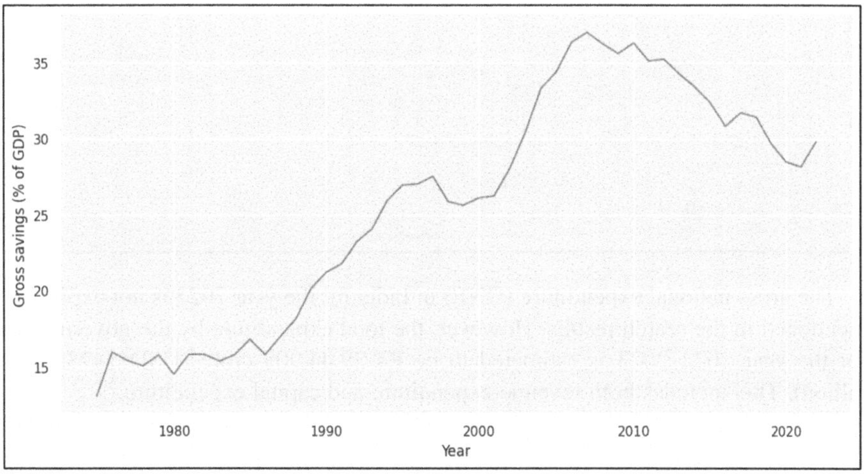

Gross national expenditure (% of GDP)

Step 9: Gross National Expenditure (% of GDP)

Load India's gross national expenditure (% of GDP).[49]

GROSSSNATEXP_df= pd.read_csv("/content/drive/My Drive/GDP-PPP/Gross national expenditure (% of GDP).csv") GROSSSNATEXP_df
GROSSSAVINGSPERGDP_df["Year"] = GROSSSAVINGSPERGDP_df["Year"].astype(int)
Plot Gross Savings

[49] Gini Index—https://data.worldbank.org/indicator/SI.POV.GINI?locations=IN

Output

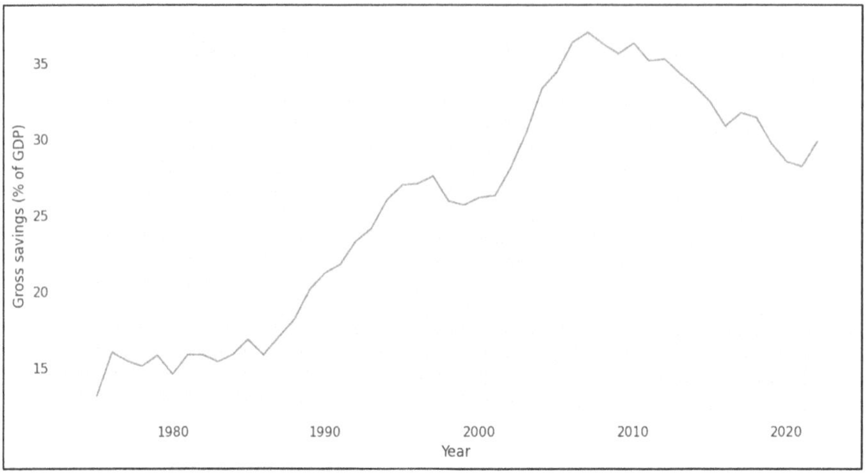

The gross national expenditure (GNE) in India for the year 2023 is not explicitly mentioned in the search results. However, the total expenditure by the government for the year 2022–2023 is estimated to be Rs 39,44,909 crore (USD $4732.229 billion). This includes both revenue expenditure and capital expenditure.[50]

As for poverty rates, according to the NITI Aayog's National Multidimensional Poverty Index (MPI) for 2023, the population living in multidimensional poverty in India is 14.96%. This is a significant reduction from the 24.8% reported for the period from 2015–2016[3]. This improvement has been attributed to advancements in various sectors including nutrition, education, sanitation, and the availability of subsidized cooking fuel[3]. It's also worth noting that around 135 million Indians managed to escape poverty between 2016[51] and 2021[2] [52].

Step 10: General Election Data

General elections in India often have a significant influence on poverty and poverty alleviation strategies. General election year is unique in India in that many governmental incentives and subsidies be introduced. The influence of this very variable on poverty levels can be observed.

[50]Union Budget 2023–24—https://prsindia.org/budgets/parliament/union-budget-2022-23-analysis

[51]NATIONAL MULTIDIMENSIONAL POVERTY INDEX: A PROGRESS REVIEW 2023—https://www.undp.org/india/national-multidimensional-poverty-index-progress-review-2023

```
GeneralElectionYear_df= pd.read_csv(GeneralElectionsInIndia.csv")
GeneralElectionYear_df
```
```
GeneralElectionYear_df["Year"] = GeneralElectionYear_df["Year"].astype(int)
```
```
Plot General Election Binary Data
```

Output

Since 1977, there have been several years when general elections[52] were held, including 1977, 1980, 1984, 1989, 1991, 1996, 1998, 1999, 2004, 2009, 2014, and 2019. To analyze this data, a data frame can be created and a binary variable can be set.

Step 11: Droughts in India

Droughts increase poverty rates in India as it disrupts agricultural outputs. Load India's drought year binary variable data.

```
DroughtYears_df= pd.read_csv("/DroughtsInIndia.csv")
DroughtYears_df
```
```
DroughtYears_df["Year"] = DroughtYears_df["Year"].astype(int)
```
```
DroughtYears_df.dtypes
```

Output

Years 1979, 1982, 1984, 1987, 2000, and 2002 were drought[53] years since 1977.

Step 12: Economic Boom and Bust Years in India

Economic events shape poverty levels. Load past economic events data to see the influence on poverty rates.

```
MacroEvents_df= pd.read_csv("EconomicEvents.csv")
```
```
MacroEvents_df
```
```
MacroEvents_df["Year"] = MacroEvents_df["Year"].astype(int)
```
```
MacroEvents_df.dtypes
```

[52] General Election Years—http://www.smetimes.in/smetimes/general-elections-2009/miscellaneous/2009/Mar/23/history-of-lok-sabha-elections5584.html/

[53] Drought Years—https://www.nature.com/articles/srep44552

Output

Years 1979, 1980, 1990, 1991, 1999, 2001, 2002, 2005, 2006, 2008, 2011, 2016, 2019, 2020, and 2021 were significant economic years. Some of the years include global[54] events too. The years 2020 and 2021 are COVID-19 years when more than 75 million Indians plunged back into poverty, setting back half a decade of gains [53].

Step 13: Public Distribution System: Procurement, Off-Take, and Stocks from the Reserve Bank of India Handbook of Statistics on Indian Economy

The Handbook of Statistics includes information on the food subsidy, specifically on the procurement, off-take, and stocks[55] of the public Distribution System.

```
PDSOFFTAKE_df= pd.read_csv("/content/drive/My Drive/GDP-PPP/PDS-Off-take.csv")
PDSOFFTAKE_df
```

```
PDSOFFTAKE_df["Year"] = PDSOFFTAKE_df["Year"].astype(int)
```

```
PDSOFFTAKE_df.dtypes
```

Output

The above code loads PDS from 1960 to 2021.

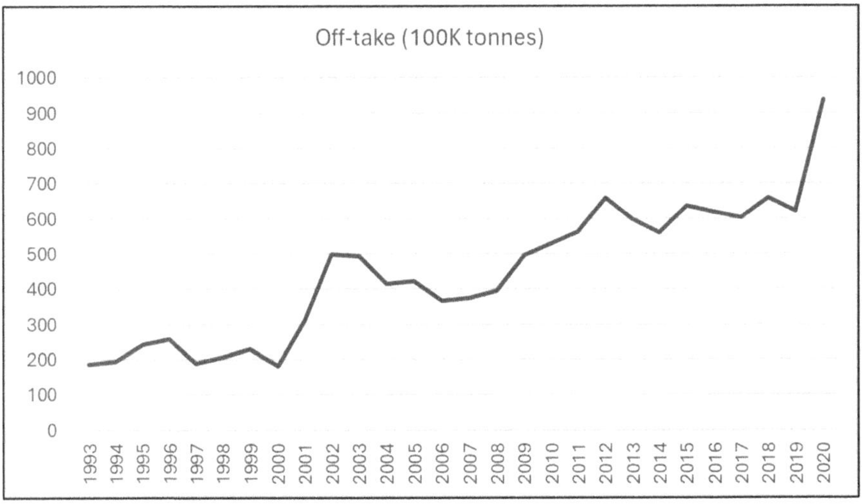

[54]What Dates Are Used for the U.S. Recession Bars?—https://fredhelp.stlouisfed.org/fred/data/understanding-the-data/recession-bars/

[55]Table 27: Public Distribution System—Procurement, Off-Take and Stocks—https://rbi.org.in/Scripts/PublicationsView.aspx?id=21833

```
PDSStocks_df= pd.read_csv("/content/drive/My Drive/GDP-PPP/PDS-Stocks.csv")
PDSStocks_df
```

```
PDSStocks_df["Year"] = PDSStocks_df["Year"].astype(int)
```

```
PDSStocks_df.dtypes
```

Output

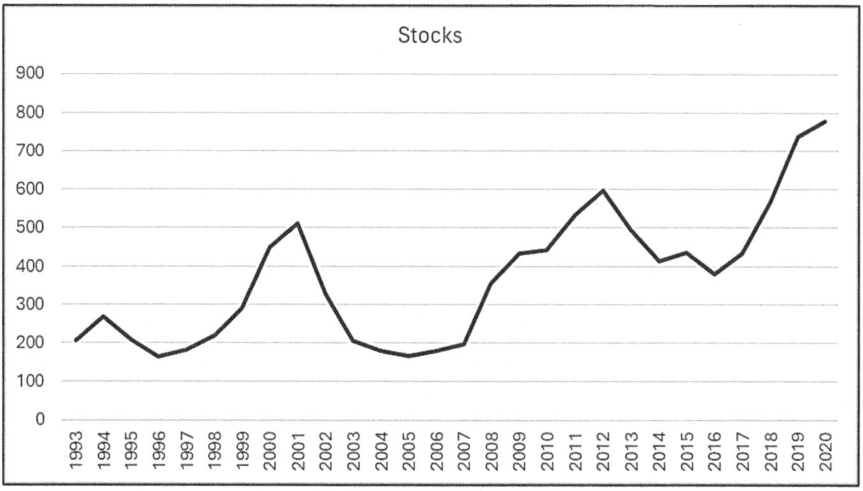

Step 14: Life Expectancy at the Birth Data

Add life expectancy at the birth data.[56]

```
LIFEEXPECATBIRTH_df= pd.read_csv("Life expectancy at birth, total (years).csv")
LIFEEXPECATBIRTH_df
```

```
LIFEEXPECATBIRTH_df["Year"] = LIFEEXPECATBIRTH_df["Year"].astype(int)
```

```
LIFEEXPECATBIRTH_df.dtypes
```

[56]Life expectancy at birth, total (years)—India—https://data.worldbank.org/indicator/SP.DYN.LE00.IN?locations=IN

Output

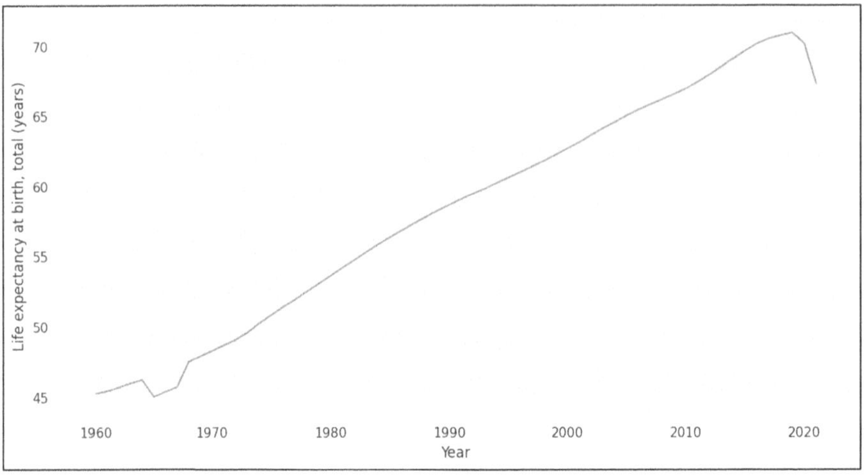

The life expectancy at birth in India for the year 2023 is 70.42 years. This is a slight increase from previous years, indicating an improvement in health and living conditions. As for poverty rates, according to the NITI Aayog's National Multidimensional Poverty Index (MPI) for 2023, the population living in multidimensional poverty in India is 14.96%. This is a significant reduction from the 24.8% reported for the period from 2015 to 2016. This improvement has been attributed to advancements in various sectors including nutrition, education, sanitation, and the availability of subsidized cooking fuel[3]. It's also worth noting that around 135 million Indians managed to escape poverty between 2016 and 2021.

Step 15: The Minimum Support Prices (MSPs) for Food Grains

The MSP is a form of market intervention by the Indian government to insure agricultural producers against any sharp fall in farm prices. The MSPs are announced by the Indian government at the beginning of the sowing season for certain crops (Parry, Maize, and Wheat) [54].

```
MSP_df= pd.read_csv("Minimum Support Price for Foodgrains According to Crop Year (Fair Average Quality).csv")
MSP_df
```

```
MSP_df["Year"] = MSP_df["Year"].astype(int)
```

```
Plot MSP
```

Output

	Year	MSP-Paddy common	MSP-Maize	MSP-Wheat
0	1990	205	180	225
1	1991	230	205	280
2	1992	270	240	330
3	1993	310	260	350
4	1994	340	280	360
5	1995	360	300	380
6	1996	380	310	475
7	1997	415	360	510

The influence of MSP on poverty in India can be understood in the following ways:

- Agricultural productivity and remunerative farm prices: the major factors contributing to poverty reduction in rural areas have been agricultural productivity, remunerative farm prices, and nonfarm employment with better wages. MSP ensures remunerative prices to farmers for their produce, which can help in reducing poverty [54].
- Income scheme for farmers: MSP was initially an incentive for farmers to adopt technology with an aim of increasing the productivity of agricultural land in the 1960s; however, in the 2000s, it is seen as a market intervention and farmer income scheme. By guaranteeing a minimum price for crops, the MSP scheme can help in increasing the income of farmers, thereby reducing poverty.
- Food security: the MSP scheme also plays a crucial role in achieving food security in the country. The food grains procured by the government through the MSP scheme are distributed to the poor at subsidized rates through the public distribution system (PDS). This not only ensures food security for the poor but also provides them with some relief from the high food prices.

It's also important to note that while MSP can help in reducing poverty, it is not the only solution. Other factors such as improving agricultural productivity, diversifying into nonfarm sectors, improving rural infrastructure, etc. also play a crucial role in poverty reduction.

Step 16: Population Growth (Annual Percentage)

The annual population growth rate of India in 2023 is 0.81%, which is a slight increase from 0.68% in 2022 [54].

```
POPGROWTH_df= pd.read_csv("Population growth (annual %).csv")
POPGROWTH_df
```

```
POPGROWTH_df["Year"] = POPGROWTH_df["Year"].astype(int)
```

```
POPGROWTH_df.dtypes
```

Output

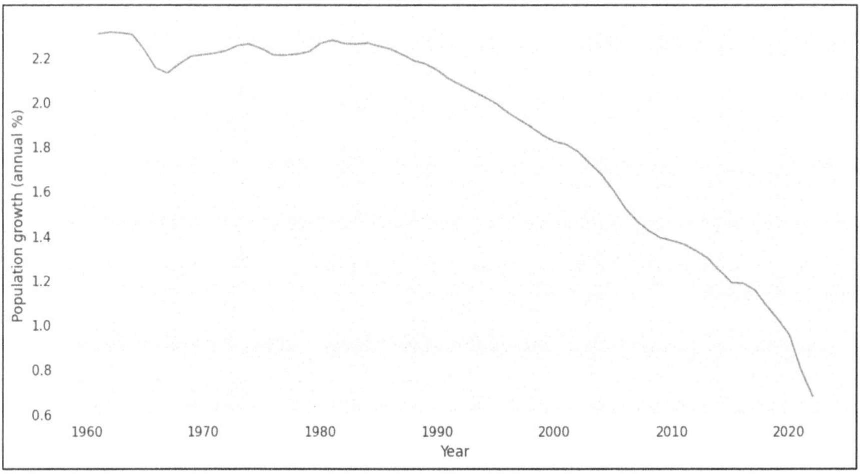

According to the World Bank, extreme poverty in India was as low as 0.8% in 2019. However, the poverty rate based on a $2.00 per day income was 10.2% in 2019. A more comprehensive measure, the Multidimensional Poverty Index, which considers health, education, and standard of living, indicates that 14.96% of India's population is considered to be in a state of multidimensional poverty [52].

Step 17: Deployment of Bank Credit by Major Sectors

India's bank lending as a share of GDP went up from 25% in 1991–1992 to 49.65% (bank credit to commercial sector (crore): 13,506,243; GDP[57] (crore) in 2022: 27,200,000) in 2022–2023. However, bank lending is still insufficient for some

[57] GDP (current LCU)—India—https://data.worldbank.org/indicator/NY.GDP.MKTP.CN?locations=IN

key sectors such as manufacturing, transport, and trading, which have more economic potential than what they get credit for, as the Karnataka Economic Survey 2021–2022 shows. Karnataka has a relatively low rate of bank loan penetration, which means that many borrowers such as small businesses, self-employed, and farmers are limited by credit and rely on informal financial institutions like cooperative credit societies or local money lenders. Too much reliance on informal finance hinders development. Therefore, it is important to increase the access to formal banking finance to improve the performance of all states in India [55]. Load deployment of bank credit by major sectors.[58]

```
BANKCREDITCOMMERCIALSEC_df= pd.read_csv("Bank Credit to Commercial Sector.csv")
BANKCREDITCOMMERCIALSEC_df
```

```
BANKCREDITCOMMERCIALSEC_df["Year"] =
BANKCREDITCOMMERCIALSEC_df["Year"].astype(int)
```

```
BANKCREDITCOMMERCIALSEC_df.dtypes
```

Output

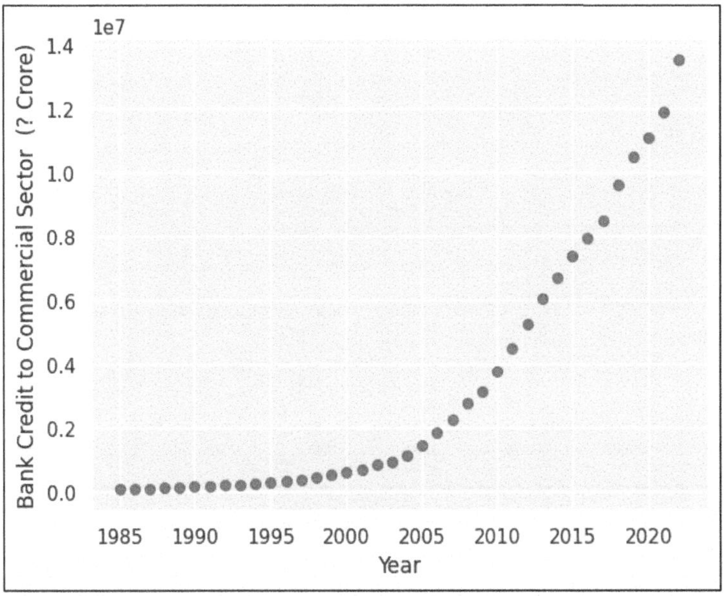

[58]Table 170: Deployment of Bank Credit by Major Sectors—https://rbi.org.in/Scripts/PublicationsView.aspx?id=21976

Step 18: Institutional Credit for Agriculture and Allied Activities

Access to credit for farming has a big effect on poverty levels in India. It helps farmers, especially small and marginal farmers who do not have the money to get the input needed for their crops, to get the capital they need to improve their farm output, income, and well-being. The study shows that only 33% of farmers can get credit facilities and that older farmers and farmers with bigger farms are more likely to get credit facilities, while farmers from disadvantaged castes are the least likely to get credit access [56].

```
CRAALoansIssued_df= pd.read_csv("CREDIT FOR AGRICULTURE AND ALLIED ACTIVITIES LoansIssued.csv")
CRAALoansIssued_df
```

```
CRAALoansIssued_df["Year"] = CRAALoansIssued_df["Year"].astype(int)
```

Output

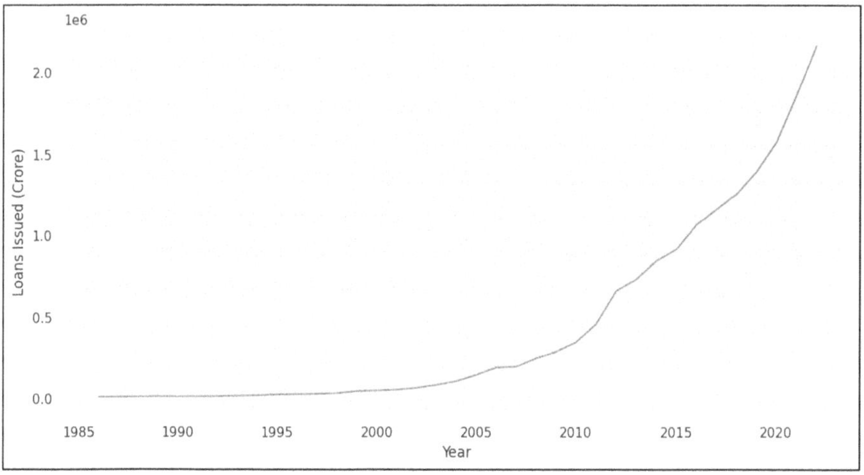


```
CRAALoansOutstanding_df= pd.read_csv("CREDIT FOR AGRICULTURE AND ALLIED ACTIVITIES Loans Outstanding.csv")
CRAALoansOutstanding_df
```

```
CRAALoansIssued_df["Year"] = CRAALoansIssued_df["Year"].astype(int)
```

```
CRAALoansOutstanding_df["Year"] = CRAALoansOutstanding_df["Year"].astype(int)
```

Output

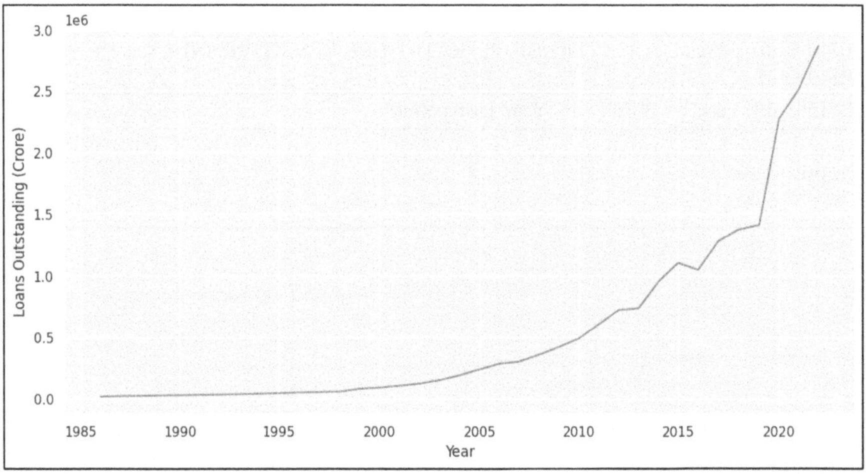

Credit availability has a significant positive impact on farm investments, such as land improvement, livestock, and equipment. It also influences positively the farm revenue expenditure, which covers the spending on seeds, equipment, labor, irrigation, plant protection chemicals, and livestock inputs. Therefore, credit availability increases farm income per hectare, livestock income, and monthly consumption expenditure. Estimates show that a 1% increase in the number of rural banked locations reduced rural poverty by about 0.4% and increased total output by 0.30%. This shows the vital role of institutional credit in reducing poverty and boosting economic growth in rural areas. However, more than 60% of farmers do not have access to credit, implying that the credit distribution system is not efficient. This requires inclusive and comprehensive policy interventions to include all farmers in the credit system, especially offering term loans to small and marginal farmers [56, 57].

Step 19: Price to Earnings Ratio of the Bombay Stock Exchange

The BSE PE ratio, or the price to earnings ratio of the Bombay Stock Exchange, is a way of measuring how much investors pay for a share in relation to the yearly net income or profit that the company makes per share. It is a financial ratio that helps to value a company: a higher PE ratio means that investors are paying more for each unit of net income, so the share is more costly compared to one with a lower PE ratio. However, the BSE PE ratio and poverty in India are not related to each other. The BSE PE ratio is a stock market indicator, while poverty is a social and economic problem. In the model, we will see if the market has any effect on the overall poverty rate.

```
BSEPE_df= pd.read_csv("AVERAGE_PE_RATIO_BSE_SENSITIVE_INDEX.csv")
BSEPE_df
```

```
BSEPE_df["Year"] = BSEPE_df["Year"].astype(int)
```

Output

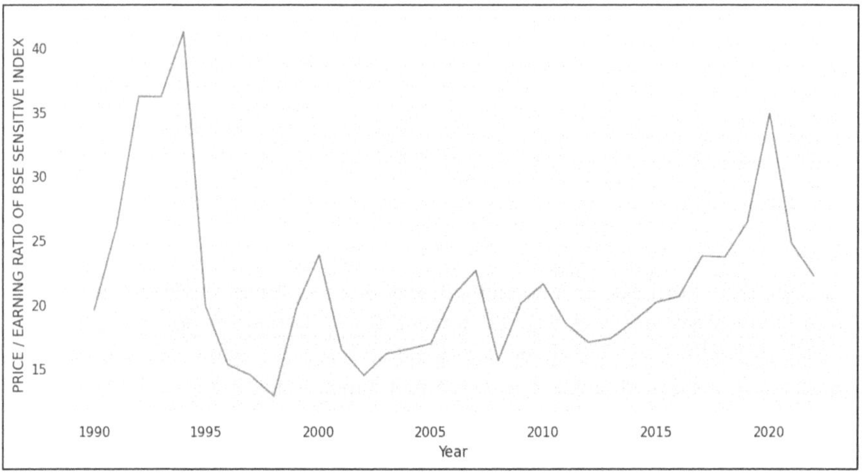

Step 20: Linear Regressor Model

Let's construct linear regression model to see the influence of identified variables on the poverty rate.

```
X = merged_df[['GDP growth (annual %)',
        'Employment in agriculture (% of total employment)',

    #'Employment in industry (% of total employment)',
    'Employment in services (% of total employment) ',
        'Inflation, consumer prices (annual %)',
        'Current account balance (% of GDP)',
        'EconomicEvents',
        'GeneralElectionYear',
        'DroughtYear',
        'Wholesale Prices_Industry Aggregates_Manufactured Products',
        'Production: Industry: Total Industry Excluding Construction ',
        'Gross national expenditure (% of GDP)',
        'Gross savings (% of GDP)',
        'Population growth (annual %)',
        # 'PRICE / EARNING RATIO OF BSE SENSITIVE INDEX',
        'MSP-Wheat'
        #'MSP-Paddy common'
        ]]

y= merged_df['Poverty Rate(<$1.90) ']

X_train,X_test, y_train , y_test = train_test_split(X ,y, test_size = 0.2,random_state=0 )

regressor = LinearRegression()
regressor.fit(X_train,y_train)

y_pred = regressor.predict(X_test)
y_pred

r2_score(y_test,y_pred)
```

The above code constructs linear regression model.

Output
0.954872 or 95.48%

Step 21: Construct Linear Regression Model Metrics

Let's evaluate the linear regression model metrics.

```
import pandas as pd
from sklearn import metrics
from sklearn.metrics import mean_squared_error, r2_score

mse = mean_squared_error(y_test, y_pred)
print("MSE:", mse)

rmse = np.sqrt(mse)
print("RMSE:", rmse)

r2 = r2_score(y_test, y_pred)
print("R2:", r2)
```

Output

```
MSE: 4.05670545504837
RMSE: 2.014126474442052
R2: 0.9548728967980639
```

Step 22: Construct Model Explainability

Let's construct model explainability to see the dominant variable.

```
eli5.explain_weights(regressor, feature_names = list(X.columns))
```

Output

y top features

Weight?	Feature
+25.307	Population growth (annual %)
+3.232	Employment in services (% of total employment)
+3.015	Employment in agriculture (% of total employment)
+1.917	EconomicEvents
+1.277	Gross savings (% of GDP)
+0.936	Current account balance (% of GDP)
+0.532	DroughtYear
+0.458	GDP growth (annual %)
+0.400	Inflation, consumer prices (annual %)
+0.364	GeneralElectionYear
+0.010	MSP-Wheat
-0.019	Production: Industry: Total Industry Excluding Construction
-0.106	Gross national expenditure (% of GDP)
-1.639	Wholesale Prices_Industry Aggregates_Manufactured Products
-296.286	<BIAS>

Increases in population growth increase poverty rates.

Step 22: Gradient Boosting Regression Model

Let's construct gradient boosting regression model to see the influence of identified variables on the poverty rate.

Prepare independent variable list:

```
numeric_features=[ 'GDP growth (annual %)',
          'Employment in agriculture (% of total employment)',
          'Gini Coefficient ',
     #'Employment in industry (% of total employment)',
     'Employment in services (% of total employment) ',
          'Inflation, consumer prices (annual %)',
          'Current account balance (% of GDP)',
          'EconomicEvents',
          'GeneralElectionYear',
          'DroughtYear',
          'Wholesale Prices_Industry Aggregates_Manufactured Products',
          'Production: Industry: Total Industry Excluding Construction ',
          'Gross national expenditure (% of GDP)', 'Gross savings (% of GDP)',
          'Population growth (annual %)',
            'PRICE / EARNING RATIO OF BSE SENSITIVE INDEX',#'Procurement','Stocks',
            'Off-take','MSP-Paddy common', 'MSP-Maize', 'MSP-Wheat',
            'Bank Credit to Commercial Sector  (? Crore)',
            'Loans Issued (Crore)',
     'Loans Outstanding (Crore)'
            ]
```

Set independent variable list to develop other ML models.

```
# high accuracy and low bias
from sklearn.ensemble import GradientBoostingRegressor
X = merged_df[numeric_features]

y= merged_df['Poverty Rate(<$1.90) ']

X_train,X_test, y_train , y_test = train_test_split(X ,y, test_size = 0.2,random_state=0 )

modelGradientBoostingRegressor          =          GradientBoostingRegressor(n_estimators=5000,
random_state=0,loss='squared_error')
modelGradientBoostingRegressor.fit(X_train,y_train)

y_pred = modelGradientBoostingRegressor.predict(X_test)
y_pred

r2_score(y_test,y_pred)
```

Output

0.995839 or 99.583%

```
import pandas as pd
from sklearn import metrics
from sklearn.metrics import mean_squared_error, r2_score

mse = mean_squared_error(y_test, y_pred)
print("MSE:", mse)

rmse = np.sqrt(mse)
print("RMSE:", rmse)

r2 = r2_score(y_test, y_pred)
print("R2:", r2)
```

Output

```
MSE: 0.3739683507533886
RMSE: 0.6115295174833252
R2: 0.9958399473302395
```

Model explainability

```
eli5.show_weights(modelGradientBoostingRegressor,top=100)
eli5.show_weights(modelGradientBoostingRegressor, feature_names =list(X.columns), target_names =
"Poverty Rate in India",top=100)
```

Output

Weight	Feature
0.3024 ± 0.0599	Population growth (annual %)
0.2336 ± 0.0929	Loans Issued (Crore)
0.1827 ± 0.0590	Employment in services (% of total employment)
0.0981 ± 0.0661	Employment in agriculture (% of total employment)
0.0655 ± 0.0669	Bank Credit to Commercial Sector (? Crore)
0.0354 ± 0.0796	MSP-Maize
0.0285 ± 0.0791	MSP-Paddy common
0.0142 ± 0.0456	MSP-Wheat
0.0122 ± 0.1011	Current account balance (% of GDP)
0.0102 ± 0.0719	Production: Industry: Total Industry Excluding Construction
0.0092 ± 0.0666	Off-take
0.0026 ± 0.0725	Loans Outstanding (Crore)
0.0015 ± 0.0621	Gross savings (% of GDP)
0.0010 ± 0.0517	Gross national expenditure (% of GDP)
0.0009 ± 0.0310	Gini Coefficient
0.0007 ± 0.1120	Wholesale Prices_Industry Aggregates_Manufactured Products
0.0007 ± 0.0741	PRICE / EARNING RATIO OF BSE SENSITIVE INDEX
0.0005 ± 0.0785	GDP growth (annual %)
0.0002 ± 0.0021	EconomicEvents
0.0001 ± 0.0807	Inflation, consumer prices (annual %)
0.0000 ± 0.0451	GeneralElectionYear
0.0000 ± 0.0096	DroughtYear

Step 23: Construct Random Forest Regression Model

Let's construct random forest regression model to see the influence of identified variables on the poverty rate.

```
# high accuracy and low bias
from sklearn.ensemble import RandomForestRegressor
X = merged_df[numeric_features]

y= merged_df['Poverty Rate(<$1.90) ']

X_train,X_test, y_train , y_test = train_test_split(X ,y, test_size = 0.2,random_state=0 )

modelRandomForestRegressor = RandomForestRegressor(max_depth=2, random_state=0)
modelRandomForestRegressor.fit(X_train,y_train)

y_pred = modelRandomForestRegressor.predict(X_test)
y_pred

r2_score(y_test,y_pred)
```

Output

0.988799 or 98.87%

```
import pandas as pd
from sklearn import metrics
from sklearn.metrics import mean_squared_error, r2_score

mse = mean_squared_error(y_test, y_pred)
print("MSE:", mse)

rmse = np.sqrt(mse)
print("RMSE:", rmse)

r2 = r2_score(y_test, y_pred)
print("R2:", r2)
```

Output

```
MSE: 1.0068535387049882
RMSE: 1.00342091801247
R2: 0.988799683868142
```

Model explainability

```
eli5.show_weights(modelRandomForestRegressor,top=100)
eli5.show_weights(modelRandomForestRegressor, feature_names =list(X.columns), target_names = "Poverty Rate in India",top=100)
```

Output

Weight	Feature
0.1638 ± 0.6862	Employment in agriculture (% of total employment)
0.1556 ± 0.6730	Population growth (annual %)
0.1090 ± 0.5692	Off-take
0.1054 ± 0.5584	Loans Issued (Crore)
0.0987 ± 0.5317	Production: Industry: Total Industry Excluding Construction
0.0722 ± 0.4600	MSP-Paddy common
0.0717 ± 0.4532	Employment in services (% of total employment)
0.0694 ± 0.4618	MSP-Maize
0.0534 ± 0.3949	MSP-Wheat
0.0407 ± 0.3589	Bank Credit to Commercial Sector (? Crore)
0.0321 ± 0.3008	Loans Outstanding (Crore)
0.0108 ± 0.1972	Gross national expenditure (% of GDP)
0.0045 ± 0.0423	Gross savings (% of GDP)
0.0039 ± 0.0339	PRICE / EARNING RATIO OF BSE SENSITIVE INDEX
0.0031 ± 0.0228	Current account balance (% of GDP)
0.0026 ± 0.0200	Inflation, consumer prices (annual %)
0.0022 ± 0.0194	Gini Coefficient
0.0005 ± 0.0079	GDP growth (annual %)
0.0003 ± 0.0067	EconomicEvents
0.0002 ± 0.0033	DroughtYear
0 ± 0.0000	Wholesale Prices_Industry Aggregates_Manufactured Products
0 ± 0.0000	GeneralElectionYear

Model Comparison and Summary

The following shows the model summary. Gradient boost regressors (95.79%) outperform among all the regressors, second linear regressor (94.86%), followed by random forest regressor (94.38%), a narrow higher than linear regression (please see Fig. 6.14).

The "best" metric for a regression model's accuracy depends on the specific problem and the desired outcome. Both mean squared error (MSE) and root mean squared error (RMSE) are commonly used metrics for regression models, but they have different characteristics:

MSE is sensitive to outliers because it squares the errors before they are averaged. It measures the average of the squares of the errors, that is, the average squared difference between the estimated values and the actual value (please see Fig. 6.15).

RMSE, which is the square root of MSE, combines the advantages of MSE while having the same scale as the target variable. It penalizes large errors more severely, making it sensitive to outliers (please see Fig. 6.16).

Gradient boost model performs well with respect to MSE and RMSE. Nevertheless, neither of these metrics alone can be considered the "best" for all situations. For example, during COVID-19 [58], the PDS data has experienced an exponential growth. In order to future forecast with higher accuracies under similar scenarios of COVID, RMSE would be a good candidate because it penalizes large errors more.

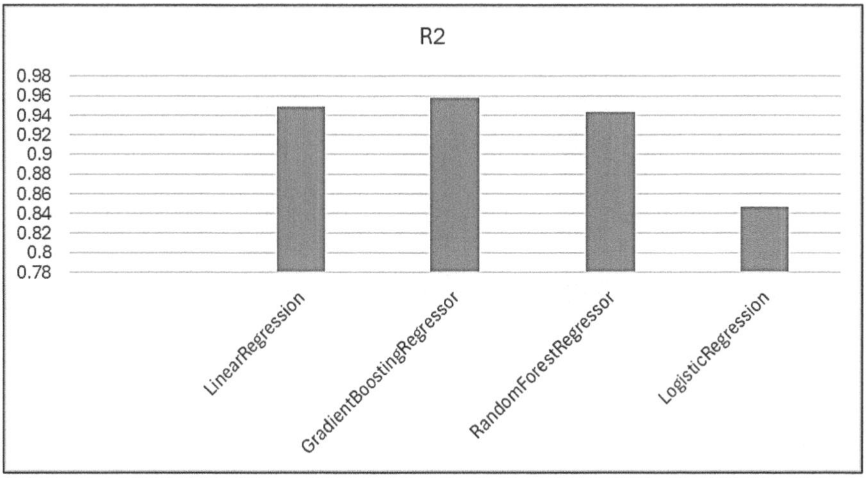

Fig. 6.14 R2

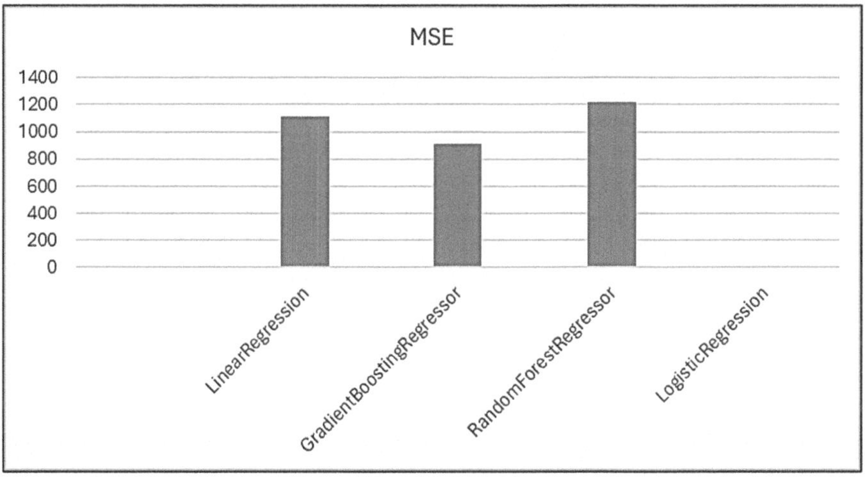

Fig. 6.15 MSE

On the other hand, if you're more interested in the average error, MSE might be more appropriate. It's also worth noting that these metrics don't tell you anything about the model's bias or variance. A model with a low MSE or RMSE might still be overfitting or underfitting the data. Therefore, it's important to use these metrics in conjunction with other diagnostic tools when evaluating your model's performance.

Increase in employment in agriculture has higher probability of increasing poverty rates. Similarly, population growth increases poverty rate. Interestingly, PDS off-take implies increased food subsidies, a direct factor of increased poverty rates.

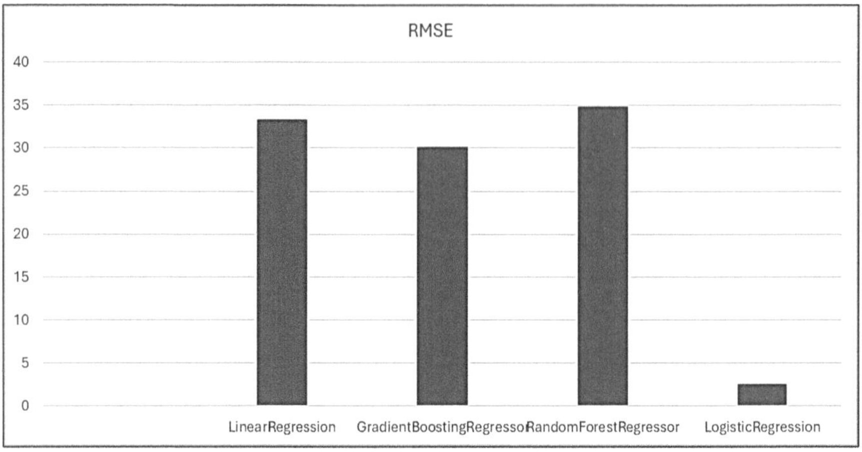

Fig. 6.16 RMSE

Loans issued also have higher influence on increasing poverty rates. Here is overall summary: some of the poverty reduction strategies include the following [10]:

- Pursuing macroeconomic policies to sustain a growth rate of at least 9% per annum over an extended period of time.
- Reducing fiscal deficit, revenue deficit, and current account deficit [59] (*current account balance (% of GDP)*).
- Strengthening mechanisms for inclusive growth and creating new job opportunities.
- Promoting credit flow to agriculture and strengthening the rural economy [24]. Expanding employment generation schemes and strengthening social security nets (*employment in agriculture (% of total employment) and employment in services (% of total employment)*). Promoting manufacturing and service sectors for job creation.
- Implementing the National Rural Employment Guarantee Scheme to generate wage employment.
- Increasing plan allocation for agriculture and implementing schemes to support farmers (*loans issued, bank credits to commercial sector*). Providing access to financial services, such as microcredit and savings accounts, can empower individuals and communities to start small businesses, invest in education, and build assets.
- Ensuring food security and maintaining affordable food prices for the poor (*inflation, CPI*)
- Improving access to quality elementary education through social programs (*public distribution system*). Providing midday meals in schools to enhance school participation and reduce hunger. Strengthening the targeted public distribution system to provide food security to those below the poverty line.

- Land reform and agricultural development: implementing land reform policies and improving agricultural productivity can help alleviate rural poverty by providing farmers with secure land tenure, access to credit, and technical support.
- Farm credit: increasing access to credit for farmers to support agricultural activities and improve their income (*loans issued and loans outstanding*) [22].
- Agricultural indebtedness: addressing the issue of farmer debt and finding solutions to alleviate their financial burden [22].
- Mission for pulses: initiatives to improve the production and productivity of pulses, including providing quality seeds and promoting related activities (*PDS off-take*, *MSP maize*, *MSP paddy*, and *MSP wheat*) [22].

Others include

- Focusing on integrated development of urban infrastructure and improving rural economies
- Ensuring safe drinking water supply and sanitation in rural areas through various programs
- Continuously working on rural sanitation programs to improve hygiene and sanitation in rural areas
- Giving highest priority to rural development and implementing various programs to improve living conditions in rural areas

A special note on BSE PE ratio: the BSE PE ratio might reflect the overall health of the companies listed on the Bombay Stock Exchange, which could indirectly impact poverty. For instance, if the companies are doing well (high earnings, low PE ratio), they might be able to create more jobs, potentially reducing unemployment and poverty. Conversely, if the companies are not doing well (low earnings, high PE ratio), they might have to lay off employees, potentially increasing unemployment and poverty. However, *many other factors influence poverty in India, such as government policies, access to education and healthcare, infrastructure, and social issues. Therefore, while the BSE PE ratio might have an indirect influence, it is not a direct or major factor in poverty in India.*

Conclusion and Policy Measures

When countries have higher national incomes, they can afford to spend more on public programs that will help children from disadvantaged backgrounds to achieve their potential (i.e., public programs will partly offset the unequal private investments). As the public programs become larger and more effective, relative mobility should increase.[59] However, this outcome is not guaranteed. It is essential that some

[59] Gross savings (% of GDP)—https://data.worldbank.org/indicator/NY.GNS.ICTR.ZS?locations=IN

of the additional fiscal space is used for public policies that improve opportunities for the most disadvantaged [60].

> Even after 52 years of independence, the provision of basic services in rural areas remains very unsatisfactory.[60] These large gaps in basic services in rural areas are not acceptable, and the government is committed to removing them rapidly.
>
> Shri Yashwant Sinha
> Minister of Finance, 2000–2001

When growth happens, poverty usually goes down regardless of whether inequality increases or decreases. This is most clearly shown in the example of developing countries like India and China. Moreover, since different growth rates have different effects on reducing poverty, we can infer that growth is beneficial for the poor, but it is not enough (necessary but not sufficient condition for poverty alleviation) [61]. Inclusive growth to marginalized and disadvantaged communities needs financial transparency and fiscal position: the fundamental tenets of transparency of financial statements and fiscal position, reflecting the government's intent, strengths, and challenges.

> Inclusive growth to marginalized and disadvantaged communities needs financial transparency and fiscal position: the fundamental tenets of transparency of financial statements and fiscal position, reflecting the government's intent, strengths, and challenges.
>
> Author

Growth alone may not be durable and lasting. It is thus important to anchor the approach of poverty reduction on fast but steady economic growth. The impact of growth on poverty reduction depends on how we define poverty, on the ability of the poor to benefit from growth, and the speed and nature of the growth [61]. In addition to economic growth, generating employment, attaining a surplus in the current account, building up foreign exchange reserves, increasing gross domestic savings, reducing interest rates on government securities, easing restrictions on capital account mobility, restructuring external debt, maintaining low inflation, maintaining a strong balance of payments position, promoting fiscal consolidation, enhancing revenue through a moderate tax administration, rationalizing expenditure, and prepaying external debt are essential to macroeconomic stability and achievement of sustainable economic and social growth [8].

The most urgent task for policymakers today is to create institutional settings and to mix pro-growth and pro-poor policies that will allow the poor to benefit from and contribute to future growth when billions of people are still in poverty [61].

Considering the crisis in the financial and real sector, global recession, rising unemployment, and inequality, future studies should not only revisit the "trickle-down" theory and especially its impact on lower income groups but also examine the

[60] Budget 2000–2001—https://www.indiabudget.gov.in/doc/bspeech/bs200001.pdf

cycle of poverty and how it relates to business cycles. However, a new distinctive poverty measure is also very important to review [61].

Policies that promote productivity growth and investment need to go hand-in-hand with social protection interventions targeting nutrition, health, and education. Social protection can help subsistence smallholders to escape poverty traps and enter into a virtuous cycle of higher productivity and income generation [62].

Poverty reduction

The policy implication is to design economic and social policies that achieve higher growth, lower inflation, sensible fiscal policy, fiscal debt reduction, subsidies, raise tax to GDP, strengthening banking system, encouragement of entrepreneurship and start-up companies are some of the macroeconomic policy results that reduce poverty and create lasting growth environment. If every developing country can keep up its economic strength, it can be a pillar of stability for the world economy and provide a secure destination for restless global capital.

Author

After reading this chapter, you should be able to calculate poverty using either the Consumer Pyramids Household Survey (approach 1) or IMF Working Paper (approach 2). You should be able to modify models in estimating poverty—a pro-poor and marginalized inclusive growth machine learning. Finally, assess the policy implications: inclusive growth to marginalized and disadvantaged communities needs financial transparency and fiscal position—the fundamental tenets of transparency of financial statements and fiscal position, reflecting the government's intent, strengths, and challenges.

References

1. Shri R.K. Shanmukham Chetty, The Budget for the Year 1947–1948, Year 1947–1948, https://www.indiabudget.gov.in/doc/bspeech/bs194748.pdf, Access Date: 11/5/2021
2. Shri Manmohan Singh, Budget 1992–93, Published Tue, 29th February 1992, https://www.indiabudget.gov.in/doc/bspeech/bs199293.pdf, Access Date: 11/23/2023
3. Prof. Madhu Dandavate, Introducing the Budget for the Year 1990–91, Published Tue, 1990, https://www.indiabudget.gov.in/doc/bspeech/bs199091.pdf, Access Date: 11/23/2023
4. Nimish Adhia, The History of Economic Development in India since Independence, Published Winter 2015, https://www.asianstudies.org/publications/eaa/archives/the-history-of-economic-development-in-india-since-independence/, Access Date: 11/11/2021
5. P. Chidambaram, Budget 2004–2005, July 8, 2004, https://www.indiabudget.gov.in/doc/bspeech/bs200405.pdf, Access Date: November 30, 2023.
6. Chandrasekar Vuppalapati, Specialty Crops for Climate Change Adaptation, Publisher Springer Cham, Hardcover ISBN 978-3-031-38398-4 Published: 15 October 2023

7. Dr. Seema Gaur and Dr. N Srinivasa Rao, Poverty Measurement in India: A Status Update, September 2020, https://rural.nic.in/sites/default/files/WorkingPaper_Poverty_DoRD_Sept_2020.pdf, Access Date: 11/13/2021
8. Shri Jaswant Singh, Budget 2003–2004, February 2003, https://www.indiabudget.gov.in/doc/bspeech/bs200304.pdf, Access Date: November 30, 2023.
9. Shri Manmohan Singh, Budget Speech 1993–1994, 27 February 1993, https://www.indiabudget.gov.in/doc/bspeech/bs199394.pdf, Access Date: 11/22/2023
10. Pranab Mukherjee, Budget 2009–2010, February 2009, https://www.indiabudget.gov.in/budget_archive/ub2009-10(I)/bs/speecha.htm, Access Date: December 05, 2023
11. K. Balchandp. Sunderarajan, You will be on BPL list if your annual income is Rs.27,000, May 21, 2011 01:45 am | Updated August 18, 2016 01:08 pm IST – New Delhi, https://www.thehindu.com/news/national/you-will-be-on-bpl-list-if-your-annual-income-is-rs27000/article2035893.ece, Access Date: 11/13/2021
12. Gaurav Datt, Poverty in India 1951–1994: Trends and Decompositions, Revised August 1997. https://documents1.worldbank.org/curated/en/253831468049773494/pdf/900360WP0Pover00Box385299B00PUBLIC0.pdf, Access Date: 11/14/2023
13. "Sinha Roy, Sutirtha; Van Der Weide, Roy. 2022. Poverty in India Has Declined over the Last Decade But Not As Much As Previously Thought. Policy Research Working Paper;9994. © World Bank, Washington, DC. http://hdl.handle.net/10986/37273 License: CC BY 3.0 IGO.", Access Date: 11/16/2023
14. The Times of India, New lines, but India still home to biggest chunk of global poor,TNN / Updated: Nov 2, 2017, 11:56 IST, https://timesofindia.indiatimes.com/india/new-lines-but-india-still-home-to-biggest-chunk-of-global-poor/articleshow/61444408.cms, Access Date: 11/13/2021
15. UNDP and OPHI (2022). Global Multidimensional Poverty Index 2022: Unpacking deprivation bundles to reduce multidimensional poverty. United Nations Development Programme (UNDP), and Oxford Poverty and Human Development Initiative (OPHI), University of Oxford.
16. Kaushal Shroff, Two New Papers on Poverty Levels Shine a Light on Need for Reliable Official Data in India, 25/Apr/2022, https://thewire.in/economy/imf-world-bank-poverty-papers-missing-data, Access Date: 11/16/2023
17. Suresh Seshadri, What is Consumer Expenditure Survey, and why was its 2017–2018 data withheld?, November 24, 2019 12:02 am | Updated December 03, 2019 07:39 pm IST, https://www.thehindu.com/business/Economy/what-is-consumer-expenditure-survey-and-why-was-its-2017-2018-data-withheld/article61611662.ece, Access Date: 11/16/2023
18. Justin Sandefur, The Great Indian Poverty Debate, 2.0, April 19 2022, https://cgdev.org/blog/great-indian-poverty-debate-20, Access Date: 11/16/2023
19. Ifeanyi Nzegwu Edochie, Samuel Freije-Rodriguez, Christoph Lakner, Laura Moreno Herrera, David Locke Newhouse, Sutirtha Sinha Roy, Nishant Yonzan, Justin Sandefur, What Do We Know about Poverty in India in 2017/18?, February 2022, https://openknowledge.worldbank.org/server/api/core/bitstreams/206cd495-76ad-53cb-9da8-5cf07eaf985e/content, Access Date: 11/16/2023
20. Bhalla, S. S., Bhasin, K. and Virmani, A. (2022). Pandemic, Poverty and Inequality: Evidence from India. https://www.elibrary.imf.org/downloadpdf/journals/001/2022/069/001.2022.issue-069-en.xml, Access Date: 11/14/2023
21. ROF. Madhu Dandavate, Introducing the budget for the year 1990–91, Published Tue, 1990, https://www.indiabudget.gov.in/doc/bspeech/bs199091.pdf, Access Date: 11/23/2023
22. P. Chidambaram, Budget 2007–2008, February 2007, https://www.indiabudget.gov.in/doc/bspeech/bs200708.pdf, Access Date: December 04, 2023
23. FAO. 2022. Credit to agriculture. Global and regional trends 2012–2020. FAOSTAT Analytical Brief Series No. 38. Rome, Access Date: 11/19/2023
24. Shri Yashwant Sinha, Budget 2000–2001, February 2000, https://www.indiabudget.gov.in/doc/bspeech/bs200001.pdf, Access Date: November 29, 2023.

25. Adam Hayes, Gini Index Explained and Gini Co-efficients Around the World, Updated May 27, 2023, https://www.investopedia.com/terms/g/gini-index.asp, Access Date: 11/14/2023
26. WorldBank, India Key conditions and challenges, Oct 2023, https://thedocs.worldbank.org/en/doc/5d1783db09a0e09d15bbcea8ef0cec0b-0500052021/related/mpo-ind.pdf, Access Date: 11/14/2021
27. Karl Usuka, An Analysis of the Determinants of Poverty in India and South Africa, Global Majority E-Journal, Vol. 10, No. 2 (December 2019), pp. 100–115, https://www.american.edu/cas/economics/ejournal/upload/usuka_accessible.pdf, Access Date: 11/12/2021
28. Lauren Frayer and Raksha Kumar, It's a mystery: Women in India drop out of the workforce even as the economy grows, Updated January 16, 2023, 7:32 AM ET, https://www.npr.org/sections/goatsandsoda/2023/01/04/1146953384/why-women-in-india-are-dropping-out-the-workforce-even-as-the-economy-grows#:~:text=As%20household%20incomes%20rise%2C%20women,because%20they%20can%20afford%20to, Access Date: December 02, 2023
29. P. Chidambaram, Budget 2006–2007, February 2006, https://www.indiabudget.gov.in/doc/bspeech/bs200607.pdf, Access Date: December 02, 2023
30. Kumar Vikram, Rural India spends most on mobiles: NSSO, New Delhi,UPDATED: Jul 9, 2016 08:54 IST, https://www.indiatoday.in/mail-today/story/rural-india-technonology-mobile-phones-national-survey-328188-2016-07-08?onetap=true, Access Date: 11/13/2021
31. India – Achievements and challenges in reducing poverty (English). A World Bank country study Washington, D.C. : World Bank Group. http://documents.worldbank.org/curated/en/232941468756917265/India-Achievements-and-challenges-in-reducing-poverty, Access Date: 11/13/2021
32. Rakesh Kochhar, Despite Poverty's Plunge, Middle-Class Status Remains Out of Reach for Many, July 8, 2015, https://www.pewresearch.org/global/2015/07/08/despite-povertys-plunge-middle-class-status-remains-out-of-reach-for-many/, Access Date: 11/16/2023
33. Shri Yashwant Sinha, Budget 2001–2002, February 2001, https://www.indiabudget.gov.in/doc/bspeech/bs200102.pdf, Access Date: November 28, 2023.
34. Nirmala Sitharaman, Budget Speech 2020–2021, February 2020, https://www.indiabudget.gov.in/doc/bspeech/bs202021.pdf, Access Date: 11/22/2023
35. Paul Cheung, Handbook on poverty statistics: concepts, methods and policy use, Dec 2005, https://unstats.un.org/unsd/methods/poverty/pdf/un_book%20final%2030%20dec%2005.pdf, Access Date: November 25, 2023.
36. Alkire, S., Foster, J. E., Seth, S., Santos, M. E., Roche, J. M., and Ballon, P. (2015). Multidimensional Poverty Measurement and Analysis. Oxford: Oxford University Press, ch. 10., Access Date: 11/17/2023
37. Ramandeep Kaur, Poverty and its Causes in India, September 2 2019, https://www.mapsofindia.com/my-india/society/poverty-in-india-and-its-causes, Access Date: 11/20/2023
38. Rakesh Kochhar, In the pandemic, India's middle class shrinks and poverty spreads while China sees smaller changes, March 18 2021, https://www.pewresearch.org/short-reads/2021/,03/18/in-the-pandemic-indias-middle-class-shrinks-and-poverty-spreads-while-china-sees-smaller-changes/, Access Date: 11/20/2023
39. Marta Schoch, Nishant Yonzan, Ruth Hill, Christoph Lakner, To end extreme poverty, getting back to pre-COVID-19 reduction rates is not enough, November 30 2022, https://blogs.worldbank.org/developmenttalk/end-extreme-poverty-getting-back-pre-covid-19-reduction-rates-not-enough, Access Date: 11/20/2023
40. Indermit Gill and Peter Nagle, Inflation could wreak vengeance on the world's poor, March 18 2022, https://www.brookings.edu/articles/inflation-could-wreak-vengeance-on-the-worlds-poor/, Access Date: 11/20/2023
41. Padma Balaji, How Inflation in India Threatens Poverty Gains, January 9 2023, https://borgenproject.org/inflation-in-india/, Access Date: 11/20/2023
42. Santosh Mehrotra and Jajati Keshari Parida, Poverty in India is on the rise again, August 04, 2021, https://www.impriindia.com/insights/poverty-in-india-on-rise-again/, Access Date: 11/20/2023

43. Rounak Jain, India's gross savings and investments edge higher in FY22 despite inflation, says SBI Research report, MAR 7, 2023, 16:33 IST, https://www.businessinsider.in/india/news/indias-gross-savings-and-investments-edge-higher-in-fy22-despite-inflation-says-sbi-research-report/articleshow/98444014.cms, Access Date: 11/20/2023
44. Singh, T. Economic growth and the state of poverty in India: sectoral and provincial perspectives. Econ Change Restruct 55, 1251–1302 (2022). https://doi.org/10.1007/s10644-021-09345-5, Access Date: 11/20/2023
45. Gaurav Datt, Poverty and growth in India, Business Insight 26th Oct 2016, https://www.monash.edu/business/cdes/cdes-research/showcase/show-all/growth,-poverty-and-inequality/poverty-and-growth-in-india, Access Date: 11/20/2023
46. INSIGHTIAS, Poverty, 2021, https://www.insightsonindia.com/indian-economy-3/poverty-and-unemployment/poverty/, Access Date: 11/20/2023
47. Arvind Panagariya Megha Mukim, A Comprehensive Analysis of Poverty in India, December 2013, https://documents1.worldbank.org/curated/en/389241468283507057/pdf/WPS6714.pdf, Access Date: 11/20/2023
48. Ouechtati, I.. (2020). The Contribution of Financial Inclusion in Reducing Poverty and Income Inequality in Developing Countries. Asian Economic and Financial Review, 10(9), 1051–1061. https://doi.org/10.18488/journal.aefr.2020.109.1051.1061, Access Date: 11/14/2023
49. WorldBank, Macro Poverty Outlook for South Asia, 2023, https://www.worldbank.org/en/publication/macro-poverty-outlook/mpo_sar, Access Date: 11/14/2023
50. Zhu Y, Bashir S, Marie M. Assessing the Relationship between Poverty and Economic Growth: Does Sustainable Development Goal Can be Achieved? Environ Sci Pollut Res Int. 2022 Apr;29(19):27613–27623. https://doi.org/10.1007/s11356-021-18240-5. Epub 2022 Jan 4. PMID: 34984611; PMCID: PMC8726512, Access Date: 11/14/2023
51. William Easterly and Stanley Fischer, Inflation and the Poor, Journal of Money, Credit and Banking, May, 2001, Vol. 33, No. 2, Part 1 (May, 2001), pp. 160–178, https://economics.mit.edu/sites/default/files/2023-05/fischer_inflation_poor.pdf, Access Date: November 25, 2023.
52. UNDP, National Multidimensional Poverty Index- A Progress Review 2023, 2023, https://www.undp.org/india/national-multidimensional-poverty-index-progress-review-2023, Access Date: December 06, 2023
53. Nikhil Inamdar and Aparna Alluri, India economy: Seven years of Modi in seven charts, 21st June 2021, 04:01 PDT, https://www.bbc.com/news/world-asia-india-57437944, Access Date: December 06, 2023
54. Seema Bathla, Pramod Kumar Joshi and Anjani Kumar, Agricultural Growth and Rural Poverty Reduction in India, 2020, https://www.rbi.org.in/Scripts/bs_viewcontent.aspx?Id=4002, Access Date: December 06, 2023
55. Karnataka Economic Survey 2021–22, Role of banks and financial institutions in the economic growth and development of Karnataka, 2021–22, https://planning.karnataka.gov.in/storage/pdf-files/Economic%20Survey/Chapter%20Eng%205.pdf, Access Date: December 06, 2023
56. Ramasamy, P., Malaiarasan, U. Agricultural credit in India: determinants and effects. Ind. Econ. Rev. 58, 169–195 (2023). https://doi.org/10.1007/s41775-023-00187-8, Access Date: December 06, 2023
57. Abhiman Das, Manjusha Senapati, Joice John, Impact of Agricultural Credit on Agriculture Production: An Empirical Analysis in India, Vol. 30, No.2, Monsoon 2009, https://rbi.org.in/scripts/bs_viewcontent.aspx?Id=2240, Access Date: December 06, 2023
58. Elif C Arbatli Saxegaard; Mattia Coppo; Nasser Khalil; Shinya Kotera; Filiz D Unsal, Inequality and Poverty in India: Impact of COVID-19 Pandemic and Policy Response, July 14, 2023, https://www.imf.org/en/Publications/WP/Issues/2023/07/20/Inequality-and-Poverty-in-India-Impact-of-COVID-19-Pandemic-and-Policy-Response-535908, Access Date: November 26, 2023.
59. Shri Manmohan Singh, Budget Speech 1991–1992, 24th July, 1991, https://www.indiabudget.gov.in/doc/bspeech/bs199192.pdf, Access Date: November 26, 2023.

60. Daniel Gerszon, Mahlerroy Van Der Weidechristoph, Laknerambar Narayanrakesh Gupta, October 18, 2021, A Kuznets curve for intergenerational mobility, https://blogs.worldbank.org/developmenttalk/kuznets-curve-intergenerational-mobility, Access Date: 11/17/2023
61. Skare Marinko and Pržiklas Družeta Romina, Poverty and economic growth: a review, 2016/01/02, https://www.researchgate.net/publication/291388021_Poverty_and_economic_growth_a_review, Access Date: 11/17/2023
62. George Rapsomanikis, The economic lives of smallholder farmers, 2015, https://www.fao.org/3/i5251e/i5251e.pdf, Access Date: 11/20/2023

Chapter 7
Conclusion

This book uses large language models to extract key information from budget speeches and understand their policy implications. These implications indicate that some of the macroeconomic policy outcomes that can contribute to higher growth, lower inflation, prudent fiscal policy, fiscal debt reduction, subsidies, increased tax to GDP, strengthened banking system, and entrepreneurship and start-up company support are crucial for designing economic and social policies that reduce poverty and create economic and social sustainability policies.

This book demonstrates that the economic and social sustainability pillar relies on three things based on the models it developed: (1) increasing farmers' income and supporting agriculture and farmers' welfare; (2) enhancing rural employment and infrastructure development, providing healthcare services, and implementing health protection schemes; and (3) advancing education and skills and creating jobs to build a productive society. The models in the book suggest that as a policy implication, increasing institutional credit for agriculture and related activities through loans, bank credit to the commercial sector, and appealing time deposits to households and firms will have a positive effect on India's long term economic growth. Moreover, higher foreign investment plays a significant role in creating jobs and economic activities that help the poor and marginalized communities escape poverty and ensure overall sustainability.

To conclude, using large language models to extract statements from budget speeches is a method of linking the past and future in the same temporal frame to evaluate the policy impact for the survival of disadvantaged communities. It's an evaluation of government policies that offer a useful perspective to improve the effectiveness of democracies and create a future where the gap between the rich and poor could be reduced with data science based intuitions. Lastly, using large language models to evaluate how well budget speeches achieve policy goals is a democracy movement for the twenty-second century. It is as basic as voting rights, but with data science.

© The Author(s), under exclusive license to Springer Nature Switzerland AG 2024

C. Vuppalapati, *Assessing Policy Effectiveness using AI and Language Models*,
International Series in Operations Research & Management Science 354,
https://doi.org/10.1007/978-3-031-56097-2_7

Appendix

Appendix A: Economic Sustainability Statements

Budget Speech 1990–1991[1] Professor Madhu Dandavate

1. We have taken some resources from the rich and used them to give some relief to the poor and the common man.
2. We have begun a process to restrain the budgetary deficit and contain the inflationary pressure.
3. We have tilted the balance of planning and investment toward the rural areas and in favor of employment.
4. Our emphasis on rural development and decentralization will allow us to integrate environmental considerations into the design of development.
5. A long-term solution to the problem of unemployed youth has to come from a shift to a more employment-oriented growth strategy.
6. We have decided to give a boost to measures which will assist the youth of this country to acquire the skills that will improve their prospects for gainful employment.
7. The union government has decided that all the facilities and privileges that were available to the Scheduled Castes will be available to them from the Union Government even after their conversion to Buddhism.
8. A strategy for greater absorption of labor in agriculture has to go hand in hand with faster growth of industry and balanced development of infrastructure.
9. This government will give priority to accelerating industrial growth in a competitive and non-monopolistic environment.

[1] Budget Speech 1990–1991—https://www.indiabudget.gov.in/doc/bspeech/bs199091.pdf

© The Editor(s) (if applicable) and The Author(s), under exclusive license to Springer Nature Switzerland AG 2024

C. Vuppalapati, *Assessing Policy Effectiveness using AI and Language Models*, International Series in Operations Research & Management Science 354, https://doi.org/10.1007/978-3-031-56097-2

10. We will aim at utilizing the talent of our scientists and engineers toward the development of appropriate technologies for agriculture, nonconventional and renewable energy, and other employment-intensive activities.
11. The government will continue to provide special facilities for Indians settled abroad to invest their savings in this country.
12. The government is of the view that the amount of compensation to be paid to the victims of the Bhopal gas tragedy under court order is too meager for the magnitude of sufferings of these innocent people.

Budget Speech 1991–1992[2] *Shri Manmohan Singh*

1. Our economy would return to a path of high sustained growth with reasonable price stability and greater social equity.
2. The starting point of our strategy should be credible fiscal adjustment and macroeconomic stabilization.
3. It should be our objective to progressively reduce the fiscal deficit of the central government, move toward a significant reduction of the revenue deficit, and reduce the current account deficit in the balance of payments.
4. We must make a conscious effort to reduce the internal debt of the government and the external debt of the nation, so that we rely more and more on our own resources to finance the process of development.
5. During the period of transition, it shall be our endeavor to minimize the burden of adjustment on the poor. We are committed to adjustment with a human face.
6. The thrust of the reform process would be to increase efficiency and international competitiveness of industrial production, utilize foreign investment and foreign technology to a greater degree, and improve the performance of the public sector.
7. We must act fast and act boldly to introduce the needed correctives, as the existing situation can only retard growth, induce recession, and fuel inflation, which would hurt the economy further and impose a far greater burden on the poor.
8. Our government attaches the highest priority to agriculture and aims to provide massive thrust to food processing and other agro-based industries to increase the income of farmers, create employment opportunities, diversify the rural economy, and foster rural industrialization.
9. We must encourage accumulation of capital and develop a new attitude toward wealth, holding it as a trust and using it in the interest of society, particularly the underprivileged and without means.

[2] Budget Speech 1991–1992—https://www.indiabudget.gov.in/doc/bspeech/bs199192.pdf

10. Our approach to development has to combine efficiency with austerity, focusing on the needs of the majority and investing in basic necessities such as drinking water, education, health, and shelter.

Budget Speech 1992–1993[3] Shri Manmohan Singh

1. Our economy, polity, and society have to be extraordinarily resilient and alert if we are to take full advantage of the opportunities and to minimize the risks associated with the increasing globalization of economic processes.
2. We have to accept the need for reform if we are to avoid an increasing marginalization of India in the evolving world economy.
3. The economic policy changes brought about by our government in the last 8 months are inspired by this vision.
4. Our longer term objective is to evolve a pattern of production which is labor intensive and generates larger employment opportunities in productive higher income jobs and reduces the disparities in income and wealth between rural and urban areas.
5. Agriculture is the foundation of our national prosperity, and no strategy of economic development can succeed in our country if it does not ensure rapid growth of production and employment in agriculture.
6. Our people cannot afford to copy the soulless consumerism and the wasteful lifestyles of the affluent countries of the West. Conspicuous consumption has to be actively discouraged. The virtues of thrift have to be emphasized.
7. The owners of wealth, as Gandhiji used to say, must learn to regard themselves as trustees of society.
8. We cannot postpone structural reform and adjustment, but we must ensure that the burden of adjustment on the poorer and weaker sections of our society is ameliorated to the maximum possible extent.

Budget Speech 1993–1994[4] Shri Manmohan Singh

1. Our reforms are inspired precisely by the vision of India's natural and human resources entitling us to become a major powerhouse of the world economy.
2. The policy initiatives we have taken do not in any way reduce our commitment to take care of the poor and the disadvantaged.
3. We have taken steps to minimize the burden of adjustment on the poor and working classes.

---

[3] Budget Speech 1992–1993—https://www.indiabudget.gov.in/doc/bspeech/bs199293.pdf

[4] Budget Speech 1993–1994—https://www.indiabudget.gov.in/doc/bspeech/bs199394.pdf

4. We have disproved those professional prophets of gloom who were predicting millions of people becoming unemployed.
5. Inflation is down and production is beginning to recover.
6. Fears of being swamped by imports as a consequence of liberalization have proven to be grossly exaggerated.
7. The investment climate has improved considerably.
8. Foreign investors are showing active interest in investment in many sectors, including critical infrastructure sectors such as power and petroleum.
9. The approvals given for foreign investment proposals indicate a substantial potential for larger investment inflows in the future.
10. We have tried to maintain strict control over expenditure.
11. The room for fiscal maneuver gained by restraining expenditure over the past 2 years must be used to give a strong fillip to development expenditure.
12. We must ensure that our economic strategy gives full support to agriculture and agro-processing industries.
13. Exports must be made a truly high priority national endeavor.
14. No economic strategy can succeed in our country, which does not recognize the central role of agriculture in supporting broad-based and equitable development.
15. The government is firmly committed to ensuring that agriculture, agro-processing, and rural development are given top priority in the design of economic policy.

Budget Speech 1994–1994[5] Shri Manmohan Singh

1. The consequence of postponing the tax reform, or of imposing additional taxes to offset the revenue loss as conventionally calculated, would be wholly unproductive in a situation where our economy is characterized by sizeable unutilized industrial capacity, record food stocks, and comfortable foreign exchange reserves.
2. The tax structure now proposed goes a long way toward the kind of modern tax system and moderate tax rates and an emphasis on compliance, which is the hallmark of all successful countries.
3. I am confident that it will provide a strong stimulus for new investments, economic revival, and international competitiveness which is what the economy needs today.
4. We are on the threshold of a new century, indeed a new millennium. There are tremendous opportunities, provided we have the wisdom and foresight to seize them.
5. There are also immense dangers if we falter or appear indecisive.

[5] Budget Speech 1993–1994—https://www.indiabudget.gov.in/doc/bspeech/bs199394.pdf

6. In pursuit of this mandate, we embarked on a far-reaching program of economic restructuring and reform.
7. The economic situation has shown substantial improvement.
8. Our foreign currency reserves, which were a little over $1 billion in June, 1991, are now close to $13 billion.
9. Our exports have increased by a remarkable 21% in dollar terms in the first 10 months of 1993–1994.
10. Despite all the fears that liberalization would lead to a flood of imports, the dollar value of our imports during April to January 1993–1994 was less than 1% higher than imports during the corresponding period of 1992–1993.

Budget Speech 1999–2000[6]Shri Manmohan Singh

1. The proposal to reduce interest rates for small borrowers in towns where the Urban Land (Ceiling and Regulation) Act is not applicable demonstrates a focus on supporting economic sustainability by providing necessary support for housing development.
2. The plan to accord tax-free status to a limited amount of municipal bonds issued each year aims to encourage municipal bodies to improve their creditworthiness and financial stability, thereby promoting economic sustainability.
3. The announcement of tax initiatives to facilitate corporate mergers and amalgamations shows a commitment to supporting industrial restructuring and adaptation to new challenges, which is crucial for long-term economic sustainability.
4. The decision to review and amend the Industries (Development and Regulation) Act with a focus on industry development rather than regulation reflects a recognition of the importance of fostering a conducive environment for industrial growth and sustainability.
5. The plan to appoint a committee to examine and propose a modern Competition Law suitable for Indian conditions indicates a shift toward promoting competition and fostering a competitive market environment, which is essential for economic sustainability.
6. The establishment of a National Statistical Commission to revamp the statistical system demonstrates a commitment to ensuring timely and reliable data for policy formulation and planning, which is crucial for effective economic management and sustainability.
7. The emphasis on strengthening the research and development capabilities of the pharmaceutical industry and promoting foreign direct investment in the sector highlights a focus on knowledge-based industries and their potential contribution to long-term economic sustainability.

[6]Budget Speech 1993–1994—https://www.indiabudget.gov.in/doc/bspeech/bs199394.pdf

8. The broad strategy of the budget, which includes measures to reduce fiscal and revenue deficits, promote productivity and employment, deepen economic reforms, safeguard the economy from external shocks, and revitalize public programs for human development, reflects a comprehensive approach to fostering economic sustainability in various sectors.

Budget Speech 2000–2001[7]Shri Yashwant Sinha

1. Our foreign exchange reserves have attained new record levels.
2. Our stock markets have soared to new heights.
3. We have strengthened our agricultural sector.
4. We have energized our financial markets.
5. We have laid the foundations of an exciting new economy.
6. We propose to put India on a sustained, equitable, and job-creating growth path of 7–8% per year.
7. *Our strategy includes strengthening the foundations of our rural economy, nurturing knowledge-based industries, modernizing traditional industries, addressing infrastructure bottlenecks, prioritizing human resource development, strengthening our role in the world economy, and establishingfiscal discipline.*
8. We are pursuing economic reforms to unlock the creative energies of our people and reap the gains of productivity growth.
9. We are implementing second-generation reforms guided by compassion and justice.
10. *We are committed to ensuring that the fruits of economic reforms are shared by all sections of society, especially those living inrural areasandmarginalizedcommunities.*
11. We are focused on improving social and economic infrastructure in rural areas, including health, education, drinking water, housing, and roads.
12. We are working toward universalization of elementary education and increasing literacy rates.
13. We are committed to the speedy economic development of the northeastern states and Sikkim.
14. We are prioritizing the development of infrastructure in the North-East, including airports, railways, power, and national highways.
15. We are encouraging agricultural and horticultural development in the North-East through schemes for minor irrigation and horticulture.
16. We have established a new Department for Disinvestment to establish a systematic policy approach to disinvestment and privatization.

[7] Budget Speech 2000–2001—https://www.indiabudget.gov.in/doc/bspeech/bs200001.pdf

17. We are restructuring and reviving potentially viable PSUs and closing down PSUs that cannot be revived.
18. We are reducing government equity in nonstrategic PSUs and fully protecting the interests of workers.
19. We are committed to fiscal discipline and putting our fiscal house in order.
20. We are increasing budget support to the plan for sustained growth and social justice.
21. We are enhancing the quality of our defense preparedness and modernizing our forces.
22. We are curbing built-in expenditure growth and bringing about structural changes in the composition of our expenditure.
23. We are reviewing all subsidies with a view to bringing in cost-based user charges wherever feasible.
24. We are reassessing the manpower requirements of government departments and limiting fresh recruitment to minimum essential needs.
25. We are encouraging autonomous institutions to maximize generation of internal resources and reviewing budgetary support to these institutions.

Budget Speech 2001–2002[8] Shri Yashwant Sinha

1. The Indian economy has continued to exhibit both growth and resilience that have characterized its performance in the past few years.
2. Overall economic growth this year is expected to be about 6% despite a series of unexpected setbacks.
3. The prices of essential commodities, and of manufactured products as a whole, have remained stable.
4. Inflation, excluding energy, was around 4% during the year.
5. The economy remained secure with record levels of foreign exchange reserves and public food stocks.
6. The agricultural sector continues to be constrained by the existence of a number of inhibiting controls and regulations.
7. The government proposes to review the operation of the Essential Commodities Act, 1955, and remove many of the restrictions that have been imposed on the free interstate movement of food grains and agricultural produce and also on the storage and stocking of such commodities.
8. The challenge is to achieve a consensus on the imposition of appropriate user charges in such a manner that the poor are protected while those who can pay are made to do so.

[8] Budget Speech 2001–2002—https://www.indiabudget.gov.in/doc/bspeech/bs200102.pdf

9. The government is accelerating the program of reforms in State Electricity Boards (SEBs) to restore their financial viability.
10. Policy initiatives designed to increase private sector participation in ports have been successfully implemented.
11. A great deal of progress has been made over the last few years in pursuing reforms in the financial sector and capital markets.
12. Efforts are being made to develop and deepen the debt market, which will benefit small investors and institutional investors alike.
13. Measures are being taken to ensure smooth and quick movement of funds through the implementation of electronic fund transfer (EFT) and real-time gross settlement (RTGS) systems.

Budget Speech 2003–2004[9] *Shri Jaswant Singh*

1. The country's macroeconomic circumstances have never been better for attaining developmental objectives of enhanced and sustainable growth, poverty eradication, employment generation, and improving the quality of life.
2. Gross domestic savings, as a proportion of GDP, have improved and stand at around 24%.
3. The government has focused on fiscal consolidation as the basis for sustained growth.
4. The Tenth Five Year Plan aims to promote balanced and equitable regional development and advance policy and administrative reforms at the state level.
5. The government has implemented additional initiatives to promote infrastructure development through leveraging public money.
6. The country's current account turned into a surplus in 2001–2002 and continued to be in surplus during the first two quarters of the current year.
7. Foreign exchange reserves reached the highest ever in a single year, with reserves crossing $75.5 billion in the third week of February.
8. Interest rates on government securities have rapidly decreased from 12% to around 7%, setting the stage for investment growth.
9. The government is focused on releasing national creativity and unleashing the entrepreneurial character and creative genius of its citizens.
10. The budget prioritizes the eradication of poverty and reduction of social and economic inequalities to ensure a wider distributive spread of national wealth and greater spending power in the hands of all citizens.

[9]Budget Speech 2003–2004—https://www.indiabudget.gov.in/doc/bspeech/bs200304.pdf

Budget Speech 2004–2005[10]Shri P. Chidambaram

1. The economic fundamentals appear strong and the balance of payments is robust.
2. Growth will be sustained by increased production and value addition in agriculture.
3. There is a marked improvement in industrial production.
4. The services sector is performing well.
5. The government is fully alert to short-term *pressures on prices*.
6. The outlook for the year is benign.
7. The government is committed to achieving rapid growth with stability and equity.
8. The government is committed to removing inadequacies in infrastructure facilities.
9. The government is committed to providing insurance cover to farming and livestock.
10. The government is implementing the Rajiv Gandhi Drinking *Water* Mission to provide adequate drinking water.
11. The government is allocating funds for the Accelerated Rural Water Supply Program.
12. The government is implementing the Urban Water Supply Program in urban areas.
13. The government is investing in desalination plants to address water scarcity.
14. The government is working on the Sethusamudram Shipping Canal Project.
15. The government is facilitating the construction of an International Container Transshipment Terminal.
16. The government is providing housing through the Indira Awas Yojana and the Golden Jubilee Rural Housing Finance Scheme.
17. The government is making efforts to make the environment in India attractive for investors.
18. The government is considering gender budgeting to highlight the gender sensitivities of budgetary allocations.
19. The government is working on targeting subsidies sharply at the poor and truly needy.
20. The government is supporting the states' finances through transfers and loans.
21. The government is empowering the states through devolution of larger resources.
22. The government is providing special economic packages for certain regions like Bihar and Jammu and Kashmir.

[10]Budget Speech 2004–2004—https://www.indiabudget.gov.in/doc/bspeech/bs200405.pdf

Budget Speech 2005–2006[11]Shri P. Chidambaram

1. We must use the current phase of high growth to improve the fiscal health of the country.
2. We must increase revenues and reorient expenditure to pay for more outlays on education, health, and infrastructure.
3. Outlays do not necessarily mean outcomes. The focus should be on improving the quality of implementation and enhancing the efficiency and accountability of the delivery mechanism.
4. *Subsidies, such as food, fertilizer, and petroleum, provide a measure of protection for the poor and will continue to be provided. However, the subsidyregime needs to be restructured in a cautious manner and after thorough discussion.*
5. *The government is committed tofiscal consolidationand reducing revenue andfiscal deficits.*
6. *The agriculture sector, which employs a significant portion of the population, needs urgent attention. There will be a focus on agricultural diversification and support for sectors like horticulture and plantation.*
7. The government will promote savings and channel them into productive investment through right policies and prudent taxes.
8. Investments in infrastructure, manufacturing, and the financial sector will be crucial for sustaining high economic growth.
9. The government will prioritize poverty reduction, employment generation, healthcare, education, and empowerment of marginalized communities.
10. The government will work toward improving the efficiency and accountability of public programs and engage in a healthy debate with civil society on the delivery mechanism.

Budget Speech 2006–2007[12]Shri P. Chidambaram

1. The growth rate in 2004–2005 was 7.5%, with the manufacturing sector growing at 8.1%.
2. Gross domestic savings increased to 29.1% of GDP, and the rate of gross capital formation increased to 30.1% of GDP.
3. GDP growth is likely to be 8.1% this year, with the manufacturing sector expected to grow at 9.4%.
4. The investment rate has increased steadily from 25.3% in 2002–2003 to 30.1% in 2004–2005.
5. Several indicators point to continued buoyancy of capital formation in the economy.
6. 5083 MW of capacity will be added to power generation in 2005–2006, and during the Tenth Plan period, the total addition is estimated at 34,000 MW.

[11] Budget Speech 2005–2006—https://www.indiabudget.gov.in/doc/bspeech/bs200506.pdf

[12] Budget Speech 2006–2007—https://www.indiabudget.gov.in/doc/bspeech/bs200607.pdf

7. Contracts have been placed for projects spanning 95 districts and covering 41,461 unelectrified and 9379 electrified villages.
8. Work is on at full steam on the Golden Quadrilateral (GQ) and the North-South and East-West Corridor.
9. The government will actively promote the establishment of new towns, preferably focused on a specific industry, for example, information technology.
10. Last year, I introduced a statement highlighting the gender sensitivities of the budgetary allocations.
11. The government is committed to the welfare of Scheduled Castes (SCs) and Scheduled Tribes (STs).

Budget Speech 2008–2009[13]Shri P. Chidambaram

1. Our economic policy must aim at plenty of and equitable distribution of wealth.
2. We must produce wealth and then divide it equitably.
3. Our work in the government is a discovery of the path to reach our goals: full employment, abolition of poverty, and elimination of inequality.
4. We must increase national income to achieve our goals of inclusive growth.
5. The UPA government is committed to inclusive growth and has taken measures like the National Rural Employment Guarantee Scheme, Midday Meal Scheme, National Rural Health Mission, and Kasturba Gandhi Balika Vidyalaya Scheme.
6. The Eleventh Five Year Plan focuses on consolidation, monitoring implementation, enforcing accountability, and measuring outcomes to ensure sustainable economic growth.
7. The allocation for ongoing programs in the Eleventh Plan will receive ample funds.
8. The UPA government has doubled agricultural credit in the first 2 years and aims to reach a level of Rs. 240,000 crore by March 2008.
9. The UPA government aims to regain India's position as a major global economic power.

Budget Speech 2009–2010 (Interim)[14]Shri Pranab Mukherjee

Here are some statements related to economic sustainability mentioned in the given context:

---

[13]Budget Speech 2008–2008—https://www.indiabudget.gov.in/budget_archive/ub2008-09/bs/speecha.htm

[14]Budget Speech 2009–2010—https://www.indiabudget.gov.in/budget_archive/ub2009-10(I)/bs/speecha.htm

1. To be equitable, economic growth has to be sustainable.
2. The Eleventh Five Year Plan provides a comprehensive framework and strategy for making growth both faster and more inclusive.
3. Our government increased the plan allocation for agriculture by 300%.
4. The Rashtriya Krishi Vikas Yojana was launched to increase the growth rate of agriculture and allied sector.
5. *Our government had announced a package for doubling the flow of credit to agriculture.*
6. The tax system has been reformed to improve its efficiency and equity.
7. The rates of union excise duties and services tax have been rationalized to enable eventual shift to the goods and services tax.
8. The government has undertaken modernization of the business processes of the tax administration through extensive use of information technology.
9. The second Administrative Reforms Commission has suggested measures to achieve a proactive, responsive, accountable, sustainable, and efficient administration.
10. The enactment of the Right to Information Act has ushered in greater accountability of the public servants.
11. The government has undertaken comprehensive reforms of the tax system to improve its efficiency and equity.
12. The government has facilitated the introduction of the state level VAT.
13. The National Rural Employment Guarantee Scheme has generated employment and enhanced wage earnings.
14. Sarva Shiksha Abhiyan has provided access to and infrastructure for elementary education.
15. *The national program of midday meals in schools has contributed to enhancement of school participation and reduction in classroom hunger.*
16. The UPA government has expanded and extended the Integrated Child Development Scheme to cover more habitations.
17. The Jawaharlal Nehru National Urban Renewal Mission has focused on integrated development for urban infrastructure and services.
18. The Rajiv Gandhi Rural Drinking Water Mission aims to supply safe drinking water to uncovered habitations.
19. The Total Rural Sanitation Program is a continuous process.

Budget Speech 2012–2013[15] Shri Pranab Mukherjee

1. We have to improve our macroeconomic environment and strengthen domestic growth drivers to sustain high growth in the medium term.
2. Focus on domestic demand-driven growth recovery.

[15] Budget Speech Budget 2012–2013—https://www.indiabudget.gov.in/doc/bspeech/bs201213.pdf

3. Create conditions for rapid revival of high growth in private investment.
4. Address supply bottlenecks in agriculture, energy, and transport sectors, particularly in coal, power, national highways, railways, and civil aviation.
5. Intervene decisively to address the problem of malnutrition, especially in the 200 high-burden districts.
6. Expedite coordinated implementation of decisions being taken to improve delivery systems, governance, and transparency; and address the problem of black money and corruption in public life.
7. *If India can continue to build on its economic strength, it can be a source of stability for the world economy and provide a safe destination forrestlessglobal capital.*
8. Strengthening investment environment: it is time to fast-track policy decisions and ensure on-time implementation of major projects.
9. Reforms in the financial sector have been pursued with the objective of more efficient market intermediation between savers and investors.
10. Encourage flow of savings in financial instruments and improve the depth of domestic capital market.
11. Encourage investment in infrastructure sector.
12. Raise FII investment limit in long-term infrastructure bonds, corporate bonds, and government securities.
13. Raise the limit on external commercial borrowings (ECB).
14. Qualify foreign investors to invest in specified Indian mutual funds and directly in equities.

Budget Speech 2014–2015[16]Shri Arun Jaitley

1. The aim is to lay down a broad policy indicator of the direction in which we wish to take this country toward sustained growth of 7–8% or above within the next 3–4 years.
2. The steps announced in this budget are the beginning of a journey toward sustained growth and macroeconomic stabilization.
3. The budget aims to achieve higher growth, lower inflation, sustained level of external sector balance, and a prudent policy stance.
4. The development strategy of the government is focused on the mandate of "Sab ka Saath Sab ka Vikas" (Together with all, development for all).
5. The government is committed to creating a vibrant and strong India, addressing the growing aspirations of the people.
6. Fiscal prudence and consolidation are of paramount importance to ensure intergenerational equity and avoid leaving behind a legacy of debt for future generations.

[16]Budget Speech Budget 2014–2015—https://www.indiabudget.gov.in/doc/bspeech/bs201415.pdf

7. There is an urgent need to generate more resources to fuel the economy, improving the tax to GDP ratio and increasing nontax revenues.
8. The government aims to promote infrastructure development and manufacturing to raise adequate resources for developmental needs.
9. The budget emphasizes the importance of targeted anti-poverty programs to ensure the well-being of the poor and address their needs.
10. The government is committed to rationalizing coal linkages, promoting renewable energy, and exploiting natural resources in a sustainable manner.
11. The budget highlights the need for good infrastructure, public transport, solid waste disposal, and sewerage treatment in urban areas.
12. The government aims to promote affordable housing and increase the flow of cheaper credit for urban poor and low-income groups.
13. The budget emphasizes the importance of export promotion and engaging with states to increase India's exports.
14. The government is focused on promoting sustainable mining practices and resolving the impasse in the mining sector.
15. *The budget highlights the importance of disseminating real-time information to farmers regarding new farming techniques, water conservation, and organic farming.*

Budget Speech 2014–2015 (Interim)[17] Shri Pranab Mukherjee

1. The fiscal deficit for 2013–2014 will be contained at 4.6% of GDP.
2. The current account deficit will be contained at USD 45 billion.
3. Efforts have been made to stabilize inflation, with WPI inflation at 5.05% and core inflation at 3.0%.
4. The agriculture sector has shown stellar performance, with increased food grain production and agricultural GDP growth estimated at 4.6%.
5. The savings rate and investment rate have remained relatively stable.
6. The economy has shown signs of stability, with declining fiscal deficit, contained current account deficit, moderated inflation, and increasing quarterly growth rate.
7. The UPA government has implemented measures to unblock projects and promote economic growth.
8. The UPA government's record of growth in various sectors such as food grains, power capacity, coal production, rural roads, education, and health expenditure is unparalleled.
9. The UPA government has delivered above the trend growth rate for the Indian economy.

[17] Budget Speech Budget 2014–2015—https://www.indiabudget.gov.in/doc/bspeech/bs.pdf

Budget Speech 2016–2017[18] *Shri Arun Jaitley*

1. We have to bring more transparency and efficiency in government procurement of goods and services.
2. To remove the difficulties and impediments to ease of doing business, we will introduce a bill to amend the Companies Act, 2013.
3. Monitoring of prices of essential commodities is a key element of good governance.
4. We will do this through mutual agreement with participating states and districts.
5. Our external situation is robust. The current account deficit has declined. Our foreign exchange reserves are at the highest ever level.
6. Our initiatives in the last 21 months have not only placed the economy on a faster growth trajectory but have bridged the trust deficit.
7. We believe in the principle that money with the government belongs to the people, and we have the sacred responsibility to spend it prudently and wisely.
8. We must strengthen our firewalls against these risks by ensuring macroeconomic stability and prudent fiscal management.
9. We see these challenges as opportunities.
10. We wish to enhance expenditure in the farm and rural sector, the social sector, and the infrastructure sector and provide for healthcare, education, and job creation.
11. A new credit rating system for infrastructure projects will be developed.
12. The duty drawback scheme has been widened and deepened to include more products and countries.
13. One hundred percent FDI will be allowed through FIPB route in marketing of food products produced and manufactured in India.
14. A new policy for management of government investment in public sector enterprises, including disinvestment and strategic sale, has been approved.
15. We will adopt a comprehensive approach for efficient management of government investment in CPSEs.

Budget Speech 2020–2021[19] *Smt. Nirmala Sitharaman*

1. Containment of inflation: the government has implemented measures to control inflation, ensuring price stability in the economy.
2. Cleaning up of accumulated loans: banks have undergone a thorough cleaning up of past decade's accumulated loans, improving their financial health.

[18] Budget Speech Budget 2016–2017—https://www.indiabudget.gov.in/doc/bspeech/bs201617.pdf
[19] Budget Speech 2020–2021—https://www.indiabudget.gov.in/doc/bspeech/bs202021.pdf

3. Recapitalization of banks: to strengthen the banking sector, recapitalization measures have been taken to infuse capital into banks, enabling them to lend more and support economic growth.
4. Formalization of the economy: steps have been taken to promote formalization of the economy, reducing the size of the informal sector and increasing tax compliance.
5. Implementation of goods and services tax (GST): the GST, a historic structural reform, has been implemented to streamline indirect taxation and create a unified tax system in the country.
6. Efficiency gains in logistic and transport sectors: the implementation of GST has resulted in efficiency gains in the logistic and transport sectors, reducing turnaround time for trucks and eliminating inspector raj.
7. Reduction in tax incidence: several rate reductions have been made under GST, resulting in a reduction in overall tax incidence and providing significant benefits to consumers.
8. Enhanced threshold and composition limits for MSMEs: the MSME sector has been provided with enhanced threshold and composition limits, reducing their tax burden and promoting their growth.
9. Macroeconomic stability measures in budget estimates: the government has estimated a fiscal deficit of 3.8% for the financial year 2019–2020 and 3.5% for the financial year 2020–2021, showcasing its commitment to maintaining macroeconomic stability.

Budget Speech 2022–2023[20] *Smt. Nirmala Sitharaman*

- Supportive policies, light-touch regulations, facilitative actions to build domestic capacities, and promotion of research and development will guide the government's approach.
- The low carbon development strategy as enunciated in the "panchamrit" that he announced is an important reflection of our government's strong commitment toward sustainable development.
- The circular economy transition is expected to help in productivity enhancement as well as creating large opportunities for new businesses and jobs.
- Five to 7% biomass pellets will be co-fired in thermal power plants resulting in CO2 savings of 38 MMT annually.
- Energy efficiency and savings measures will be promoted, particularly in large commercial buildings through the Energy Service Company (ESCO) business model.
- The policies and required legislative changes to promote agroforestry and private forestry will be brought in.

[20] Budget Speech 2022–2023—https://www.indiabudget.gov.in/doc/bspeech/bs202223.pdf

- The outlay for capital expenditure in the Union Budget is being stepped up sharply by 35.4% to support speedy and sustained economic revival and consolidation.
- Promoting digital economy and fintech, technology-enabled development, energy transition, and climate action.
- The initiatives of the last year's budget have seen significant progress and have been provided with adequate allocations in this budget as well.

The Strengthening of Health Infrastructure, Speedy Implementation of the Vaccination Program, and the Nationwide Resilient Response to the Current Wave of the Pandemic

Budget Speech 2023–2024[21] *Smt. Nirmala Sitharaman*

- Sustainable cities of tomorrow: encouraging urban planning reforms and actions to transform cities into sustainable cities of tomorrow
- Making cities ready for municipal bonds: incentivizing cities to improve their credit worthiness for municipal bonds through property tax governance reforms and ring-fencing user charges on urban infrastructure
- Urban infrastructure development fund: establishing an urban infrastructure development fund (UIDF) to create urban infrastructure in Tier 2 and Tier 3 cities, leveraging resources from the grants of the 15th Finance Commission
- Green credit program: notifying a green credit program under the Environment (Protection) Act to incentivize environmentally sustainable and responsive actions by companies, individuals, and local bodies
- GOBARdhan scheme: establishing 500 new "waste-to-wealth" plants under the GOBARdhan scheme for promoting circular economy and introducing a 5% compressed biogas (CBG) mandate
- Green growth: implementing programs and policies for green fuel, green energy, green farming, green mobility, green buildings, and green equipment, to reduce carbon intensity and provide green job opportunities
- Digital public infrastructure for agriculture: building digital public infrastructure for agriculture as an open-source, open standard, and interoperable public good

[21] Budget Speech 2023–2024—https://www.indiabudget.gov.in/

Appendix B: Social Sustainability Statements

Budget Speech 2020–2021[22] *Smt. Nirmala Sitharaman*

1. Our government is committed to the goal of doubling farmers' incomes by 2022, ensuring prosperity for farmers and making farming competitive.
2. We have provided energy sovereignty through the KUSUM scheme and input sovereignty through the Paramparagat Krishi Vikas Yojana, promoting sustainable agricultural practices.
3. We have provided resilience for 6.11 crore farmers insured under the PM Fasal Bima Yojana, protecting them from crop losses and ensuring their financial stability.
4. Focus on the cultivation of pulses and expansion of micro-irrigation through the Krishi Sinchai Yojana has raised the self-reliance of the country and promoted sustainable agricultural practices.
5. We have implemented the Model Agricultural Land Leasing Act, 2016, and the Model Agricultural Produce and Livestock Marketing (Promotion and Facilitation) Act, 2017, to liberalize farm and livestock markets and remove distortions.
6. We are promoting the use of renewable energy in agriculture through the PM-KUSUM scheme, which removes farmers' dependence on diesel and kerosene and links pump sets to solar energy.
7. We are encouraging the balanced use of all kinds of fertilizers, including traditional organic and other innovative fertilizers, to reduce the excessive use of chemical fertilizers and promote sustainable farming practices.
8. We are implementing the Village Storage Scheme, run by self-help groups, to provide farmers with good holding capacity and reduce their logistics costs, promoting sustainable agriculture practices.
9. We are promoting the development of a seamless national cold supply chain for perishables, including milk, meat, and fish, through the setting up of "Kisan Rail" and the inclusion of refrigerated coaches in express and freight trains.
10. We are expanding integrated farming systems in rainfed areas, promoting multitier cropping, beekeeping, solar pumps, solar energy production in non-cropping seasons, and zero-budget natural farming.
11. We are promoting the use of negotiable warehousing receipts (e-NWR) integrated with e-NAM to facilitate financing in the agriculture sector and provide farmers with better access to credit.
12. We are taking steps to eliminate diseases in livestock, such as foot and mouth disease, brucellosis, and peste des petits ruminants (PPR), by 2025, ensuring the health and well-being of animals and improving the livelihoods of farmers.
13. We are promoting the development, management, and conservation of marine fishery resources, ensuring the sustainability of our blue economy.

[22] Budget Speech 2020–2021—https://www.indiabudget.gov.in/doc/bspeech/bs202021.pdf

14. We are providing viability gap funding for the setting up of efficient warehouses at the block/taluk level, promoting the efficient storage and management of agricultural produce.
15. We are working toward water stress management in 100 water-stressed districts, ensuring the sustainable use and conservation of water resources.
16. We are implementing various schemes and programs to improve the well-being and livelihoods of farmers, promoting social sustainability in rural areas.

Budget Speech 1991–1992[23]Shri Manmohan Singh

Based on the given context

1. Our economy would return to a path of high sustained growth with reasonable price stability and greater social equity.
2. There has been inadequate emphasis on reduction of costs, upgradation of technology, and improvement of living standards.
3. This task has become imperative as we prepare ourselves to be an internationally competitive economy.
4. The main objective of the National Renewal Fund will be to ensure that the cost of technical change and modernization does not devolve on the workers.
5. The fund will not merely provide ameliorative measures for the workers affected in the course of technical change but, more importantly, provide retraining to them.
6. The basic challenge of our times is to ensure that wealth creation is not only tempered by equity and justice but is harnessed to the goal of removal of poverty and development for all.
7. All wealth is a social product, and those who create and own it have to hold it as a trust and use it in the interest of society, particularly the underprivileged and without means.
8. In highlighting the significance of reform, my purpose is not to give a fillip to mindless and heartless consumerism we have borrowed from the affluent societies of the West.
9. Our approach to development has to combine efficiency with austerity, holding our society together in pursuit of the noble goal of banishing poverty, hunger, and disease.
10. The allocation for social welfare programs, such as the welfare of backward classes, rehabilitation and retraining of scavengers, and education, is being increased to ensure inclusive development.

[23]Budget Speech 1992–1993—https://www.indiabudget.gov.in/doc/bspeech/bs199293.pdf

Budget Speech 1993–1994[24]*Shri Manmohan Singh*

1. The policy initiatives we have taken do not in any way reduce our commitment to take care of the poor and the disadvantaged.
2. We have taken steps to minimize the burden of adjustment on the poor and working classes.
3. The priorities for economic policy, at this critical stage of our economic restructuring, are very clear: rural development and the vital social services such as education and health.
4. The government is firmly committed to ensuring that agriculture, agro-processing, and rural development are given top priority in the design of economic policy.

Budget Speech 1994–1995[25]*Shri Manmohan Singh*

Based on the provided context, there are several banking policies and measures that have been implemented to support poor and marginalized communities. These include the following:

1. Reduction in lending rates: the government and financial institutions have reduced the minimum lending rates to stimulate investment in the economy. This helps to make credit more accessible and affordable for poor and marginalized communities.
2. Restructuring of regional rural banks (RRBs): the weak condition of RRBs has been a major concern in rural credit. Comprehensive restructuring measures, including cleaning up their balance sheets and infusion of fresh capital, are being undertaken for 50 out of 196 RRBs to make them more viable and effective instruments of rural banking.
3. Strengthening cooperative credit structure: measures are being taken to strengthen the cooperative credit structure, which has played a significant role in rural development. This includes initiating steps to revamp the cooperative system and improve its viability, along with increasing budget provisions for rural development.
4. Enhanced rural credit system: efforts are being made to strengthen the overall rural credit system. This includes improving the recovery of banks' dues, setting up the Board for Financial Supervision to supervise banks and financial institutions, and implementing measures to encourage greater discipline among borrowers.
5. Modernization of capital markets: the government is committed to modernizing the capital markets to ensure transparency, speed of settlements, and adequate

[24] Budget Speech 1993–1994—https://www.indiabudget.gov.in/doc/bspeech/bs199394.pdf

[25] Budget Speech 1994–1995—https://www.indiabudget.gov.in/doc/bspeech/bs199394.pdf

investor protection. This includes establishing a model national stock exchange with screen-based trading and introducing a depository system for scripless trading.

It is important to note that the provided context does not cover all banking policies for poor and marginalized communities. There may be additional policies and measures that have been implemented or are being considered.

Budget Speech 1999–2000[26] *Shri Manmohan Singh*

Social sustainability and Banking policies:

1. The provision of credit for agriculture and rural development is a priority sector lending for banks. This ensures that farmers and rural communities have access to financial services and credit for agricultural activities.
2. The Rural Infrastructure Development Fund (RIDF) provides financing for rural infrastructure projects of the state governments. This includes funding for village-level infrastructure projects implemented by Gram Panchayats and self-help groups.
3. The Kisan Credit Card Scheme provides timely and cost-effective credit to farmers. This scheme aims to simplify the credit delivery system and provide flexible credit options to farmers.
4. The National Program for Rural Industrialization (NPRI) promotes the development of microenterprises, especially in rural areas. It focuses on providing credit and support to self-help groups and NGOs for the promotion of microenterprises.
5. The National Movement of Watershed Development aims to foster community infrastructure and income generation opportunities for the landless and poor, particularly those belonging to Scheduled Castes, Scheduled Tribes, and Other Backward Classes.
6. The Samagra Awas Yojana is a comprehensive housing scheme that aims to provide shelter, sanitation, and drinking water to the poor and marginalized communities. This scheme focuses on upgrading kutcha dwellings and providing shelter to all shelterless households.
7. The Swaran Jayanti Gram Swa-Rozgar Yojana aims to merge and streamline self-employment programs for the rural poor. This initiative focuses on greater participation of Gram Panchayats and flexibility in execution to meet the needs of local people.
8. The National Human Development Initiative empowers the weakest sections of the population and aims to minimize rural-urban disparities. The Gram

[26]Budget Speech 1999–2000—https://www.indiabudget.gov.in/doc/bspeech/bs199394.pdf

Panchayat plays a pivotal role in implementing various components of this initiative.

9. The priority sector lending includes lending to nonbanking finance companies or other financial intermediaries for on-lending to the tiny sector. This ensures increased outreach of banks to the tiny sector, providing them with access to credit.
10. The National Housing Bank implements schemes like the Golden Jubilee Rural Housing Finance Scheme, which provides financing for housing development in rural areas and targets the housing needs of the marginalized communities.

Budget Speech 2000–2001[27] *Shri Yashwant Sinha*

1. Our government stands fully committed to ensure that the fruits of economic reforms are shared by all sections of society, especially those living in rural areas and more particularly the Scheduled Castes, Scheduled Tribes, and Other Backward Classes.
2. Five elements of social and economic infrastructure are critical to the quality of life especially in rural areas: health, education, drinking water, housing, and roads.
3. *Even after 52 years of independence, the provision of basic services inrural areasremains very unsatisfactory. These large gaps in basic services inrural areasare not acceptable, and the government is committed to removing them rapidly.*
4. Universalization of elementary education is one of our key objectives. A new Department of Elementary Education and Literacy has already been created under the Ministry of Human Resources Development to give a new thrust and focus to these efforts.
5. The plan allocation for elementary education has been increased to ensure better access to education.
6. *A new Department of Drinking Water Supply in the Ministry of Rural Development has been set up to intensify the efforts and accelerate the pace of coverage. Our objective is to provide drinking water facilities in all rural habitations in the next 5 years.*
7. The Reproductive and Child Health program will receive increased funding to improve healthcare in rural areas.
8. Various housing schemes are being implemented to provide affordable housing to the rural population.
9. A social security scheme, "Janashree Bima Yojana," has been introduced to provide insurance cover to the poorest sections of society.

[27] Budget Speech 2000–2001—https://www.indiabudget.gov.in/doc/bspeech/bs200001.pdf

10. A task force will be set up to review existing legislation and government schemes pertaining to the role of women in the national economy, with the aim of empowering women.
11. Efforts will be made to improve access to national resources for women and ensure their rightful place in the mainstream of economic development.
12. Measures will be taken to preserve and nurture forests and the environment for the benefit of the weakest sections of society.
13. Policies and schemes will be implemented to support and promote the small-scale industry sector, which plays a vital role in industrial production, employment generation, and exports.

Budget Speech 2001–2002[28]Shri Yashwant Sinha

1. Recognizing the need for increasing investments in social sectors, the plan allocation for the Ministry of Health and Family Welfare has been stepped up from Rs 4920 crore to Rs 5780 crore. This includes an allocation of Rs 180 crore for HIV/Aids Control Program.
2. Our small-scale entrepreneurs have proved their competitiveness in providing over 35% of national exports. To enable further new investment and technology upgradation in some of the key export-oriented sectors, it is now proposed to deserve another 14 items related to leather goods, shoes, and toys.
3. The government has recently announced a New Textile Policy aimed at preparing the industry for the new challenges of global competition. A scheme for setting up integrated apparel parks is being initiated. This will enable the deserved readymade garment industry to set up modern units with the best infrastructure.
4. The government has set up a separate National Scheduled Tribes Finance and Development Corporation with an authorized share capital of Rs 500 crore to improve the welfare of scheduled tribes. The allocation for the schemes for welfare of scheduled tribes in the Ministry of Tribal Affairs has been enhanced from Rs 786 crore to Rs 986 crore in the coming year.
5. The government has launched the "Janashree Bima Yojana" to extend social security cover to the poorest sections of society. A special scheme for landless agricultural laborers, the Khetihar Mazdoor Bima Yojana, will provide benefits of insurance cover and a pension to beneficiaries. Additionally, a Shiksha Sahyog Yojana will provide an education allowance to children of parents living below the poverty line.
6. The government proposes to give an enlarged role to the state governments in both procurement and distribution of food grains for the public distribution system (PDS) in their respective states. Financial assistance will be provided

[28] Budget Speech 2001–2002—https://www.indiabudget.gov.in/doc/bspeech/bs200102.pdf

to enable them to procure and distribute food grains to BPL families at subsidized rates.
7. Policy initiatives have been implemented to increase private sector participation in ports. Capacity in Indian major ports is expected to increase, along with substantial capacity addition in minor ports. The formation of the Tariff Authority for Major Ports ensures transparent and fair tariff levels.
8. Progress has been made in pursuing reforms in the financial sector and capital markets. Measures will be taken to develop and deepen the debt market, which will benefit small investors and institutional investors alike. A clearing corporation will be set up under the active encouragement of the RBI, and electronic systems will be implemented to facilitate transparent trading and settlement of government securities.
9. The government is implementing schemes to provide credit support, improve the quality of power supply, and augment resources for rural electrification, benefiting dalit bastis, households of scheduled tribes, and other weaker sections of society.
10. The government is launching schemes for women's empowerment, providing microcredit to poor assetless women, and assisting women in difficult circumstances.

Budget Speech 2003–2004[29] *Shri Jaswant Singh*

1. Eradication of poverty is crucial, as it is both a moral and economic issue of our times.
2. Renewed progress on the front of agriculture, as it is the nation's lifeblood.
3. Development of physical and social infrastructure to ensure rapid and efficient growth.
4. Enhancing growth, improving incomes, generating employment, and promoting exports through specific manufacturing sectors.
5. Fiscal consolidation to sustain growth and eliminate budgetary drags.
6. Addressing reduction of social and economic inequalities by implementing critical reforms.
7. Releasing national creativity and promoting economic and social progress.
8. Diversification and value chain development in agriculture to respond to second-generation issues.
9. Promoting a vibrant small-scale industry to contribute to industrial and export growth and provide sustained income and employment.
10. Accessing global markets with quality consumer goods produced under flexible conditions.

[29] Budget Speech 2003–2004—https://www.indiabudget.gov.in/doc/bspeech/bs200304.pdf

11. Promoting India as a production center and investment destination through the India Development Initiative.
12. Accelerating disinvestment to unlock the productive potential of undertakings and reorient the government toward governance.
13. Increasing the cap on foreign direct investment in banking companies to facilitate the setting up of subsidiaries and attract investment in private banks.

Budget Speech 2004–2005[30]Shri P. Chidambaram

1. Government is committed to providing insurance cover to farming and livestock, ensuring the sustainability of the agricultural sector.
2. Government is dedicated to removing the inadequacies in infrastructure facilities through a mix of policy and fiscal measures, promoting sustainable growth.
3. The Rajiv Gandhi Drinking Water Mission aims to bring all drinking water schemes under its umbrella, ensuring access to adequate drinking water for all.
4. The Accelerated Rural Water Supply Program focuses on renewing water sources and serving uncovered and partially covered habitations, promoting sustainable water supply in rural areas.
5. The city of Chennai and other cities will have desalination plants installed, addressing the acute scarcity of drinking water and promoting sustainable water management.
6. The Fund for the Regeneration of Traditional Industries aims to support and revive traditional industries, promoting their growth, exports, and cultural heritage.
7. The implementation of value added tax (VAT) will create a modern and efficient trade tax system, promoting sustainable revenue generation.
8. The introduction of a "defined contribution" pension scheme for central government employees ensures a sustainable pension system.
9. The National Employment Guarantee Act aims to guarantee 100 days of employment in a year to one able-bodied person in every poor household, addressing unemployment and poverty.
10. Special attention will be given to the welfare, especially education, of minorities, promoting their social sustainability.
11. Microfinance initiatives, such as the self-help group (SHG)-bank linkage program, aim to provide cost-effective banking services to the poor, promoting their financial sustainability.
12. The government follows a 5-year roadmap to achieve rapid growth with stability and equity, ensuring sustainable development.

[30]Budget Speech 2004–2004—https://www.indiabudget.gov.in/doc/bspeech/bs200405.pdf

Budget Speech 2005–2006[31] *Shri P. Chidambaram*

1. We must increase our revenues and reorient expenditure to pay for more outlays on education, health, and infrastructure.
2. The Prime Minister has repeatedly emphasized the need to improve the quality of implementation and enhance the efficiency and accountability of the delivery mechanism.
3. Civil society should also engage the government in a healthy debate on the efficiency of the delivery mechanism.
4. Subsidies provide a measure of protection for the poor and we shall continue to provide subsidies.
5. The Ministry of Agriculture intends to make procurement of food grains more cost-effective through decentralized procurement.
6. *The National Horticulture Mission will ensure an end-to-end approach having backward and forward linkages covering research, production, postharvest management, processing, and marketing.*
7. Government will, therefore, through a mix of right policies and prudent taxes, promote savings and devise ways and means to channel these savings into productive investment.
8. Skills development, especially for the youth who have only minimal formal education, is an area that can no longer be ignored.
9. We shall build on the growing external strengths of the economy and further liberalize trade policy.
10. The incipient investment boom in infrastructure, industry, and services needs to be nurtured through further reforms in the financial sector

Budget Speech 2006–2007[32] *Shri P. Chidambaram*

1. Old age pensions are granted under the National Social Assistance Programme (NSAP) to destitute persons above the age of 65 years at Rs. 75 per month. The pension amount will be increased to Rs. 200 per month.
2. The government will actively promote the establishment of new towns focused on specific industries or themes, such as information technology, education, or health. Projects are being planned in West Bengal and Karnataka.
3. Gender budgeting is being implemented, with schemes where 100% of the allocation is for the benefit of women or where at least 30% of the allocation is targeted toward women. This covers 24 demands for grants in various ministries and departments.

[31] Budget Speech 2005–2006—https://www.indiabudget.gov.in/doc/bspeech/bs200506.pdf

[32] Budget Speech 2006–2007—https://www.indiabudget.gov.in/doc/bspeech/bs200607.pdf

4. The welfare and development of Scheduled Castes (SCs) and Scheduled Tribes (STs) are prioritized. Allocations for schemes benefiting only SCs and STs have been enhanced by 14.5%, and schemes with at least 20% allocation for SCs and STs have been enhanced by 13.9%.
5. The government is committed to safeguarding and preserving oral and intangible heritage, such as Kuttiyattam, Vedic Chanting, and Ramlila. An initial provision of Rs. 5 crore has been made for this purpose.
6. The government is working toward evolving a consensus on the issue of subsidies, with a focus on targeting subsidies at the poor and the truly needy.
7. The government is implementing a strategy of enhanced revenue mobilization through better tax administration, which is yielding tangible results.
8. *Gross domestic savings and gross capital formation have increased, indicating positive economic growth and development.*

Please note that these statements are based on the provided context and may not encompass all social sustainability statements in the budget speech.

Budget Speech 2007–2008[33] Shri P. Chidambaram

From the given

1. The health insurance scheme has so far covered 300,000 weavers and will be extended to more weavers (Handlooms).
2. Government proposes to take up an additional 100–150 clusters in 2007–2008 (Handlooms).
3. Following the credit policy for small and medium enterprises (SMEs) announced in August 2005, outstanding credit to the SME sector increased from Rs. 135,200 crore at end of December 2005 to Rs. 173,460 crore at end of December 2006 (small and medium enterprises).
4. The coir industry provides employment to a large number as well as earns valuable foreign exchange. I am happy to announce a scheme for the modernization and technology upgradation of the coir industry with special emphasis on major coir producing states (coir industry).
5. In allocating resources, school education must have primacy. Hence, I propose to increase the allocation for school education by about 35% (Sarva Shiksha Abhiyan and Midday Meal Scheme).
6. Government will reward the employer once the physically challenged employee is regularized and is enrolled under the Employees Provident Fund (EPF) and the Employees State Insurance (ESI) (employment for the physically challenged).

[33] Budget Speech 2007–2008—https://www.indiabudget.gov.in/doc/bspeech/bs200708.pdf

Please note that these statements are just a selection from the provided context and may not cover all aspects of social sustainability.

Budget Speech 2008–2009[34]*Shri P. Chidambaram*

1. The National Rural Employment Guarantee Scheme has proven to be a historic measure of empowerment of Scheduled Castes and Scheduled Tribes and, especially, of women.
2. *The Midday Meal Scheme is the largest school lunch program in the world covering 11.4 crore children.*
3. The National Rural Health Mission has taken improved healthcare to rural India by strengthening the primary health centers, of which 8756 have been made 24 × 7.
4. The Kasturba Gandhi Balika Vidyalaya Scheme has enrolled 182,000 girls in residential schools, thus helping to bridge the gender gap in education.
5. Bharat Nirman has made impressive progress in providing drinking water, all-weather roads, telephones, and electricity to rural areas.
6. The allocation for the Central Plan has increased by 16% over the previous year, ensuring ample funds for ongoing programs.
7. The government has allocated funds for the development of information technology and information technology-enabled services, including schemes for establishing broadband Internet-enabled common service centers in rural areas and statewide area networks.
8. The Ministry of Textiles has continued schemes for the development of integrated textile parks and technology upgradation in the textile sector.
9. Efforts have been made to support micro-, small, and medium enterprises, including a secular rise in the number of registered units, production, employment, and exports.
10. Special grants have been provided to institutions of excellence in agriculture, education, and research.
11. Funds have been allocated to the Indian Council of Cultural Relations to promote India's soft power through music, literature, dance, art, cuisine, and films.
12. A one-time grant has been provided to the National Tiger Conservation Authority to protect and conserve tigers.
13. The government is implementing a Central Plan Scheme Monitoring System to effectively monitor the implementation and outcomes of various schemes and strengthen evaluation through independent evaluations conducted by research institutions.

[34] Budget Speech 2008–2008—https://www.indiabudget.gov.in/budget_archive/ub2008-09/bs/speecha.htm

Budget Speech 2009–2010 (Interim)[35] *Shri Pranab Mukherjee*

- To be equitable, economic growth has to be sustainable.
- To be sustainable, economic growth has in turn to be all inclusive.
- In pursuant of that vision, the UPA government in the National Common Minimum Programme had declared its intention to make growth more inclusive.
- The Eleventh Five Year Plan provides a comprehensive framework and strategy for making growth both faster and more inclusive.
- Focused attention has been given by our government to the agriculture sector.
- Employment generation schemes have to be expanded, and social security nets have to be strengthened to protect the vulnerable sections of our society.
- National Rural Employment Guarantee Scheme has resulted in increased wage employment, enhanced wage earnings, and improved equity with significant benefits flowing to SC/ST and women.
- Sarva Shiksha Abhiyan has made significant contribution in providing access to and infrastructure for elementary education.
- The national program of midday meals in schools has contributed to enhancement of school participation, reduction in classroom hunger, and fostering of social and gender parity.
- In our commitment to reduce the malnutrition levels in the country, the UPA government has recently adopted the New WHO Child Growth Standards for monitoring growth of children under ICDS.
- Jawaharlal Nehru National Urban Renewal Mission was launched to give focused attention to integrated development for urban infrastructure and services in mission mode, in identified cities.
- Rajiv Gandhi Rural Drinking Water Mission is envisaged to supply safe drinking water to uncovered habitations and slipped back habitations.
- Efforts of our government and the financing institutions have led to a rapid growth of credit-linked women self-help groups.
- The Rashtriya Mahila Kosh will be strengthened by enhancing its authorized capital.
- "Priyadarshini Project," which is a rural women's empowerment and livelihood program, was launched with the assistance of IFAD.
- A revised and modified scheme named "Indira Gandhi National Old Age Pension Scheme" was launched.
- Two new schemes—Indira Gandhi National Widow Pension Scheme and Indira Gandhi National Disability Pension Scheme—are being launched in the current year.
- To enable young widows to become self-sufficient, they will be given priority in admissions to I, WoITIs, and National/Regional institutes.

[35] Budget Speech 2009–2010—https://www.indiabudget.gov.in/budget_archive/ub2009-10(I)/bs/speecha.htm

Budget Speech 2012–2013[36]*Shri Pranab Mukherjee*

- Address supply bottlenecks in agriculture, energy, and transport sectors, particularly in coal, power, national highways, railways, and civil aviation.
- Intervene decisively to address the problem of malnutrition especially in the 200 high-burden districts.
- Expedite coordinated implementation of decisions being taken to improve delivery systems, governance, and transparency; and address the problem of black money and corruption in public life.
- The government has further evolved its approach to divestment of central public sector enterprises (CPSEs). The CPSEs are being given a level playinlis-a-vis the private sector with regard to practices like buybacks and listing at stock exchange.
- Our government has taken definite steps to create food security at the household level by making food a legal entitlement for all targeted people, especially for the poor and vulnerable segments of our population.
- Following the decision taken in the Prime Minister's National Council on India's nutritional challenges, a multi-sectoral program to address maternal and child malnutrition in selected 200 high-burden districts is being rolled out during 2012–2013.
- Integrated Child Development Services (ICDS) scheme is being strengthened and restructured.
- National program of midday meals in schools has enhanced enrollment, retention, and attendance and also helped in improving nutrition levels among children.
- Rajiv Gandhi Scheme for Empowerment of Adolescent Girls, SABLA, has been introduced last year with a view to address the nutritional needs and other educational and skills development initiatives for self-development of adolescent girls in the age group of 11 to 18 years.

Budget Speech 2014–2015 (Interim)[37]*Shri Pranab Mukherjee*

1. The Ministry of MSME will create the "India Inclusive Innovation Fund" to promote grassroots innovations with social returns to support enterprises in the MSME sector.
2. In order to promote entrepreneurship among the Scheduled Castes and to provide concessional finance to them, IFCI will set up a venture capital fund for Scheduled Castes.
3. The restructured ICDS has been implemented in 400 districts and will be rolled out in the remaining districts from January 4, 2014.

[36] Budget Speech Budget 2012–2013—https://www.indiabudget.gov.in/doc/bspeech/bs201213.pdf
[37] Budget Speech Budget 2014–2015—https://www.indiabudget.gov.in/doc/bspeech/bs.pdf

4. Government has approved the National Agro-Forestry Policy 2014 which has multiple objectives including employment, productivity, conservation, and adaptation.
5. A mechanism for marketing minor forest produce has been introduced, and the budget has allocated 444.59 crore to continue the scheme in 2014–2015.
6. A new plan scheme with an allocation of 100 crore has been approved to promote community radio stations.
7. The Ministry of Health and Family Welfare has delivered new technologies to the people: the JE vaccine, a diagnostic test for thalassemia, and a Magnivisualizer for detection of cervical cancer.

Budget Speech 2014–2015[38]Shri Arun Jaitley

1. The government is committed to providing safe drinking water in rural areas affected by contaminants such as fluoride, arsenic, and pesticides.
2. The government aims to provide 24 × 7 uninterrupted power supply to all homes in rural areas.
3. The government plans to launch the Shyama Prasad Mukherji Rurban Mission to deliver integrated infrastructure and economic development in rural areas.
4. The government proposes to allocate funds for the welfare of Scheduled Castes and Scheduled Tribes.
5. The government intends to extend tax incentives on home loans to encourage people, especially the young, to own houses.
6. The government plans to allocate funds for the National Livelihood Mission to eliminate rural poverty through sustainable livelihood options.
7. The government aims to support the development of low-cost affordable housing for the urban poor.
8. The government proposes to include slum development in the list of corporate social responsibility (CSR) activities.
9. The government plans to launch the Deendayal Upadhyaya Gram Jyoti Yojana to augment power supply to rural areas and strengthen distribution systems.
10. The government intends to allocate funds for the development of rural housing through the National Housing Bank.

Budget Speech 2016–2017[39]Shri Arun Jaitley

1. No amount of politics would be of any avail until the masses in India are well educated, well fed, and well cared for.

---

[38] Budget Speech Budget 2014–2015—https://www.indiabudget.gov.in/doc/bspeech/bs201415.pdf

[39] Budget Speech Budget 2016–2017—https://www.indiabudget.gov.in/doc/bspeech/bs201617.pdf

2. The government is also launching a new initiative to ensure that the BPL families are provided with a cooking gas connection, supported by a government subsidy. This will significantly improve the health of women and those BPL families who suffer adversely from the ill effects of Chulha cooking.
3. The Pradhan Mantri Fasal Bima Yojana has already been announced to protect the farmer from the adverse consequences of nature. The farmer will pay a nominal amount of insurance premium and get the highest ever compensation in the event of any loss suffered.
4. We have decided to embark upon a massive mission to provide LPG connection in the name of women members of poor households. This will benefit about 1 crore 50 lakh households below the poverty line in 2016–2017.
5. The National Land Record Modernization Programme has been revamped under the Digital India Initiative and will be implemented as a central sector scheme with effect from April 1, 2016. The revamped program will build an integrated land information management system.
6. It is proposed to launch a new restructured scheme, namely, Rashtriya Gram Swaraj Abhiyan, for which 655 crore is being set apart in 2016–2017.
7. The government shall also endeavor to continue with the ongoing reform program and ensure the passage of the constitutional amendments to enable the implementation of the goods and services tax and the passage of insolvency and bankruptcy law and other important reform measures which are pending before the Parliament.
8. The government will undertake three major schemes to help the weaker sections of society: Pradhan Mantri Fasal Bima Yojana, a health insurance scheme, and a cooking gas connection scheme for BPL families.
9. Our agenda for the next year is, therefore, to "Transform India" in this direction.
10. Since its launch, the National Skills Development Mission has created an elaborate skilling ecosystem and imparted training to 76 lakh youth. We want to bring entrepreneurship to the doorsteps of the youth through Pradhan Mantri Kaushal Vikas Yojana (PMKVY).

Budget Speech 2022–2023[40]Shri Pranab Mukherjee

No social sustainability measures.

[40]Budget Speech 2022–2023—https://www.indiabudget.gov.in/budget_archive/ub2009-10(I)/bs/speecha.htm

Budget Speech 2022–2023[41] Shri Pranab Mukherjee

No social sustainability measures.

Budget Speech 2023–2023[42] Shri Pranab Mukherjee

Based on the given context, here are some statements related to social sustainability:

- "Making cities ready for municipal bonds"—this statement emphasizes the importance of improving urban infrastructure and governance to enhance the creditworthiness of cities, which can lead to sustainable development.
- "Urban sanitation"—this statement highlights the focus on achieving 100% mechanical desludging of septic tanks and sewers, along with the scientific management of waste. It aims to improve public health and hygiene in cities.
- "Inclusive development"—this priority focuses on the government's philosophy of inclusive growth, covering various sections of society, including farmers, women, youth, OBCs, Scheduled Castes, Scheduled Tribes, divyangjan, and economically weaker sections.
- "Digital public infrastructure for agriculture"—this statement highlights the plan to build digital public infrastructure for agriculture, which will provide open-source and interoperable tools to farmers. It aims to promote technology adoption and improve agricultural practices.
- *"PM-PRANAM"—this program aims to incentivize states and union territories to promote alternativefertilizersand balanced use of chemicalfertilizers, contributing to sustainable agricultural practices.*
- "GOBARdhan scheme"—this scheme aims to establish waste-to-wealth plants, including compressed biogas plants, to promote the circular economy. It focuses on the efficient utilization of organic waste resources.

These statements reflect the government's commitment to social sustainability by addressing issues related to urban development, sanitation, inclusive growth, agricultural practices, and waste management.

[41] Budget Speech 2023–2024—https://www.indiabudget.gov.in/budget_archive/ub2009-10(I)/bs/speecha.htm

[42] Budget Speech 2023–2024—https://www.indiabudget.gov.in/budget_archive/ub2009-10(I)/bs/speecha.htm

Appendix C: Poverty Reduction Strategies

Budget Speech 1991–1992[43] *Shri Manmohan Singh*

Based on the given context, here are some poverty reduction strategies mentioned:

5. Encouraging wealth creation and accumulation of capital
6. Implementing a regime of austerity to remove stumbling blocks for wealth creation
7. Developing a new attitude toward wealth, considering it as a social product and using it for the benefit of society, especially the underprivileged
8. Practicing and promoting the philosophy of trusteeship, where wealth owners hold it as a trust and use it in the interest of society
9. Combining efficiency with austerity in the approach to development
10. Addressing fiscal adjustments and macroeconomic stabilizations to improve the resources position
11. Ensuring prompt payment of dues owed by State Electricity Boards to national entities
12. Improving fiscal performance and streamlining the working of public enterprises at the state level
13. Minimizing the burden of adjustment on the poor during the period of transition
14. Providing enhanced outlays in the social sectors for poverty eradication and employment creation
15. Stepping up the plan outlay for the Ministry of Rural Development
16. Reducing fiscal deficit, revenue deficit, and current account deficit
17. Increasing efficiency and international competitiveness of industrial production
18. Utilizing foreign investment and technology to increase productivity of investment
19. Modernizing the financial sector and improving the performance of the public sector
20. Accelerating the program for the conversion of dry latrines and increasing allocation for the rehabilitation and retraining of scavengers.
21. Stepping up the allocation for the programs of the Ministry of Welfare, Department of Women and Child Development, and Health and Family Welfare
22. Investing in education, particularly elementary education, to develop human resources and provide quality education to all, especially children from marginalized groups

Please note that these strategies are specific to the given context and may not encompass all poverty reduction strategies in general.

---

[43] Budget Speech 1991–1992—https://www.indiabudget.gov.in/doc/bspeech/bs199192.pdf

Budget Speech 1993–1994[44] *Shri Manmohan Singh*

1. Reform and modernize the tax system.
2. Correct the slippage in the fiscal deficit.
3. Build on the success achieved in the external sector.
4. Provide a stimulus for a strong industrial recovery.
5. Reorient development policies and programs to address poverty, unemployment, and social deprivation.
6. Consolidate and deepen progress in restoring the health of the banking system.
7. Move toward convertibility on the current account.
8. Reduce reliance on foreign borrowing through foreign direct and portfolio investment.
9. Retire high-cost debt and allow companies to prepay past foreign loans.
10. Undertake comprehensive restructuring of regional rural banks (RRBs) to make them more viable and sustainable.
11. It is important to note that these are specific strategies mentioned in the provided context, and there may be other poverty reduction strategies not explicitly mentioned.

Budget Speech 1999–2000[45] *Shri Yashwant Sinha*

Based on the provided context, here are some poverty reduction strategies mentioned:

1. Downsizing government and reforming the public sector to reduce nondevelopmental expenditure
2. Implementing the National Program for Rural Industrialization (NPRI) to boost rural industrialization and reduce rural-urban disparities
3. Launching the National Human Development Initiative (NHDI) to empower vulnerable groups by providing access to food, healthcare, education, employment, and shelter
4. Strengthening the targeted public distribution system to provide food security to those below the poverty line.
5. Introducing the "Annapurna" scheme to provide free food grains to indigent senior citizens
6. Expanding and improving health infrastructure and services
7. Promoting self-help groups and NGOs as channels for funds to microenterprises
8. Providing priority sector lending to food and agro-processing industries
9. Implementing a credit-linked capital subsidy scheme for the construction of cold storages and godowns

[44] Budget Speech 1993–1994—https://www.indiabudget.gov.in/doc/bspeech/bs199394.pdf

[45] Budget Speech 1999–2000—https://www.indiabudget.gov.in/doc/bspeech/bs19992000.pdf

10. Providing special financial assistance to states that undertake land reform
11. Experimenting with incentive discounts for farmers for lifting fertilizer during lean months
12. Reorienting schemes for the development of degraded and wastelands to involve local self-help groups and landless poor
13. Implementing a multipronged program to strengthen the rural economy, including measures for agriculture and rural development

Please note that this is not an exhaustive list and there may be other poverty reduction strategies mentioned elsewhere in the provided context.

Budget Speech 2000–2001[46] *Shri Yashwant Sinha*

Based on the information given, the poverty reduction strategies mentioned include the following:

1. Providing housing support through schemes like Indira Awas Yojana and credit-cum-subsidy scheme
2. Refinancing for construction of houses under the Golden Jubilee Rural Housing Finance Scheme
3. Providing equity support to HUDCO for finance of rural housing construction
4. Support from cooperative sector and voluntary agencies for housing construction
5. Introducing a new scheme of group insurance, "Janashree Bima Yojana," for social security cover
6. Setting up a task force to review existing legislation and schemes for women's empowerment
7. Launching the Pradhan Mantri Gramodaya Yojana to fulfill critical needs of rural people, including rural roads and connectivity
8. Focusing on universalization of elementary education through schemes like Sarva Shiksha Abhiyan and District Primary Education Program.
9. Enhancing efforts for drinking water supply in rural areas
10. Allocating funds for reproductive and child health programs
11. Improving the financial position of state governments through fiscal reforms
12. *Promoting credit flow to agriculture and strengthening the rural economy*
13. Financing rural infrastructure projects through the Rural Infrastructure Development Fund (RIDF)
14. Supporting microfinance initiatives for poverty alleviation

[46] Budget Speech 2000–2001—https://www.indiabudget.gov.in/doc/bspeech/bs200001.pdf

Budget Speech 2001–2002[47] Shri Yashwant Sinha

Based on the given context, the following poverty reduction strategies can be inferred:

1. Speeding up agricultural sector reforms and better management of the food economy
2. Intensifying infrastructure investment and deepening structural reforms
3. Improving educational opportunities and programs of social security for human development
4. Stringent expenditure control of nonproductive expenditure, rationalization of subsidies, and improvement in the quality of government expenditure
5. Accelerating the privatization process and restructuring of public enterprises
6. Widening the tax base and administering a fair and equitable tax regime for revenue enhancement
7. Downsizing government establishments and rightsizing the ministries/departments to improve efficiency
8. Implementing zero-based budgeting and retaining only efficient and essential schemes
9. Reviewing and removing restrictions on the movement and stocking of food grains and agricultural produce
10. *Reducing the fiscal deficitand completely eliminating the revenue deficitover the next 5 years*
11. Enhancing the maintenance and quality of government services by revising user charges and postal rates
12. Scrutinizing recruitment requirements and limiting fresh recruitment to reduce manpower
13. Streamlining the surplus pool to redeploy and retrain surplus staff and offering attractive VRS packages
14. Enhancing the standard license fee (rent) on government accommodation
15. Suspending the facility of LTC (leave travel concession) for a certain period
16. Maximizing the use of information technology in government activities to promote efficiency and computerizing various operations

[47] Budget Speech 2001–2002—https://www.indiabudget.gov.in/doc/bspeech/bs200102.pdf

Budget Speech 2003–2004[48] *Shri Jaswant Singh*

Based on the given context, some poverty reduction strategies mentioned include the following:

1. Promoting infrastructure through public-private partnerships
2. Providing hand pumps in water-scarcity areas and schools
3. Rejuvenating traditional water sources in villages
4. Research and development support in pharmaceuticals, wind, and solar energy
5. Eradicating poverty as a moral and economic issue
6. Progress in agriculture through a second revolution
7. Rapid and efficient development of physical and social infrastructure
8. Encouraging specific manufacturing sectors and knowledge-based industries
9. Fiscal consolidation and elimination of budgetary drags
10. Addressing social and economic inequalities
11. Implementing reforms to promote growth
12. Deregulation and de-bureaucratization of systems
13. Releasing national creativity and promoting entrepreneurship
14. Providing food security and diversifying agriculture
15. Considering debt relief for heavily indebted poor countries
16. Utilizing grants or project assistance for developing countries
17. Restructuring and reorganizing the Ministry of Finance
18. Introducing tax reforms, including the implementation of a value added tax (VAT) and integrating services into the tax net

Budget Speech 2004–2005[49] *Shri P. Chidambaram*

Based on the given context, here are some poverty reduction strategies mentioned:

1. Empowering the poor with access to education and healthcare
2. Facilitating their full participation in the growth process through gainful employment
3. Providing basic education for children and ensuring they remain in school for at least 8 years
4. Ensuring access to drinking water in every habitation
5. Strengthening the public health system to provide basic medical care
6. Increasing investments and targeted interventions to create job opportunities for the poor
7. Implementing programs like Food for Work, Sarva Shiksha Abhiyan, and Midday Cooked-Meal Scheme

---

[48] Budget Speech 2003–2004—https://www.indiabudget.gov.in/doc/bspeech/bs200304.pdf

[49] Budget Speech 2004–2004—https://www.indiabudget.gov.in/doc/bspeech/bs200405.pdf

8. Expanding the Antyodaya Anna Yojana to provide subsidized food grains to more families
9. Launching the National Employment Guarantee Act to guarantee 100 days of employment in a year
10. Allocating funds specifically for the welfare of Scheduled Castes and Scheduled Tribes.
11. Promoting microfinance initiatives through self-help groups to provide banking services to the poor
12. Setting goals to improve agricultural product markets, provide electricity and telecommunication facilities for all, and enhance access to housing and healthcare
13. Increasing investment in industry, both public and private, to create new jobs

Budget Speech 2005–2006[50]*Shri P. Chidambaram*

1. Direct cash transfers or social safety net programs: these programs provide cash or in-kind transfers to individuals or households living in poverty, helping them meet their basic needs and improve their living conditions.
2. Employment generation and job creation: promoting economic growth and creating employment opportunities are crucial in reducing poverty. This can be done through investments in infrastructure, agriculture, and small and medium-sized enterprises.
3. Education and skills development: providing access to quality education and skills development programs can help individuals acquire the necessary knowledge and skills to escape poverty and secure better job opportunities.
4. Health and nutrition programs: improving access to healthcare services and nutrition programs, especially for vulnerable populations such as children and pregnant women, can help reduce poverty by improving overall health and well-being.
5. *Microfinanceand financial inclusion: providing access to financial services, such as microcredit and savings accounts, can empower individuals and communities to start small businesses, invest ineducation, and build assets.*
6. *Land reform and agricultural development: implementing land reform policies and improving agricultural productivity can help alleviate rural poverty by providing farmers with secure land tenure, access to credit, and technical support.*
7. Gender equality and women's empowerment: addressing gender inequalities and promoting women's empowerment are crucial in reducing poverty. This can be done through policies and programs that promote women's education, access to healthcare, and economic opportunities.

[50]Budget Speech 2005–2006—https://www.indiabudget.gov.in/doc/bspeech/bs200506.pdf

8. Social inclusion and empowerment: ensuring that marginalized and vulnerable groups have equal access to resources, services, and opportunities is essential in poverty reduction. This includes addressing issues of discrimination, social exclusion, and inequality.
9. Sustainable development and environmental protection: promoting sustainable development practices and protecting the environment can help alleviate poverty by ensuring long-term access to natural resources and improving resilience to climate change impacts.
10. Good governance and institutional reforms: strengthening governance systems, promoting transparency, and combating corruption are essential in creating an enabling environment for poverty reduction strategies to be effective.

Budget Speech 2006–2007[51] *Shri P. Chidambaram*

Based on the given context, here are some poverty reduction strategies mentioned:

1. Planned urbanization to promote growth, employment, and better quality of life
2. Promotion of new towns focused on specific industries or themes
3. Increase in old age pensions under the National Social Assistance Programme
4. Gender budgeting and allocation of funds for women-centric schemes
5. Allocation of funds for the welfare and development of Scheduled Castes and Scheduled Tribes
6. Improvement of access to safe drinking water and sanitation
7. Enhancement of the National Rural Health Mission
8. Expansion of the Integrated Child Development Services scheme
9. Implementation of the National Rural Employment Guarantee Scheme
10. Launch of the Jawaharlal Nehru National Urban Renewal Mission
11. *Promotion ofmanufacturingand services sectors for job creation*
12. Support for the textile industry through schemes like Technology Upgradation Fund and Integrated Textiles Parks
13. Focus on handloom sector development and cluster development approach
14. Priority treatment for the food processing industry in terms of bank credit
15. Development of PC&P investment regions to attract investment and employment
16. Promotion of information technology and IT-enabled services to attract high-tech manufacturing
17. Support for small and medium enterprises through policy changes and credit availability

Please note that these are not an exhaustive list of poverty reduction strategies, but rather the ones mentioned in the given context.

---

[51] Budget Speech 2006–2007—https://www.indiabudget.gov.in/doc/bspeech/bs200607.pdf

Budget Speech 2007–2008[52] Shri P. Chidambaram

Based on the given context, some poverty reduction strategies mentioned include the following:

1. Rajiv Gandhi Grameen Vidyutikaran Yojana: providing electricity to rural areas to improve living conditions and promote economic activities
2. Sarva Shiksha Abhiyan and Midday Meal Scheme: focusing on education by improving access to schools and providing nutritious meals to children.
3. Bharat Nirman: a flagship program aimed at improving rural infrastructure, such as roads, housing, irrigation, and water supply
4. *Farm credit: Increasing access to credit for farmers to support agricultural activities and improve their income*
5. *Agricultural indebtedness: addressing the issue of farmer debt and finding solutions to alleviate their financial burden*
6. *Mission for pulses: initiatives to improve the production and productivity of pulses, including providing quality seeds and promoting related activities*
7. National Rural Employment Guarantee Scheme (NREGS): a demand-driven scheme that provides legal guarantees of employment in rural areas
8. National Health Mission: focusing on improving healthcare in rural areas, including the recruitment and training of healthcare workers and increasing access to healthcare services
9. HIV/AIDS control program: implementing measures to control the spread of HIV/AIDS, including prevention, treatment, and support for affected individuals
10. Integrated Child Development Services (ICDS): expanding the program to cover all habitations and settlements, providing support to pregnant women, lactating mothers, and children below the age of 6

Please note that this is not an exhaustive list of poverty reduction strategies, and there may be other strategies mentioned in the full context that are not included here.

Budget Speech 2008–2009[53] Shri P. Chidambaram

Some poverty reduction strategies include the following:

1. Promoting economic growth and job creation: this involves implementing policies that stimulate economic growth and create employment opportunities, such

[52] Budget Speech 2007–2008—https://www.indiabudget.gov.in/doc/bspeech/bs200708.pdf

[53] Budget Speech 2008–2008—https://www.indiabudget.gov.in/budget_archive/ub2008-09/bs/speecha.htm

as investing in infrastructure development, supporting entrepreneurship, and attracting foreign direct investment.
2. Investing in education and skills development: providing access to quality education and vocational training can equip individuals with the necessary skills and knowledge to secure better job opportunities and improve their income levels.
3. Improving access to basic services: ensuring access to essential services like healthcare, clean water, sanitation, and electricity can improve living conditions and reduce poverty.
4. *Enhancing agricultural productivity and rural development: supporting smallholder farmers and investing in agricultural infrastructure, technology, and research can increase agricultural productivity, improvefood security, and liftrural communitiesout of poverty.*
5. *Social protection programs: implementing social safety nets, such as conditional cash transfers, foodsubsidies, and pension schemes, can provide a basic level of income and support to vulnerable populations, helping them escape poverty and improve their well-being.*
6. Empowering women and promoting gender equality: promoting women's rights, providing access to education and healthcare, and ensuring equal opportunities for women in the workforce can help reduce gender disparities and alleviate poverty.
7. *Enhancing financial inclusion: expanding access to financial services, such asmicrofinanceand mobile banking, can empower individuals and small businesses to save, invest, and access credit, enabling them to build assets and improve their economic prospects.*
8. *Promoting inclusive andsustainable development: ensuring that development policies and programs benefit all segments of society, includingmarginalizedgroups, and promoting sustainable practices that protect the environment can contribute to poverty reduction and long-term economic stability.*

It is important to note that poverty reduction strategies may vary depending on the specific context and challenges faced by a country or region.

Budget Speech 2009–2010 (Interim)[54]Shri Pranab Mukherjee

Some poverty reduction strategies mentioned in the given context include the following:

- *Pursuing macroeconomic policies to sustain a growth rate of at least 9% per annum over an extended period of time*

[54]Budget Speech 2009–2010—https://www.indiabudget.gov.in/budget_archive/ub2009-10(I)/bs/speecha.htm

- Strengthening mechanisms for inclusive growth and creating new job opportunities
- Reducing the proportion of people living below the poverty line by half by 2014
- *Expanding employment generation schemes and strengthening social security nets*
- Increasing plan allocation for agriculture and implementing schemes to support farmers
- Ensuring food security and maintaining affordable food prices for the poor
- Giving highest priority to rural development and implementing various programs to improve living conditions in rural areas
- Implementing the National Rural Employment Guarantee Scheme to generate wage employment
- Improving access to quality elementary education through the Sarva Shiksha Abhiyan program
- Providing midday meals in schools to enhance school participation and reduce hunger
- Universalizing the Integrated Child Development Scheme to reduce malnutrition levels
- Focusing on integrated development of urban infrastructure and services through the Jawaharlal Nehru National Urban Renewal Mission
- Ensuring safe drinking water supply and sanitation in rural areas through various programs
- Continuously working on rural sanitation programs to improve hygiene and sanitation in rural areas

Budget Speech 2012–2013[55] *Shri Pranab Mukherjee*

Based on the given pieces of context, some poverty reduction strategies mentioned include the following:

1. Improving the targeting and delivery of subsidies to ensure they reach the intended beneficiaries and reduce leakage
2. Implementing direct transfer of subsidies to the retailer and eventually to the farmer, reducing misuse of fertilizers
3. Enhancing transparency and customer service in public sector oil marketing companies to reduce leakage
4. Increasing budgetary allocation for rural drinking water and sanitation
5. Strengthening Panchayats through the Rajiv Gandhi Panchayat Sashaktikaran Abhiyan (RGPSA)

[55] Budget Speech Budget 2012–2013—https://www.indiabudget.gov.in/doc/bspeech/bs201213.pdf

6. Implementing the Backward Regions Grant Fund scheme to support development projects in backward areas
7. Expanding the Pradhan Mantri Gram Sadak Yojana (PMGSY) to accelerate connectivity in rural areas
8. Increasing budgetary allocation for education, including the Right to Education (RTE) Act and Rashtriya Madhyamik Shiksha Abhiyan (RMSA)
9. Implementing a scheme for education loans and setting up a credit guarantee fund to ensure better flow of credit to deserving students
10. Enhancing healthcare through the National Rural Health Mission (NRHM) and National Urban Health Mission
11. Improving employment opportunities through the Mahatma Gandhi National Rural Employment Guarantee Scheme (MGNREGS) and the National Rural Livelihood Mission (NRLM)

Budget Speech 2014–2015[56]Shri Arun Jaitley

Some poverty reduction strategies include the following:

1. Promoting inclusive economic growth: this involves creating job opportunities, increasing productivity, and improving access to markets for the poor.
2. Enhancing social protection: this includes implementing social safety nets, such as cash transfer programs or subsidies for basic needs, to provide a safety net for vulnerable populations.
3. Improving access to basic services: this includes ensuring access to education, healthcare, clean water, and sanitation for all individuals, regardless of their socioeconomic status.
4. Empowering women and marginalized groups: this involves promoting gender equality and providing opportunities for women and marginalized groups to participate in decision-making processes, access resources, and earn income.
5. Investing in human capital development: this includes investing in education, skills training, and healthcare to equip individuals with the necessary knowledge and capabilities to improve their livelihoods.
6. Strengthening rural development: this involves supporting agriculture, rural infrastructure development, and access to credit and markets for rural populations.
7. Promoting sustainable livelihoods: this includes supporting entrepreneurship, microfinance, and income-generating activities to help individuals lift themselves out of poverty.

[56]Budget Speech Budget 2014–2015—https://www.indiabudget.gov.in/doc/bspeech/bs201415.pdf

8. Enhancing governance and institutional capacity: this involves improving transparency, accountability, and the rule of law to ensure effective implementation of poverty reduction programs and efficient use of resources.
9. Addressing social exclusion and discrimination: this includes combating discrimination based on gender, ethnicity, disability, or other factors that contribute to poverty and inequality.
10. Promoting international cooperation: this involves working collaboratively with other countries, international organizations, and civil society to address global poverty issues and promote sustainable development.

Budget Speech 2016–2017[57] *Shri Arun Jaitley*

Based on the given context, the following poverty reduction strategies can be identified:

1. Increasing farmers' income and promoting agriculture and farmers' welfare
2. Enhancing rural employment and infrastructure development
3. Providing healthcare services and implementing health protection schemes
4. Improving education, skills, and job creation to create a productive society
5. Enhancing infrastructure and investment to improve the quality of life
6. Implementing financial sector reforms to bring transparency and stability
7. Improving governance and ease of doing business to enable people to realize their full potential
8. Ensuring fiscal discipline and prudent management of government finances
9. Implementing tax reforms to reduce compliance burden and promote citizen participation
10. Promoting entrepreneurship among marginalized sections of society
11. Providing social sector programs and schemes to support vulnerable sections of society
12. Implementing schemes like Pradhan Mantri Fasal Bima Yojana and health insurance schemes to protect farmers and provide financial support during emergencies
13. Providing cooking gas connections to below poverty line families to improve their health and reduce the adverse effects of traditional cooking methods

Please note that this list is based on the given context and may not be exhaustive.

[57] Budget Speech Budget 2016–2017—https://www.indiabudget.gov.in/doc/bspeech/bs201617.pdf

Budget Speech 2020–2021[58] *Smt. Nirmala Sitharaman*

17. Our government is committed to the goal of doubling farmers' incomes by 2022, ensuring prosperity for farmers and making farming competitive.
18. We have provided energy sovereignty through the KUSUM scheme and input sovereignty through the Paramparagat Krishi Vikas Yojana, promoting sustainable agricultural practices.
19. We have provided resilience for 6.11 crore farmers insured under the PM Fasal Bima Yojana, protecting them from crop losses and ensuring their financial stability.
20. Focus on the cultivation of pulses and expansion of micro-irrigation through the Krishi Sinchai Yojana has raised the self-reliance of the country and promoted sustainable agricultural practices.
21. We have implemented the Model Agricultural Land Leasing Act, 2016, and the Model Agricultural Produce and Livestock Marketing (Promotion and Facilitation) Act, 2017, to liberalize farm and livestock markets and remove distortions.
22. We are promoting the use of renewable energy in agriculture through the PM-KUSUM scheme, which removes farmers' dependence on diesel and kerosene and links pump sets to solar energy.
23. We are encouraging the balanced use of all kinds of fertilizers, including traditional organic and other innovative fertilizers, to reduce the excessive use of chemical fertilizers and promote sustainable farming practices.
24. We are implementing the Village Storage Scheme, run by self-help groups, to provide farmers with good holding capacity and reduce their logistics costs, promoting sustainable agriculture practices.
25. We are promoting the development of a seamless national cold supply chain for perishables, including milk, meat, and fish, through the setting up of "Kisan Rail" and the inclusion of refrigerated coaches in express and freight trains.
26. We are expanding integrated farming systems in rainfed areas, promoting multitier cropping, beekeeping, solar pumps, solar energy production in non-cropping seasons, and zero-budget natural farming.
27. We are promoting the use of negotiable warehousing receipts (e-NWR) integrated with e-NAM to facilitate financing in the agriculture sector and provide farmers with better access to credit.
28. We are taking steps to eliminate diseases in livestock, such as foot and mouth disease, brucellosis, and peste des petits ruminants (PPR), by 2025, ensuring the health and well-being of animals and improving the livelihoods of farmers.
29. We are promoting the development, management, and conservation of marine fishery resources, ensuring the sustainability of our blue economy.
30. We are providing viability gap funding for the setting up of efficient warehouses at the block/taluk level, promoting the efficient storage and management of agricultural produce.

[58] Budget Speech 2020–2021—https://www.indiabudget.gov.in/doc/bspeech/bs202021.pdf

31. We are working toward water stress management in 100 water-stressed districts, ensuring the sustainable use and conservation of water resources.
32. We are implementing various schemes and programs to improve the well-being and livelihoods of farmers, promoting social sustainability in rural areas.

Appendix D: Public Distribution System (PDS)

Budget Speech 2020–2021[59] *Smt. Nirmala Sitharaman*

- The public distribution system (PDS) in India provides essential food items to the eligible beneficiaries at subsidized rates. The items included in the PDS vary from state to state, but generally, the common items provided through the PDS include the following:

    1. Rice
    2. Wheat
    3. Sugar
    4. Pulses (such as lentils and chickpeas)
    5. Edible oils
    6. Salt
    7. Fortified food products (such as fortified wheat flour and fortified rice)
    8. Kerosene (in some states)

Budget Speech 2007–2008[60] *Shri P. Chidambaram*

The public distribution system (PDS) in India provides essential commodities to the public at subsidized rates. The items typically distributed through the PDS include the following:

1. Rice
2. Wheat
3. Sugar
4. Pulses (such as lentils, chickpeas)
5. Edible oils (such as mustard oil, palm oil)
6. Salt
7. Kerosene oil (used for cooking and lighting)
8. Fair price shop (FPS) items (vary from state to state)

Please note that the exact items provided through the PDS may vary from state to state.

[59] Budget Speech 2020–2021—https://www.indiabudget.gov.in/doc/bspeech/bs202021.pdf
[60] Budget Speech 2007–2008—https://www.indiabudget.gov.in/doc/bspeech/bs200708.pdf

Budget Speech 2008–2009[61] Shri P. Chidambaram

The public distribution system (PDS) provides various essential food items at subsidized rates to eligible beneficiaries. The specific items included in the PDS may vary from region to region, but generally, the following items are distributed:

1. Rice
2. Wheat
3. Sugar
4. Pulses (such as lentils, chickpeas, etc.)
5. Edible oils (such as mustard oil, soybean oil, etc.)
6. Salt
7. Fortified food products (such as fortified wheat flour, fortified rice, etc.)
8. Kerosene oil (used for cooking and lighting purposes)

Please note that the availability and quantity of these items may vary depending on the region and the government's policies.

Appendix E: Macroeconomic Stability Measures

Budget Speech 1990–1991[62] Professor Madhu Dandavate

Based on the context provided, the following macroeconomic stability measures can be identified:

1. Containment of inflation: the government formed a cabinet committee on prices and took effective steps to increase the supply of essential commodities, break the inflationary psychology, and contain inflation.
2. Increasing food grain stocks: the government gave high priority to stepping up procurement efforts and rebuilding stocks of food grains to maintain price stability and economic security.
3. Priority on exports: to combat pressure on the balance of payments and ensure a viable situation, the government has provided top priority to exports. The Import Export Policy focused on promoting exports that earn high net foreign exchange.
4. Reduction of import bill: the government aimed to reverse the trend of increasing import bills, particularly for bulk items like oil, which were causing a significant outflow of foreign exchange.

---

[61] Budget Speech 2008–2008—https://www.indiabudget.gov.in/budget_archive/ub2008-09/bs/speecha.htm

[62] Budget Speech 1990–1991—https://www.indiabudget.gov.in/doc/bspeech/bs199091.pdf

5. Fiscal discipline: the government recognized that the fiscal imbalance was the root cause of inflation and other economic problems. Efforts were made to reduce the burden of administrative expenditure and consider the effectiveness of subsidies.
6. Tax compliance and reduction of black money: the government aimed to improve tax compliance by combining reasonable tax rates, simpler tax laws, effective tax administration, and strong deterrents against tax evasion. Measures were also taken to curb the generation of black money.
7. Strengthening financial institutions: the government aimed to strengthen the role of public financial institutions by granting them functional autonomy while ensuring their accountability. Guidelines were to be framed to ensure objective actions in the capital market.
8. Enhancing credit delivery: the government expressed concern over the low credit deposit ratio in certain regions and urged the Reserve Bank of India to improve credit delivery in those areas while maintaining financial discipline.
9. Improving banking services: the government acknowledged the need for greater competition, operational flexibility, and responsiveness to the needs of the public in the banking sector. A committee was proposed to recommend measures for enhancing public satisfaction with banking services.
10. Formation of Securities Exchange Board of India (SEBI): the government aimed to introduce legislation to give statutory authority to SEBI, which had been announced by the previous government but not yet implemented.

Please note that these measures are specific to the context provided and may not encompass all possible macroeconomic stability measures.

Budget Speech 1991–1992[63]*Shri Manmohan Singh*

Based on the provided context, the macroeconomic stability measures mentioned include the following:

1. Credible fiscal adjustment and macroeconomic stabilization
2. Progressive reduction of fiscal deficit
3. Significant reduction of revenue deficit
4. Reduction of current account deficit in the balance of payments
5. Curbing exponential growth in internal and external debt
6. Reduction of internal debt and external debt reliance
7. Essential reforms in economic policy and economic management
8. Increase in efficiency and international competitiveness of industrial production
9. Utilization of foreign investment and foreign technology
10. Increase in productivity of investment

[63]Budget Speech 1991–1992—https://www.indiabudget.gov.in/doc/bspeech/bs199192.pdf

11. Rapid modernization of India's financial sector
12. Improvement in the performance of the public sector
13. Reduction of transmission and distribution line losses
14. Ensuring prompt payment of dues owed by State Electricity Boards
15. Improvement of fiscal performance and streamlining of public enterprises by state governments
16. Protection of the poor during the period of transition
17. Control of inflation and price rise
18. Enhanced outlays in social sectors for the protection of the poor
19. Upliftment of the weakest and most vulnerable sections of society
20. Stepping up plan outlay for rural development
21. Reduction in nominal and real interest rates
22. Establishment of a separate board for the regulation and functioning of stock exchanges and the securities industry
23. Giving statutory powers to the Securities and Exchange Board of India
24. Comprehensive reforms relating to trading on stock exchanges
25. Promotion and development of mutual funds, including tax incentives
26. Review of policies and procedures for nonresident Indian investments
27. Relaxations and removal of procedural difficulties for NRIs in setting up industrial and other ventures

Budget Speech 1993–1994[64] Shri Manmohan Singh

Based on the given context, the following macroeconomic stability measures can be identified:

1. Fiscal correction: reducing the fiscal deficit both at the center and in the states as a percentage of GDP to curb inflationary expectations
2. Development expenditure: using the fiscal maneuver gained from restraining expenditure to boost development expenditure, particularly for poverty alleviation programs
3. Exchange rate unification: eliminating the dual exchange rate arrangement and allowing all exporters and foreign exchange earners to convert 100% of their earnings at the market rate
4. Export promotion: reviewing and removing restrictions on exports, ensuring adequate credit for exports, and taking steps to stimulate exports
5. Poverty alleviation: taking steps to minimize the burden of adjustment on the poor and working classes
6. Removal of bureaucratic interference: removing unnecessary bureaucratic interference in economic activity to create an environment conducive to innovation, production, and growth

---

[64] Budget Speech 1993–1994—https://www.indiabudget.gov.in/doc/bspeech/bs199394.pdf

7. Review of laws and procedures: constituting a special review group in each ministry to review existing laws and procedures and identify necessary changes to align them with the new economic environment
8. Control over expenditure: maintaining strict control over expenditure while making unavoidable increases for subsidies and other necessary provisions

It is important to note that these measures are specific to the given context and may not encompass all possible macroeconomic stability measures.

Budget Speech 1994–1995[65] *Shri Manmohan Singh*

Based on the provided context, the following macroeconomic stability measures can be identified:

1. Accelerating the reform and modernization of the tax system
2. Improving tax administration to ensure fair and firm enforcement of tax laws
3. Implementing tax reforms to address tax evasion and promote compliance
4. Maintaining fiscal discipline to correct slippage in the fiscal deficit
5. Focusing on external sector performance and phased integration with the world economy
6. Providing stimulus for a strong industrial recovery, especially for investment and capital goods production
7. Reorienting development policies and programs to address poverty, unemployment, and social deprivation
8. Consolidating and deepening progress in restoring the health of the banking system
9. Moving toward convertibility on the current account to liberalize access to foreign exchange
10. Reviewing and making necessary changes to the Foreign Exchange Regulation Act, 1973
11. Encouraging foreign direct and portfolio investment through policies and bilateral investment treaties
12. Managing external debt prudently and retiring high-cost debt through prepayment

It is worth noting that these measures are specific to the provided context and may not represent an exhaustive list of all macroeconomic stability measures.

[65] Budget Speech 1994–1995—https://www.indiabudget.gov.in/doc/bspeech/bs199495.pdf

Budget Speech 1999–2000[66] Shri Yashwant Sinha

Based on the given context, the macroeconomic stability measures mentioned are as follows:

1. Downsizing government to control nondevelopmental expenditure
2. Revenue and fiscal deficit reduction to free up resources for investment and growth
3. Reform of indirect taxes to promote productivity and employment
4. Deepening and widening economic reforms in major sectors
5. Safeguarding the economy from external shocks and reviving export
6. Strengthening knowledge-based industries
7. Revitalizing and redirecting public programs for human development, with a focus on empowering the poor and marginalized community
8. Strengthening the rural economy through a multipronged program
9. Revamping the export credit scheme to simplify procedures and provide competitive rates
10. Reducing transaction costs related to foreign trade licensing, tax procedures, and the banking system
11. Expanding the list of automatic approvals for foreign direct investment (FDI) in important sectors
12. Creating a Foreign Investment Implementation Authority (FIIA) to ensure quick implementation of FDI approvals
13. Introducing a Gold Deposit Scheme to mobilize idle gold assets, reduce dependence on imported gold, and provide a regular source of income for depositors.

Budget Speech 2000–2001[67] Shri Yashwant Sinha

Based on the given context, here are some macroeconomic stability measures mentioned:

1. Downsizing government: the objective of downsizing the government is mentioned as a measure to manage the fiscal deficit and promote fiscal responsibility.
2. *Fiscal Responsibility Act: the need for a strong institutional mechanism embodied in a Fiscal Responsibility Act is mentioned to manage the fiscal deficitin the medium term.*
3. Fiscal reform at state level: the deteriorating financial position of state governments is mentioned, and the need for fiscal reform at the state level is emphasized.

[66] Budget Speech 1999–2000—https://www.indiabudget.gov.in/doc/bspeech/bs19992000.pdf

[67] Budget Speech 2000–2001—https://www.indiabudget.gov.in/doc/bspeech/bs200001.pdf

4. *Strengthening rural economy: measures to strengthen the rural economy, such as promoting credit flow to agriculture, launching initiatives to promoterural credit, and financing rural infrastructure projects, are mentioned.*
5. *Curbing expenditure growth: initiatives to curb built-in expenditure growth and bring about structural changes in expenditure composition, such as subjecting ongoing schemes to zero-based budgeting scrutiny and reassessing manpower requirements, are mentioned.*
6. Reviewing subsidies: the need to review all subsidies with the aim of bringing in cost-based user charges wherever feasible is mentioned.
7. *Establishingfiscal discipline: the importance of establishing a credible framework offiscal disciplineis emphasized as a crucial element of the overall strategy.*
8. Second-generation reforms: the need for thorough-going economic reforms to unlock the creative energies of the people and reap the gains of productivity growth is mentioned.

These are some of the macroeconomic stability measures mentioned in the given context.

Budget Speech 2001–2002[68]*Shri Yashwant Sinha*

Based on the given context, here are some strategies that can be considered:

1. *Speeding up agricultural sector reforms and better management of the food economy: this can be achieved by implementing policies that improve irrigation systems, provide better access to credit for farmers, and promote the use of advanced farming techniques.*
2. Intensification of infrastructure investment: increase public investment in infrastructure projects such as roads, bridges, ports, and power plants. This will not only create jobs but also attract private investment and stimulate economic growth.
3. Continued reform in the financial sector and capital markets: implement measures to deepen the debt market, promote transparency, and encourage participation of small investors. This will provide long-term funding options for infrastructure projects and boost investor confidence.
4. Human development through better educational opportunities and social security programs: invest in education and skills development programs to enhance the quality of the workforce. Implement social security programs to provide a safety net for vulnerable populations.

[68] Budget Speech 2001–2002—https://www.indiabudget.gov.in/doc/bspeech/bs200102.pdf

5. Stringent expenditure control and rationalization of subsidies: implement measures to control nonproductive expenditure, reduce subsidies to unaffordable levels, and ensure that subsidies reach the deserving beneficiaries.
6. Acceleration of the privatization process and restructuring of public enterprises: continue with the privatization of state-owned enterprises to improve efficiency and attract private investment. Restructure public enterprises to make them more competitive and profitable.
7. Revenue enhancement through widening the tax base and administration of a fair and equitable tax regime: implement measures to broaden the tax base, reduce tax evasion, and ensure a fair and transparent tax system. This will help increase government revenue and reduce the fiscal deficit.
8. Strengthening the management of public sector banks: provide greater autonomy to bank managements and abolish recruitment boards to improve efficiency and decision-making in public sector banks.
9. Liberalizing the capital account: further liberalize the capital account to attract foreign investment and encourage Indian companies to invest abroad. This can be done by relaxing restrictions on overseas investments and simplifying regulations.
10. Focus on power sector reforms: implement measures to reduce transmission and distribution losses, stop electricity theft, and impose economic tariffs. Restore financial viability of State Electricity Boards through reforms and milestones set in MOUs with state governments.

These strategies aim to address the concerns mentioned in the context and promote sustainable economic growth.

Budget Speech 2004–2005[69]*Shri P. Chidambaram*

1. The economic fundamentals appear strong and the balance of payments is robust.
2. Growth will be sustained by increased production and value addition in agriculture.
3. There is a marked improvement in industrial production.
4. The services sector is performing well.
5. The government is fully alert to short-term pressures on prices.
6. The outlook for the year is benign.
7. The government is committed to achieving rapid growth with stability and equity.
8. The government is committed to removing inadequacies in infrastructure facilities.

[69] Budget Speech 2004–2004—https://www.indiabudget.gov.in/doc/bspeech/bs200405.pdf

9. The government is committed to providing insurance cover to farming and livestock.
10. The government is implementing the Rajiv Gandhi Drinking Water Mission to provide adequate drinking water.
11. The government is allocating funds for the Accelerated Rural Water Supply Program.
12. The government is implementing the Urban Water Supply Program in urban areas.
13. The government is investing in desalination plants to address water scarcity.
14. The government is working on the Sethusamudram Ship Canal Project.
15. The government is facilitating the construction of an International Container Transshipment Terminal.
16. The government is providing housing through the Indira Awas Yojana and the Golden Jubilee Rural Housing Finance Scheme.
17. The government is making efforts to make the environment in India attractive for investors.
18. The government is considering gender budgeting to highlight the gender sensitivities of budgetary allocations.
19. The government is working on targeting subsidies sharply at the poor and truly needy.
20. The government is supporting the states' finances through transfers and loans.
21. The government is empowering the states through devolution of larger resources.
22. The government is providing special economic packages for certain regions like Bihar and Jammu and Kashmir.

Budget Speech 2005–2006[70]Shri P. Chidambaram

From the given context, some of the macroeconomic stabilization initiatives mentioned are as follows:

1. Improving fiscal health of the country by increasing revenues and reorienting expenditure toward education, health, and infrastructure.
2. Restructuring subsidies in a cautious manner and after thorough discussion.
3. Implementing the New Pricing Scheme for fertilizers to make procurement more cost-effective.
4. Taking appropriate decisions regarding petroleum products based on the recommendations of the Lahiri Committee.
5. Nurturing the investment boom in infrastructure, industry, and services through further reforms in the financial sector.

---

[70]Budget Speech 2005–2006—https://www.indiabudget.gov.in/doc/bspeech/bs200506.pdf

6. Introducing amendments to the Banking Regulation Act to remove bounds to the statutory liquidity ratio, allow banking companies to issue preference shares, and enable consolidated supervision of banks and their subsidiaries.
7. Introducing amendments to the Reserve Bank of India Act to remove limits of the cash reserve ratio and enable RBI to lend or borrow securities.
8. Implementing a defined contribution pension scheme for newly recruited central government employees and extending it to the unorganized sector.
9. Providing equity support and loans to central public sector enterprises.
10. Promoting savings and mobilizing financial resources through a mix of right policies and prudent taxes.
11. Preparing a roadmap for agricultural diversification and focusing on fruits, vegetables, flowers, dairy, poultry, fisheries, pulses, and oilseeds.
12. Launching the National Horticulture Mission to ensure an end-to-end approach in horticulture production and marketing.
13. Setting up an expert committee to suggest improvements to the Price Stabilization Fund for the plantation sector.
14. Considering a program for massive replantation and rejuvenation in the tea sector.

Budget Speech 2006–2007[71] Shri P. Chidambaram

Based on the given context, the following macroeconomic stability measures can be identified:

1. Fiscal prudence: the government has emphasized fiscal prudence through enhanced revenues and expenditure control.
2. Monetary stability: the government has focused on maintaining monetary stability and managing external debt.
3. High growth rate: the government has achieved a high growth rate, with GDP growth expected to be 8.1% and the manufacturing sector expected to grow at 9.4% in the current year.
4. Investment promotion: the government has enhanced investment by steadily increasing the investment rate from 25.3% in 2002–2003 to 30.1% in 2004–2005.
5. Infrastructure development: the government has focused on augmenting infrastructure, including power generation, electrification of villages, and the construction of national highways.
6. Support for small and medium enterprises (SMEs): the government has implemented various measures to support SMEs, including providing credit guarantees, reducing guarantee fees, and extending insurance cover.

[71] Budget Speech 2006–2007—https://www.indiabudget.gov.in/doc/bspeech/bs200607.pdf

7. Cluster development: the government has adopted the cluster development model to promote manufacturing, renew industrial towns, and build new industrial townships.
8. Tourism development: the Ministry of Tourism has taken initiatives to develop tourist destinations, circuits, and villages with core competencies in handicrafts, handlooms, and culture.
9. Insurance and banking reforms: the government is examining the recommendations of the K.P. Narasimhan Committee on insurance, and important bills to amend banking laws and establish a Pension Fund Regulatory and Development Authority are before Parliament.
10. Capital market development: the government has implemented measures to deepen, broaden, and strengthen the capital market, including increasing FII investment limits, allowing mutual funds to invest in overseas instruments, and setting up an investor protection fund.
11. Unified market for corporate bonds: the government has accepted the recommendations of an expert committee and will take steps to create a single, unified exchange-traded market for corporate bonds.

Budget Speech 2007–2008[72]*Shri P. Chidambaram*

20. Fiscal prudence: ensuring that government spending and revenue collection are balanced to avoid excessive borrowing and debt accumulation
21. Monetary stability: maintaining stable and low inflation through appropriate monetary policy measures, such as controlling money supply and interest rates
22. Exchange rate stability: managing the value of the domestic currency to prevent excessive fluctuations and maintain competitiveness in international trade
23. Price stability: controlling inflation and preventing excessive price increases to protect the purchasing power of the currency and maintain stable consumer prices
24. Financial sector stability: ensuring the stability and soundness of the banking and financial system through regulations and supervision to prevent financial crises and maintain confidence in the system
25. External sector stability: balancing imports and exports to avoid excessive trade deficits or surpluses and managing external debt levels to prevent vulnerability to external shocks
26. Employment stability: promoting a stable labor market with low unemployment rates to ensure income stability and social welfare
27. Economic growth: encouraging sustainable and steady economic growth to create employment opportunities and improve living standards

[72]Budget Speech 2007–2008—https://www.indiabudget.gov.in/doc/bspeech/bs200708.pdf

28. Public debt management: implementing effective debt management strategies to prevent excessive accumulation of public debt and ensure debt sustainability
29. Policy coordination: coordinating macroeconomic policies across different sectors and levels of government to ensure coherence and effectiveness in achieving macroeconomic stability

Budget Speech 2008–2009[73] *Shri P. Chidambaram*

1. Fiscal discipline: maintaining a balanced budget and controlling government spending to avoid excessive borrowing and debt accumulation
2. Monetary policy: implementing policies to manage the money supply, interest rates, and inflation to ensure price stability and control inflationary pressures
3. Exchange rate stability: managing the value of the domestic currency to prevent excessive fluctuations and promote stability in international trade and investment
4. Financial sector regulation: implementing regulations and policies to ensure the stability and soundness of the financial system, including monitoring and oversight of banks, insurance companies, and other financial institutions
5. Public debt management: developing strategies to manage and reduce public debt levels to ensure long-term fiscal sustainability
6. External balance: monitoring and managing the balance of payments to prevent excessive trade deficits or surpluses and maintain stability in international transactions
7. Structural reforms: implementing policies to improve the efficiency and competitiveness of the economy, including measures to promote investment, enhance productivity, and foster innovation
8. Employment and social protection: implementing policies to promote full employment, reduce inequality, and provide social safety nets to protect vulnerable populations
9. Transparency and accountability: promoting transparency in economic policies and decision-making processes to build trust and accountability among policymakers and stakeholders
10. Crisis management and contingency planning: developing mechanisms and policies to effectively respond to economic crises, such as financial market disruptions or natural disasters, and mitigate their impact on the economy

[73] Budget Speech 2008–2008—https://www.indiabudget.gov.in/budget_archive/ub2008-09/bs/speecha.htm

Budget Speech 2009–2010 (Interim Budget)[74]*Shri Pranab Mukherjee*

Based on the given context, the macroeconomic stability measures mentioned are:

1. Relaxing the FRBM targets: the government decided to relax the Fiscal Responsibility and Budget Management (FRBM) targets to provide a demand boost and counter the effects of the global financial meltdown.
2. Reverting to fiscal consolidation: the medium-term objective is to revert to the path of fiscal consolidation at the earliest.
3. Accelerating policy reforms: there is a need to accelerate the pace of policy reforms, including in the financial sector, to make the economy more competitive.
4. Strengthening economic regulatory and oversight systems: efforts are being made to make the economic regulatory and oversight systems more efficient and effective to bring the economy back to a 9% growth path.
5. Expansion of employment generation schemes: employment generation schemes need to be expanded to create more job opportunities.
6. Strengthening social security nets: social security nets need to be strengthened to protect the vulnerable sections of society.
7. Pursuing macroeconomic policies for sustained growth: the aim is to sustain a growth rate of at least 9% per annum over an extended period of time.
8. Reducing poverty levels: the goal is to reduce the proportion of people living below the poverty line to less than half of the current levels by 2014.
9. Making growth inclusive: the focus is on making economic growth inclusive and sustainable, with a commitment to the rise of all sections of society.
10. Increasing plan allocation for agriculture: the government has increased the plan allocation for agriculture by 300% between 2003–2004 and 2008–2009.
11. Launching the Rashtriya Krishi Vikas Yojana: a scheme was launched to increase the growth rate of agriculture and allied sectors to 4% per annum during the Eleventh Plan period.
12. Doubling the flow of credit to agriculture: measures were taken to double the flow of credit to agriculture and strengthen the short-term cooperative credit structure.

These are some of the macroeconomic stability measures mentioned in the given context.

[74]Budget Speech 2009–2010 (I)—https://www.indiabudget.gov.in/budget_archive/ub2009-10(I)/bs/speecha.htm

Budget Speech 2009–2010 (Interim)[75] *Shri Pranab Mukherjee*

Based on the given context, some macroeconomic stability measures that have been mentioned include the following:

- Relaxing the FRBM targets: this measure was taken to provide a demand boost to counter the situation created by the global financial meltdown.
- Simplifying and making guidelines consistent across sectors: this measure aims to provide an impetus to foreign investment in India.
- Pursuing macroeconomic policies to sustain a growth rate of at least 9% per annum: this measure is intended to ensure sustained economic growth over an extended period of time.
- Strengthening mechanisms for inclusive growth: this measure focuses on creating new work opportunities and reducing the proportion of people living below the poverty line.
- Accelerating the pace of policy reforms: this measure aims to make the economy more competitive, particularly in the financial sector.
- Expanding employment generation schemes and strengthening social security nets: this measure is intended to protect vulnerable sections of society during economic downturns.
- Implementing fiscal and administrative measures to ease supply side constraints: this measure was taken to address the sharp rise in global inflation and domestic prices.
- Providing substantial fiscal stimulus: this measure includes tax relief and increased expenditure on public projects to boost demand and create employment and public assets.

Budget Speech 2012–2013[76] *Shri Pranab Mukherjee*

Based on the given context, some macroeconomic stability measures mentioned are as follows:

1. Strengthening domestic growth drivers: this involves focusing on domestic demand-driven growth recovery and creating conditions for rapid revival of high growth in private investment.
2. Addressing supply bottlenecks: this includes addressing supply bottlenecks in agriculture, energy, and transport sectors, particularly in coal, power, national highways, railways, and civil aviation.

---

[75] Budget Speech 2009–2010—https://www.indiabudget.gov.in/budget_archive/ub2009-10(I)/bs/speecha.htm

[76] Budget Speech Budget 2012–2013—https://www.indiabudget.gov.in/doc/bspeech/bs201213.pdf

3. Improving delivery systems, governance, and transparency: this involves expediting coordinated implementation of decisions to improve delivery systems, governance, and transparency, as well as addressing the problem of black money and corruption in public life.
4. Fiscal consolidation: this includes efforts to raise *the tax-GDP ratio* and lower expenditure, particularly on subsidies, to improve the quality of public spending and reduce the fiscal deficit.
5. *Strengthening macroeconomic fundamentals: this involves striking a balance betweenfiscal consolidationand strengthening macroeconomic fundamentals to create adequate headroom to deal with future shocks in the global environment.*
6. Implementing the National Manufacturing Policy: this policy aims to raise the share of manufacturing in GDP to 25% and create 10 crore jobs within a decade. It encourages the setting up of National Investment and Manufacturing Zones (NIMZs) across the country.

Please note that these measures are specific to the given context and may not cover all possible macroeconomic stability measures.

Budget Speech 2014–2015 (Interim)[77]*Shri Pranab Mukherjee*

Some of the macroeconomic stabilization initiatives mentioned in the context are as follows:

1. Containing fiscal deficit at 4.6% of GDP
2. Containing current account deficit at USD 45 billion
3. Adding USD 15 billion to foreign exchange reserves
4. Focusing on fiscal stability and maintaining a target of 3% fiscal deficit of GDP by 2016–2017
5. Accepting a moderate level of inflation while aiming for high growth
6. Implementing recommendations of the Financial Sector Legislative Reforms Commission
7. Rebuilding and expanding infrastructure, adopting proven models, and increasing the use of the PPP model
8. Focusing on manufacturing, especially for export, and providing tax waivers or rebates on exported products
9. Addressing obstacles in project implementation and setting up the Cabinet Committee on Investment and Project Monitoring Group to clear the way for completing projects
10. Taking pathbreaking decisions like fully decontrolling sugar and gradually correcting diesel prices

[77]Budget Speech Budget 2014–2015—https://www.indiabudget.gov.in/doc/bspeech/bs.pdf

Budget Speech 2014–2015[78]*Shri Arun Jaitley*

1. Higher growth: the government aims to achieve sustained growth of 7–8% or above within the next 3–4 years.
2. Lower inflation: the government aims to achieve lower levels of inflation.
3. Sustained level of external sector balance: the government aims to maintain a manageable current account deficit.
4. Prudent policy stance: the government aims to introduce fiscal prudence and consolidation to ensure fiscal stability.
5. Fiscal deficit reduction: the government aims to reduce the fiscal deficit to 4.1% of GDP in the current year, with a roadmap to further reduce it to 3.6% in 2015–2016 and 3% in 2016–2017.
6. Expenditure management: the government will establish an Expenditure Management Commission to review and reform government expenditure for maximum efficiency and output.
7. Overhaul of subsidy regime: the government plans to reform the subsidy regime, including food and petroleum subsidies, to make it more targeted and efficient.
8. Increase tax to GDP ratio: the government aims to improve the tax to GDP ratio and increase nontax revenues to generate more resources for the economy.
9. Strengthening banking system: the government plans to infuse equity into banks to meet Basel-III norms and raise additional resources. It also aims to increase public ownership in banks through the sale of shares to common citizens.
10. Encouraging investment: the government aims to encourage investment in the economy, including capital investment by PSUs and the creation of a virtuous investment cycle.
11. Development of smart cities: the government plans to develop smart cities to accommodate the increasing urban population and improve living standards.
12. Revival of special economic zones (SEZs): the government aims to revive SEZs as effective instruments of industrial production, economic growth, export promotion, and employment generation.
13. Promotion of entrepreneurship and start-up companies: the government plans to establish a fund to provide equity, quasi-equity, soft loans, and other risk capital for start-up companies. It also aims to review the financial architecture for the MSME sector and facilitate forward and backward linkages.
14. E-Visa for tourism: the government plans to introduce electronic travel authorization (e-Visa) to boost tourism in India.

These are some of the macroeconomic stability initiatives mentioned in the given context.

[78]Budget Speech Budget 2014–2015—https://www.indiabudget.gov.in/doc/bspeech/bs201415.pdf

Budget Speech 2022–2023[79] *Smt. Nirmala Sitharaman*

Based on the provided context, the following macroeconomic stabilization initiatives can be identified:

- Promotion of energy efficiency and savings measures in large commercial buildings through the Energy Service Company (ESCO) business model
- Establishment of four pilot projects for coal gasification and conversion of coal into chemicals
- I*mplementation of policies and legislative changes to promote agroforestry and private forestry, with financial support provided to farmers belonging* to Scheduled Castes and Scheduled Tribes
- Stepping up public capital expenditure by 35.4% to stimulate economic revival and consolidation
- Public investment to crowd in private investment and pump priming private investment and demand
- Expansion and extension of the Emergency Credit Line Guarantee Scheme (ECLGS) to support MSMEs, particularly in the hospitality sector.
- Revamping the Credit Guarantee Trust for Micro and Small Enterprises (CGTMSE) scheme to facilitate additional credit for MSMEs
- Launching the Raising and Accelerating MSME Performance (RAMP) program to enhance the resilience, competitiveness, and efficiency of the MSME sector
- Reorienting skilling programs and partnerships with the industry to promote continuous skilling avenues and aligning the National Skill Qualification Framework (NSQF) with industry needs
- Support for research and development in sunrise opportunities, with government contributions provided
- Implementation of supportive policies, light-touch regulations, and facilitative actions to build domestic capacities and promote research and development
- Allocation of funds for the production-linked incentive for the manufacture of high-efficiency solar modules to support domestic manufacturing for the goal of 280 GW of installed solar capacity by 2030
- Emphasis on circular economy transition, with action plans in sectors such as electronic waste, end-of-life vehicles, used oil waste, and toxic and hazardous industrial waste
- Transition to a carbon-neutral economy through the co-firing of biomass pellets in thermal power plants, resulting in CO2 savings

[79] Budget Speech 2022–2023—https://www.indiabudget.gov.in/doc/bspeech/bs202223.pdf

Appendix F: General Strategies

Budget Speech 2020–2021[80] *Smt. Nirmala Sitharaman*

10. Encouraging data-driven innovation: the government can implement policies to enable the private sector to build data center parks throughout the country. This will allow firms to incorporate data in every step of their value chains, leading to more efficient and innovative processes.
11. Digital connectivity for public institutions: the government aims to provide digital connectivity to all public institutions at the Gram Panchayat level. This can be achieved by expanding the BharatNet program and providing fiber to the home (FTTH) connections. This will enable better access to digital services and improve the delivery of public services.
12. Focus on intellectual property creation and protection: to expand the base for knowledge-driven enterprises, the government can initiate national-level science schemes to create a comprehensive database. This will support intellectual property creation and protection, fostering innovation and economic growth.
13. Support for start-ups: the government proposes to provide early-stage funding, including a seed fund, to support the ideation and development of start-ups. This can be done by creating a favorable ecosystem for start-ups, providing mentorship, and facilitating access to capital and resources.
14. Investment in quantum technologies: the government plans to allocate an outlay of 8000 crore over 5 years for the National Mission on Quantum Technologies and Applications. This investment can be used to promote research, development, and commercialization of quantum technologies, leading to advancements in computing, communications, and cybersecurity.
15. Empowering women and child welfare: the government's Beti Bachao Beti Padhao initiative can be further strengthened by focusing on improving access to education, healthcare, sanitation, and financial inclusion for women and children. This can be achieved through targeted programs, awareness campaigns, and policy interventions.
16. Agricultural reforms: to ensure prosperity for farmers, the government can focus on liberalizing farm markets, removing distortions, and investing in logistics, agri-services, and farm-based activities. Additionally, sustainable cropping patterns, technology adoption, and cooperation with states can be promoted to enhance agricultural productivity and income.
17. Strengthening cold supply chain: the government can support the establishment of a seamless national cold supply chain for perishables, including milk, meat, and fish. This can be done through partnerships with the Indian Railways and the launch of initiatives like Kisan Rail and Krishi Udaan. These efforts will improve value realization and market access for farmers.

[80] Budget Speech 2020–2021—https://www.indiabudget.gov.in/doc/bspeech/bs202021.pdf

18. Cluster-based approach for horticulture: supporting states to adopt a cluster-based approach, focusing on "one product one district" in the horticulture sector, can enhance marketing and export opportunities. This can involve providing infrastructure, technology, and market linkages to promote the growth of specific horticultural products in designated districts.
19. These strategies aim to leverage technology, promote innovation, empower marginalized sections of society, and drive sustainable growth in various sectors of the economy.

Budget Speech 1990–1991[81] *Professor Madhu Dandavate*

Based on the given context, here are some strategies that can be considered:

1. Environmental conservation: implement measures to address soil erosion, salinity, and loss of tree cover. Promote sustainable practices in agriculture and forestry. Encourage the use of renewable energy sources, and reduce pollution from industries and transportation.
2. Employment generation: focus on developing an employment-oriented growth strategy. Invest in vocational training programs to enhance the skills of the youth and improve their employability. Provide credit support for self-employment initiatives.
3. Social inclusion: ensure that all facilities and privileges available to Scheduled Castes are also extended to those who have converted to Buddhism. Work toward eradicating social stigma and discrimination.
4. Agricultural development: increase investment in irrigation, land development, and soil and moisture conservation to improve agricultural productivity, especially in rainfed and semiarid areas. Promote diversification of agriculture and the development of agro-based processing industries.
5. Industrial growth: prioritize industrial growth in a competitive and non-monopolistic environment. Review and simplify the Industrial Licensing Policy to encourage competition and prevent monopolies.
6. Rural development: allocate a significant portion of investible resources for agriculture and rural development. Increase budgetary support for the rural sector. Provide debt relief and fertilizersubsidies to benefit rural areas.
7. Land reforms: initiate steps to include land reforms legislation in the Constitution. Work with state governments to restructure land relations and ensure equitable distribution of land.
8. Debt relief for farmers: provide relief to farmers burdened with debt, especially those with loans up to Rs. 10,000. Implement measures to break the cycle of indebtedness and low incomes.

[81] Budget Speech 1990–1991—https://www.indiabudget.gov.in/doc/bspeech/bs199091.pdf

9. Research and development: focus on developing appropriate technologies for agriculture, nonconventional and renewable energy, and other intensive employment activities. Encourage public spending on R&D, and incentivize the use of indigenous technologies.
10. Engagement with Indians settled abroad: simplify procedures to facilitate investment of savings by Indians settled abroad. Provide special facilities and support to strengthen their intellectual, philosophical, and social links with their country of origin.

These strategies aim to address various challenges and promote sustainable development, employment generation, social inclusion, and agricultural and industrial growth.

Budget Speech 1991–1992[82] *Shri Manmohan Singh*

Based on the given context, here are some strategies that could be considered:

1. Focus on attracting substantial investment: develop policies and incentives to attract high-tech industries and foreign investment, which can provide access to advanced technology and global markets.
2. Enhance the public sector's role in economic growth: review and concentrate public sector investments in strategic areas that require high technology and have the potential for generating large surpluses.
3. Implement fiscal adjustment and macroeconomic stabilization: prioritize credible fiscal adjustment and macroeconomic stabilization to reduce fiscal deficits, revenue deficits, and current account deficits. This will help curb the exponential growth in internal and external debt and limit the burden on debt servicing.
4. Undertake essential reforms in economic policy and management: implement reforms to eliminate waste and inefficiency, increase efficiency and international competitiveness of industrial production, utilize foreign investment and technology, improve productivity of investment, modernize the financial sector, and enhance the performance of the public sector.
5. Minimize the burden of adjustment on the poor: ensure that the adjustment process does not adversely affect the poor by controlling inflation and price rise, providing enhanced outlays in social sectors, prioritizing employment creation and poverty eradication in rural areas, and uplifting the weakest and most vulnerable sections of society.
6. Increase investment in social sectors: allocate resources for investment in social sectors, particularly in education, to develop human resources and provide quality education to all, including children from disadvantaged backgrounds.

[82] Budget Speech 1992–1993—https://www.indiabudget.gov.in/doc/bspeech/bs199293.pdf

7. Accelerate programs for rural development and welfare: increase the pace of programs like the Rajiv Gandhi Technology Mission for rural development, accelerate the conversion of dry latrines, and allocate more funds for the rehabilitation and retraining of scavengers.
8. Collaborate with state governments: encourage state governments to improve their fiscal performance, streamline the working of public enterprises, and ensure prompt payment of dues to central government entities.

It is important to note that these strategies are based on the given context and may need to be adapted to specific circumstances and priorities.

Budget Speech 2000–2001[83]Shri Yashwant Sinha

Based on the given context, here are some strategies that can be inferred:

1. *Strengthening the rural economy: focus on improving agriculture and allied activities to boost the rural economy. This can be done through investment in infrastructure, technology, and providing support to farmers.*
2. Promote knowledge-based industries: nurture industries such as infotech, biotechnology, and pharmaceuticals to harness their potential for growth and job creation. This can be achieved through investment in research and development, providing incentives for innovation, and creating a conducive environment for these industries to thrive.
3. *Modernize traditional industries: strengthen and modernize traditional industries such as textiles, leather, agro-processing, and the small-scale sector. This can be done through technological upgradation, skills development, and providing access to finance and markets.*
4. Address infrastructure bottlenecks: focus on improving infrastructure in areas such as power, roads, ports, telecom, railways, and airways. This can be achieved through increased investment, public-private partnerships, and streamlining regulatory processes.
5. Prioritize human resource development: give high priority to education, health, and other social services, with a special emphasis on the poorest and weakest sections of society. This can be done through increased investment in these sectors, improving access to quality education and healthcare, and implementing targeted welfare programs.
6. *Boost exports and foreign investment: strengthen India's role in the world economy by promoting exports, attracting higher foreign investment, and managing external debt prudently. This can be achieved through trade promotion, creating a favorable investment climate, and implementing sound fiscal and monetary policies.*

[83]Budget Speech 2000–2001—https://www.indiabudget.gov.in/doc/bspeech/bs200001.pdf

7. *Establishfiscal discipline: ensure a credible framework offiscal disciplineto support the overall growth strategy. This can be done through measures such as reducing wasteful expenditure, increasing revenue generation, and implementing effective budgetary controls.*
8. Encourage economic reforms: pursue thorough-going economic reforms to unlock the creative energies of the people and promote productivity growth. These reforms should be guided by compassion and justice, taking into account the interests of all stakeholders.
9. Promote venture capital finance: liberalize the tax treatment for venture capital funds to encourage investment in knowledge-based enterprises. Simplify procedures and provide support for domestic and overseas venture capital funds.
10. Strengthen the capital market: ensure investor protection and promote the growth of the capital market. This can be done through measures such as establishing investor protection funds, enhancing regulatory mechanisms, and encouraging transparency and accountability.
11. Restructure weak public sector banks: address the issue of nonperforming assets (NPAs) in public sector banks by implementing comprehensive restructuring programs, recapitalizing weak banks, and improving mechanisms for recover bank dues.
12. Enhance operational flexibility of the RBI: amend relevant legislation to provide greater operational flexibility to the Reserve Bank of India (RBI) for conducting monetary policy and regulating the financial system.

These strategies aim to promote sustainable and inclusive economic growth, job creation, and poverty reduction in India.

Budget Speech 1993–1994[84]Shri Manmohan Singh

1. Continue fiscal discipline: the government should focus on reducing the fiscal deficit further to curb inflationary expectations and maintain stability in the economy.
2. Increase development expenditure: utilize the fiscal maneuver gained from restraining expenditure to boost development expenditure, especially for poverty alleviation programs and rural development.
3. Simplify tax system: implement tax reforms to create a simpler tax system with moderate rates and a greater emphasis on compliance. This will help in promoting economic growth and reducing tax evasion.
4. Support agriculture and agro-processing industries: give priority to agriculture and agro-processing industries as they have the potential to increase

[84]Budget Speech 1993–1994—https://www.indiabudget.gov.in/doc/bspeech/bs199293.pdf

employment and income in rural areas. This can be achieved through policies that provide support, incentives, and investment in these sectors.

5. Promote exports: make exports a high priority national endeavor to reduce the dependence on foreign financing and achieve self-reliance. This can be done by implementing policies that support and incentivize export-oriented industries.
6. Ensure adequate credit availability: address the issue of inadequate credit for the commercial sector by reducing the statutory liquidity ratio (SLR) to release more resources for commercial lending. This will help in converting weak industrial recovery into a strong revival.
7. Lower interest rates: with the reduction in inflation, there is scope for further reduction in interest rates. The government should work with the Reserve Bank of India to lower interest rates, which will stimulate economic recovery and investment.
8. Focus on structural reforms: continue with structural reforms aimed at increasing efficiency in resource use and improving international competitiveness. This includes trade and industrial policy restructuring, promoting domestic competition, encouraging foreign investment, and integrating the Indian economy with the world economy.
9. Prioritize social services: ensure that rural development, education, and healthcare receive adequate support and funding. These vital social services are crucial for the well-being and livelihood of the majority of the population.
10. Maintain commitment to the poor and disadvantaged: despite the economic reforms, the government should continue to take care of the poor and disadvantaged sections of society. Steps should be taken to minimize the burden of adjustment on them and ensure their well-being is not compromised.

Budget Speech 1999–2000[85]*Shri Yashwant Sinha*

Based on the given context, here are some strategies that can be derived:

1. Revenue and fiscal deficit reduction: begin a medium-term process of reducing revenue and fiscal deficits to free up more resources for productive investment and growth, as well as to contain inflation.
2. Indirect tax reform: undertake a major reform of indirect taxes to promote productivity and employment.
3. Economic reforms: deepen and widen economic reforms in all major sectors and accelerate internal liberalization to release the productive energies and creativity of farmers, manufacturers, traders, and service providers.
4. Safeguard the economy: take measures to safeguard the economy from external shocks, revive exports, and stimulate domestic engines for growth revival.

[85]Budget Speech 1999–2000—https://www.indiabudget.gov.in/doc/bspeech/bs19992000.pdf

5. Strengthen knowledge-based industries: focus on strengthening knowledge-based industries to prepare for the challenges of the new millennium.
6. Revitalize public programs for human development: revitalize and redirect public programs for human development, with a focus on empowering the poor and vulnerable sections of society, such as Scheduled Castes, Scheduled Tribes, and Other Backward Classes.
7. Strengthen rural economy: implement a multipronged program to further strengthen the rural economy, recognizing the critical importance of agriculture.
8. Public sector reform and disinvestment: continue with a judicious mix of strengthening strategic units, privatizing nonstrategic ones through gradual disinvestment or strategic sale, and devising viable rehabilitation strategies for weak units.
9. Improve credit delivery to small-scale industries: take initiatives to improve the availability and delivery system of credit to small-scale industries, such as increasing the limit for composite loans, simplifying computation of working capital limits, and encouraging lending by banks to nonbanking finance companies for on-lending to the tiny sector.
10. National Program for Rural Industrialization: implement a National Program for Rural Industrialization (NPRI) to set up 100 rural clusters every year, aimed at boosting rural industrialization, benefiting rural artisans and unemployed youth, and reducing rural-urban disparities.
11. National Human Development Initiative (NHDI): implement the NHDI to empower vulnerable groups in society by providing access to food, healthcare, education, employment, and shelter. Strengthen the targeted public distribution system, expand and improve health infrastructure and services, and provide food security to indigent senior citizens through the "Annapurna" scheme.

These strategies aim to address various challenges and promote inclusive and sustainable economic growth, poverty reduction, and human development.

Budget Speech 2000–2001[86]*Shri Yashwant Sinha*

In the given context, the following strategies have been mentioned to support micro-, small, and medium enterprises (MSMEs):

1. Providing credit support: measures have been taken to improve credit flow to MSMEs by reducing collateral requirements for small loans and increasing the composite loan limit for small borrowers.
2. Establishing specialized SSI bank branches: public sector banks are encouraged to set up specialized SSI branches in every district and SSI clusters within districts to improve banking services for MSMEs.

[86] Budget Speech 2000–2001—https://www.indiabudget.gov.in/doc/bspeech/bs200001.pdf

3. Launching a credit guarantee scheme: a central scheme has been formulated to provide credit guarantee for MSMEs. This scheme will cover loans up to Rs.10 lakhs from the banking sector.
4. Increasing equity support: the limit for equity support under the National Equity Fund Scheme, administered by SIDBI, will be raised from Rs. 15 lakh to Rs. 25 lakh.
5. Promoting technology development and modernization: the Technology Development Modernization Fund Scheme, administered by SIDBI, is being extended for another 3 years to support MSMEs in adopting modern technologies.

These are the strategies mentioned in the given context.

Budget Speech 2001–2002[87] *Shri Yashwant Sinha*

Based on the given context, here are some strategies that can be considered:

1. Speeding up agricultural sector reforms and better management of the food economy: this can be achieved by implementing policies that improve irrigation systems, provide better access to credit for farmers, and promote the use of advanced farming techniques.
2. Intensification of infrastructure investment: increase public investment in infrastructure projects such as roads, bridges, ports, and power plants. This will not only create jobs but also attract private investment and stimulate economic growth.
3. Continued reform in the financial sector and capital markets: implement measures to deepen the debt market, promote transparency, and encourage participation of small investors. This will provide long-term funding options for infrastructure projects and boost investor confidence.
4. Human development through better educational opportunities and social security programs: invest in education and skills development programs to enhance the quality of the workforce. Implement social security programs to provide a safety net for vulnerable populations.
5. Stringent expenditure control and rationalization of subsidies: implement measures to control nonproductive expenditure, reduce subsidies to unaffordable levels, and ensure that subsidies reach the deserving beneficiaries.
6. Acceleration of the privatization process and restructuring of public enterprises: continue with the privatization of state-owned enterprises to improve efficiency and attract private investment. Restructure public enterprises to make them more competitive and profitable.
7. *Revenue enhancement through widening the tax base and administration of a fair and equitable tax regime: implement measures to broaden the tax base,*

[87] Budget Speech 2001–2002—https://www.indiabudget.gov.in/doc/bspeech/bs200102.pdf

reduce tax evasion, and ensure a fair and transparent tax system. This will help increase government revenue and reduce the fiscal deficit.

8. Strengthening the management of public sector banks: provide greater autonomy to bank managements and abolish recruitment boards to improve efficiency and decision-making in public sector banks.
9. Liberalizing the capital account: further liberalize the capital account to attract foreign investment and encourage Indian companies to invest abroad. This can be done by relaxing restrictions on overseas investments and simplifying regulations.
10. Focus on power sector reforms: implement measures to reduce transmission and distribution losses, stop electricity theft, and impose economic tariffs. Restore financial viability of State Electricity Boards through reforms and milestones set in MOUs with state governments.

These strategies aim to address the concerns mentioned in the context and promote sustainable economic growth.

Budget Speech 2003–2004[88] *Shri Jaswant Singh*

Based on the given context, here are some strategies that could be considered:

1. Diversification and value chain development in agriculture: encourage farmers to diversify their crops and move up the value chain by adopting modern farming techniques, utilizing technology, and accessing market information to meet the changing demands of consumers.
2. Addressing land degradation and water logging: implement measures to combat land degradation and water logging, such as promoting sustainable farming practices, watershed management, and efficient irrigation systems.
3. Infrastructure development: focus on the development of both physical and social infrastructure, including roads, bridges, schools, healthcare facilities, and access to clean water. Encourage public-private partnerships to leverage private sector investment in infrastructure projects.
4. Promoting manufacturing sectors: identify and support specific manufacturing sectors, particularly those that rely on knowledge and technology, to enhance growth, improve incomes, generate employment, and promote exports.
5. Fiscal consolidation: emphasize the need for fiscal consolidation to eliminate budgetary drags and self-laid traps that hinder economic growth. This can be achieved through efficient allocation of resources, reducing wasteful expenditure, and increasing revenue generation.
6. Reduction of inequalities: address both social and economic inequalities by implementing reforms that promote inclusive growth and ensure a wider

[88] Budget Speech 2003–2004—https://www.indiabudget.gov.in/doc/bspeech/bs200304.pdf

distributive spread of national wealth. This may include targeted social welfare programs, skills development initiatives, and measures to enhance income levels for all citizens.

7. De-bureaucratization: streamline bureaucratic processes, and reduce red tape to facilitate ease of doing business and encourage entrepreneurship. This can be achieved through simplifying regulations, digitizing government services, and promoting a culture of efficiency and accountability.
8. Leveraging public-private partnerships: explore opportunities to leverage private sector investment in key sectors such as infrastructure, research and development, and renewable energy. This can help mobilize additional resources and expertise to accelerate development initiatives.

It is important to note that these strategies are based on the given context and may need to be further tailored and adapted to specific circumstances and priorities.

Budget Speech 2004–2005[89] Shri P. Chidambaram

Based on the provided context, here are some strategies that could be considered:

1. Empowering the poor: focus on providing universal access to education and healthcare and facilitating their full participation in the growth process through gainful employment. This can be achieved by investing in education and skills development programs, creating job opportunities, and ensuring access to affordable healthcare.
2. Poverty alleviation programs: expand and strengthen programs like the Antyodaya Anna Yojana, which provides subsidized food grains to poor families. Increase the coverage of such programs, and ensure efficient distribution of essential commodities.
3. Investment in agriculture and rural infrastructure: double agricultural credit in 3 years, accelerate the completion of irrigation projects, and invest in rural infrastructure. This will help improve agricultural productivity, create employment opportunities, and enhance the overall development of rural areas.
4. Focus on education and healthcare: levy access to generate additional funds for education, with a special emphasis on providing basic education to all children. Upgrade technical education institutes to meet industry demands and promote public-private partnerships in this sector. Also, ensure access to quality healthcare through health insurance schemes.
5. Employment generation: launch the National Employment Guarantee Act to guarantee 100 days of employment in a year to one able-bodied person in every poor household. Additionally, implement the Food for Work program in backward districts to provide employment opportunities and create durable assets.

[89] Budget Speech 2004–2004—https://www.indiabudget.gov.in/doc/bspeech/bs200405.pdf

6. Welfare of marginalized communities: allocate specific funds for the welfare of Scheduled Castes, Scheduled Tribes, and minorities. Provide reservations in various plan schemes and focus on their education and overall development.
7. Microfinance and self-help groups: promote microfinance initiatives and strengthen the self-help group (SHG)-bank linkage program. Encourage matured SHGs to transition from credit to starting microenterprises, thereby promoting entrepreneurship and economic empowerment.
8. Fiscal prudence and financial discipline: while enhancing investment in all sectors of the economy, maintain fiscal prudence and financial discipline. Consider the availability of resources and the absorptive capacity of various sectors.

These strategies aim to address the challenges of poverty, unemployment, and inequality while promoting inclusive growth and development.

Budget Speech 2005–2006[90] *Shri P. Chidambaram*

Based on the given context, here are some strategies that can be derived:

1. Strengthen and modernize agricultural universities and research institutions to enhance agricultural research and development.
2. Create a National Fund for Strategic Agricultural Research to provide financial support for agricultural research initiatives.
3. Scale up manufacturing capacities to global standards by implementing the recommendations of the Investment Commission and the National Manufacturing Competitiveness Council.
4. Launch the Manufacturing Competitiveness Programme to support small and medium enterprises in strengthening their operations and competitiveness.
5. Focus on the textile sector by continuing the Technology Upgradation Fund (TUF) scheme and introducing a 10% capital subsidy scheme for the textile processing sector.
6. Adopt a cluster development approach for the production and marketing of handloom products to support the handloom sector.
7. Expand the coverage of the life insurance scheme for handloom weavers.
8. Provide a supportive policy environment and stable tax policies for the services sector, recognizing the leading role played by the private sector.
9. Provide equity support and loans to central public sector enterprises to promote their growth and development.
10. Promote savings and channel them into productive investment through the capital market, banks, insurance companies, pension funds, and superannuation funds.

[90] Budget Speech 2005–2006—https://www.indiabudget.gov.in/doc/bspeech/bs200506.pdf

11. Address the problems faced by farmers with a sense of urgency by supporting states in their public investment in agriculture.
12. Prepare a roadmap for agricultural diversification, focusing on fruits, vegetables, flowers, dairy, poultry, fisheries, pulses, and oilseeds.
13. Launch the National Horticulture Mission to ensure an end-to-end approach in horticulture production, postharvest management, processing, and marketing.
14. Introduce a program for massive replantation and rejuvenation in the plantation sector, particularly for tea.
15. Implement gender budgeting to ensure gender sensitivities in budgetary allocations and benefit-incidence analyses.
16. Increase equity support for the National Minorities Development and Finance Corporation to bring minorities more into the development process.
17. Allocate resources for preexamination coaching of candidates belonging to minority communities.
18. Establish a special purpose vehicle (SPV) to finance infrastructure projects in sectors like roads, ports, airports, and tourism.
19. Provide "viability gap" funding for infrastructure projects to bridge the financial gap.
20. Implement the PURA (Provision of Urban Amenities in Rural Areas) concept to address issues like unemployment, isolation from markets, lack of connectivity, and migration to cities.
21. Launch the National Urban Renewal Mission to renew and develop cities to cope with rapid urbanization.

Budget Speech 2006–2007[91] Shri P. Chidambaram

Based on the provided context, here are some strategies that could be considered:

1. Increase farm credit: The government can continue to encourage banks to increase the level of credit to farmers and add more farmers to their portfolio. This can be done by providing incentives to banks and closely monitoring their progress.
2. Focus on repair and restoration of water bodies: The government can prioritize the repair, renovation, and restoration of water bodies in different states. This can be done by signing memorandums of understanding with the participating state governments and allocating funds for these projects.
3. Promote planned urbanization: the government can actively promote the establishment of new towns, preferably focused on specific industries or themes. This can attract investment, create employment opportunities, and improve the quality of life in these areas.

[91] Budget Speech 2006–2007—https://www.indiabudget.gov.in/doc/bspeech/bs200607.pdf

4. *Increase social assistance programs: the government can increase the pension amount for destitute persons above the age of 65 years and work toward establishing a system where the pension is directly credited to the beneficiary's account in a post office or bank. Additionally, the government can urge state governments to make equal contributions to increase the pension amount.*
5. *Gender-sensitive budgeting: The government can continue to prioritize gender budgeting by allocating funds to schemes that benefit women or have a significant allocation targeted toward women. Ministries and departments can also establish gender budgeting cells to ensure a gender perspective in their budgets.*
6. Welfare of Scheduled Castes and Scheduled Tribes: the government can enhance the allocations for schemes benefiting SCs and STs and increase the equity contribution to organizations like the National SC Finance and Development Corporation and the National Safai Karamchari Finance and Development Corporation.
7. Infrastructure development: the government can focus on infrastructure development, such as the construction of highways and the development of ports. This can be done by restructuring organizations like the National Highway Authority of India and increasing the plan allocation for the Department of Shipping.
8. Financial sector reforms: the government can continue to implement reforms in the banking sector, such as converting non-tradable special securities into tradable securities. This can provide banks with additional resources for lending to the productive sectors of the economy.

These strategies can help in promoting economic growth, improving the livelihoods of farmers, ensuring social welfare, and enhancing infrastructure development in the country.

Budget Speech 2007–2008[92]Shri P. Chidambaram

Based on the given context, here are some strategies that can be considered:

1. Increase farm credit: continue to focus on increasing farm credit to support farmers and improve their economic viability. Set higher targets for farm credit, and bring more farmers into the institutional credit system.
2. Implement special plans for distressed districts: allocate funds for implementing special plans in distressed districts, especially for water-related schemes and subsidiary income generation for farmers.
3. Address agricultural indebtedness: act on the recommendations of the Committee on Agricultural Indebtedness to address the issue of agricultural indebtedness and provide relief to farmers.

[92]Budget Speech 2007–2008—https://www.indiabudget.gov.in/doc/bspeech/bs200708.pdf

4. Focus on improving pulses production: address the stagnation in pulses production by expanding the availability and quality of certified seeds and implementing measures to boost production and productivity.
5. Invest in infrastructure: tap into resources available outside the budget, and leverage them for investment, especially in the infrastructure sector. This will help in improving the overall economic growth and development.
6. Increase allocation for major sectors: allocate substantial funds for major sectors like education, health, and Bharat Nirman to ensure inclusive growth and development.
7. Strengthen Sarva Shiksha Abhiyan (SSA): increase the allocation for SSA to provide quality education and improve access to education for all children.
8. Encourage investment: continue to encourage investment in the economy by providing a conducive environment for both domestic and foreign investors. This can be done through policy reforms, ease of doing business, and providing incentives for investment.
9. Focus on power sector: take new initiatives in the power sector to increase electricity generation capacity and address the growing demand for power. This can include awarding Ultra Mega Power Projects (UMPPs) and facilitating private participation in power generation and transmission projects.
10. Improve infrastructure development: focus on improving infrastructure development in areas like roads, water resources, and rural infrastructure. This will help in boosting economic growth and improving the quality of life for people in rural areas.

Budget Speech 2008–2009[93] Shri P. Chidambaram

Based on the given context, here are some strategies that can be considered:

1. Strengthening and expanding existing schemes: continue to invest in and improve the effectiveness of schemes like the National Rural Employment Guarantee Scheme, Midday Meal Scheme, National Rural Health Mission, and Kasturba Gandhi Balika Vidyalaya Scheme. This will further empower marginalized communities, bridge gender gaps, and improve access to education and healthcare.
2. Focus on infrastructure development: increase the pace of providing basic amenities like drinking water, all-weather roads, telephones, and electricity to rural areas. This will improve the quality of life and create opportunities for economic growth in these regions.
3. Enhancing skills development: establish a world-class skills development program to address the challenge of imparting skills required by a growing economy. This program should be scalable and involve collaboration between the government, private sector, and international sources to ensure adequate funding and expertise.

[93] Budget Speech 2008–2008—https://www.indiabudget.gov.in/budget_archive/ub2008-09/bs/speecha.htm

4. Upgrading vocational training institutes: allocate funds for the upgradation of industrial trainInstitutes (ITIs) to ensure that they are equipped to provide relevant and high-quality vocational training. This will help in developing a skilled workforce to support a knowledge- and technology-driven economy.
5. Improving defense forces recruitment: allocate funds for the improvement of infrastructure in Sainik Schools to attract and train future leaders of the defense forces. This will help address the issue of attrition at the officer level and ensure a strong defense system.
6. Strengthening the public distribution system: implement measures to improve the efficiency and effectiveness of the public distribution system to ensure that essential commodities reach the intended beneficiaries. This will help in reducing poverty and ensuring food security.
7. Promoting the growth of key sectors: increase allocation for sectors like coal, information technology, textiles, and micro-, small, and medium enterprises (MSMEs) to support their growth and development. This will create employment opportunities and contribute to economic growth.
8. *Encouraging investment and entrepreneurship: create a conducive environment for investment and entrepreneurship by implementing policies that promote ease of doing business, provide financial incentives, and support innovation and research and development.*
9. Strengthening governance and accountability: ensure close monitoring of program implementation, enforce accountability, and measure outcomes in terms of targets achieved and their quality. This will help in improving the efficiency and effectiveness of government programs and ensure that resources are utilized optimally.
10. Promoting inclusive growth: continuously strive to achieve inclusive growth by focusing on reducing poverty, eliminating inequality, and ensuring equitable distribution of wealth. This can be done by implementing policies that promote social justice, equal opportunities, and inclusive development.

Budget Speech 2008–2009[94]*Shri P. Chidambaram*

Based on the given context, here are some strategies that could be considered:

1. Accelerate policy reforms: focus on implementing policy reforms, especially in the financial sector, to make the economy more competitive and attract investments.
2. Strengthen economic regulatory and oversight systems: improve the efficiency and effectiveness of economic regulatory and oversight systems to bring the economy back to a 9% growth path.
3. Expand employment generation schemes: increase the scope and reach of employment generation schemes to create more job opportunities, especially for the vulnerable sections of society.

[94] Budget Speech 2008–2009—https://www.indiabudget.gov.in/budget_archive/ub2008-09/bs/speecha.htm

4. Strengthen social security nets: enhance social security nets to protect the weak and downtrodden sections of society, ensuring their financial and social well-being.
5. Increase investment in agriculture: continue to allocate significant funds for the agriculture sector, focusing on initiatives that promote growth and development in the agricultural and allied sectors.
6. Enhance credit availability for farmers: implement measures to increase the flow of credit to the agriculture sector, ensuring that farmers have access to affordable and timely credit for their farming activities
7. Focus on inclusive growth: develop and implement policies that promote inclusive growth, reducing the proportion of people living below the poverty line and creating opportunities for all sections of society.
8. Improve delivery mechanisms: strengthen the delivery mechanisms to ensure that development programs and benefits reach the intended beneficiaries efficiently and effectively.
9. Establish a unique identification system: implement a comprehensive system of unique identification for the resident population of the country to streamline the delivery of development programs and services.
10. Recapitalize public sector banks: provide necessary capital infusion to public sector banks to maintain their capital adequacy ratio and ensure continued credit growth to support economic growth.
11. Increase infrastructure and rural development expenditure: consider increasing plan expenditure, particularly in infrastructure development and rural development, to stimulate the economy and bridge existing gaps.
12. Consider contracyclical policies: in times of economic slowdown, consider implementing contracyclical policies, such as increasing expenditure in infrastructure and rural development, to counter the negative impact of the global recession.

Please note that these strategies are based on the given context and may not cover all aspects or be comprehensive.

Budget Speech 2009–2010 (Interim)[95] *Shri Pranab Mukherjee*

Based on the given context, here are some strategies that could be considered:

1. Accelerate policy reforms: focus on implementing policy reforms, especially in the financial sector, to make the economy more competitive and attract investments.

[95] Budget Speech 2009–2010—https://www.indiabudget.gov.in/budget_archive/ub2009-10(I)/bs/speecha.htm

2. Strengthen economic regulatory and oversight systems: improve the efficiency and effectiveness of economic regulatory and oversight systems to bring the economy back to a 9% growth path.
3. Expand employment generation schemes: increase the scope and reach of employment generation schemes to create more job opportunities, especially for the vulnerable sections of society.
4. Strengthen social security nets: enhance social security nets to protect the weak and downtrodden sections of society, ensuring their financial and social well-being.
5. Increase investment in agriculture: continue to allocate significant funds for the agriculture sector, focusing on initiatives that promote growth and development in the agricultural and allied sectors.
6. Enhance credit availability for farmers: implement measures to increase the flow of credit to the agriculture sector, ensuring that farmers have access to affordable and timely credit for their farming activities.
7. Focus on inclusive growth: develop and implement policies that promote inclusive growth, reducing the proportion of people living below the poverty line and creating opportunities for all sections of society.
8. Improve delivery mechanisms: strengthen the delivery mechanisms to ensure that development programs and benefits reach the intended beneficiaries efficiently and effectively.
9. Establish a unique identification system: implement a comprehensive system of unique identification for the resident population of the country to streamline the delivery of development programs and services.
10. Recapitalize public sector banks: provide necessary capital infusion to public sector banks to maintain their capital adequacy ratio and ensure continued credit growth to support economic growth.
11. Increase infrastructure and rural development expenditure: consider increasing plan expenditure, particularly in infrastructure development and rural development, to stimulate the economy and bridge existing gaps.
12. *Consider contracyclical policies: in times of economicslowdown, consider implementing contracyclical policies, such as increasing expenditure in infrastructure and rural development, to counter the negative impact of theglobal recession.*

Please note that these strategies are based on the given context and may not cover all aspects or be comprehensive.

Budget Speech 2012–2013[96]Shri Pranab Mukherjee

Based on the given context, here are some strategies that can be considered:

[96] Budget Speech Budget 2012–2013—https://www.indiabudget.gov.in/doc/bspeech/bs201213.pdf

1. Focus on domestic demand-driven growth recovery: encourage policies and initiatives that stimulate domestic consumption and investment, such as tax incentives for individuals and businesses, infrastructure development, and support for small and medium enterprises.
2. Accelerate the pace of reforms: implement reforms that improve the ease of doing business, attract foreign investment, and promote competition in key sectors. This can include streamlining regulations, reducing bureaucratic red tape, and enhancing transparency and governance.
3. Address supply bottlenecks: prioritize investments in critical sectors like coal, power, national highways, railways, and civil aviation to address infrastructure gaps and improve efficiency. This can involve public-private partnerships, attracting private investment, and implementing technology-driven solutions.
4. Improve delivery systems and governance: enhance the efficiency and effectiveness of government programs and services by leveraging technology, promoting transparency, and reducing corruption. This can include the use of mobile-based systems for subsidy transfers, Aadhaar validation for public distribution systems, and online portals for customer service.
5. Strengthen the investment environment: fast-track policy decisions, and ensure timely implementation of major projects to boost investor confidence. This can involve creating a conducive business environment, providing clarity on regulations, and engaging with state governments to build consensus on key reforms.
6. Encourage foreign direct investment: continue efforts to attract foreign investment, particularly in sectors like retail, by addressing concerns and building consensus with state governments. This can involve providing incentives, simplifying regulations, and promoting the benefits of FDI in terms of reducing costs and benefiting consumers and producers.
7. Deepen the capital market: implement measures to encourage savings and investment in the domestic capital market, such as introducing schemes like the Rajiv Gandhi Equity Savings Scheme. This can involve providing tax incentives for retail investors, increasing the investment limit for foreign investors, and expanding the range of eligible financial instruments.

It is important to note that these strategies are based on the given context and may need to be adapted to the specific circumstances and priorities of the Indian economy.

Budget Speech 2014–2015 (Interim)[97] *Shri Pranab Mukherjee*

Based on the given context, here are some strategies that can be derived:

[97] Budget Speech Budget 2014–2015—https://www.indiabudget.gov.in/doc/bspeech/bs.pdf

1. Implement minimum tariff protection: introduce a minimum tariff protection policy to incentivize manufacturing goods in India rather than importing them. This will help boost domestic production and reduce reliance on imports.
2. *Prioritizesubsidies: evaluate and prioritizesubsidiesbased on their necessity and deserving recipients. Allocatesubsidiesonly to those who truly need them, ensuring efficient use of limited resources.*
3. Address urban decay: focus on addressing the decay in cities by tapping into their wealth and resources. Implement a new model of governance to rebuild cities and make them more livable and governable.
4. *Emphasize skills development: give equal importance to skills development alongside secondary and universityeducation, total sanitation, and universal healthcare. Prioritize skills development to meet the demands of the job market and enhance employability.*
5. Share responsibility between states and center: encourage states to bear a reasonable proportion of the financial costs of implementing flagship programs. This will allow the central government to allocate more resources to sectors like defense, railways, national highways, and telecommunications.
6. Introduce GST and DTC: pass the goods and services tax (GST) laws and the Direct Taxes Code (DTC) to modernize tax laws and improve revenue generation. These reforms will help streamline the tax system and promote economic growth.
7. Establish research funding organization: set up a research funding organization to fund research projects through a competitive process. Provide tax benefits to individuals and organizations contributing to this organization. This will promote scientific research and innovation.
8. Crackdown on offshore accounts: continue investigations into illegal offshore accounts held by Indians. Take necessary actions to determine tax liabilities and impose penalties on offenders. Strengthen efforts to obtain information from countries concerned, and prosecute those involved in tax evasion.
9. *Focus on agriculture: continue supporting the agriculture sector by promoting increased food grain production, exports, and agricultural credit. Implement policies to ensure sustainable growth, and improve the livelihoods of farmers.*
10. *Encourage investment: take measures to boost investment in sectors like mining andmanufacturing. Address issues that hinder investment growth, and aim for a higher growth rate by maintaining a favorable investment climate.*
11. Provide relief for education loans: introduce a moratorium period for education loans taken up to a certain date and outstanding on a specific date. Take over the liability for outstanding interest up to a certain date, and provide relief to borrowers. Allocate funds to support this initiative.
12. Strengthen financial markets: implement measures to deepen and strengthen the Indian financial markets. This includes revamping the ADR/GDR scheme, liberalizing the rupee-denominated corporate bond market, and creating a comprehensive record for all financial assets.
13. Regulate commodity derivatives market: strengthen the regulatory framework of the commodity derivatives market to prevent payment crises, and ensure the

stability of the financial market. Amend relevant acts to address any loopholes and improve oversight.

Budget Speech 2014–2015[98] *Shri Arun Jaitley*

Based on the given context, here are some strategies that could be considered:

1. Focus on higher growth: the Finance Minister emphasizes the importance of achieving higher growth for the Indian economy. To achieve this, the government could implement policies that promote investment, innovation, and entrepreneurship. This could include providing incentives for businesses, improving ease of doing business, and investing in infrastructure development.
2. Targeted anti-poverty programs: the Finance Minister acknowledges the large population below the poverty line and the need to ensure that anti-poverty programs are well-targeted. The government could develop and implement programs that specifically address the needs of the poor, such as providing access to education, healthcare, and employment opportunities.
3. Fiscal prudence and consolidation: the Finance Minister highlights the importance of fiscal prudence and reducing fiscal deficit. The government could focus on improving tax collection, reducing wasteful expenditure, and implementing measures to increase nontax revenues. This could help generate more resources for the economy and reduce the burden of debt on future generations.
4. Strengthening the financial sector: the Finance Minister recognizes the importance of a strong and modern financial sector for economic growth. The government could work on strengthening the legislative and regulatory framework, as well as promoting the development of capital markets. This could include measures to deepen the corporate bond market, liberalize the ADR/GDR regime, and introduce new financial instruments like REITs and InvITs.
5. Promoting skills development and employment: the Finance Minister proposes the launch of a national multi-skill program called Skill India, which aims to skill the youth and provide training and support for traditional professions. The government could focus on implementing this program effectively, partnering with industries to provide vocational training, and creating an environment conducive to job creation.
6. Enhancing financial inclusion: the Finance Minister announces the launch of a Financial Inclusion Mission to provide banking services to all households in the country. The government could work on expanding access to banking services, particularly for marginalized sections of society, through initiatives like opening bank accounts for every household and providing credit facilities.
7. Encouraging private sector participation in infrastructure: the Finance Minister acknowledges the need for long-term financing for infrastructure projects and

[98] Budget Speech Budget 2014–2015—https://www.indiabudget.gov.in/doc/bspeech/bs201415.pdf

encourages larger private sector participation. The government could create a favorable environment for private investment in infrastructure by providing flexible loan structures, reducing regulatory preemption, and simplifying licensing processes for differentiated banks serving niche interests.

These strategies are based on the given context and can be further developed and tailored to specific needs and priorities.

Budget Speech 2016–2017[99]Shri Arun Jaitley

Based on the given context, here are some strategies that can be derived:

1. Strengthening the banking sector: the government should continue with the implementation of the "Plan for Revamping of Public Sector Banks" and provide necessary support to address the issue of stressed assets. This can be done through recapitalization of banks and providing additional capital if required.
2. Enhancing financial inclusion: the government should focus on promoting schemes like the Pradhan Mantri Mudra Yojana (PMMY) to provide financial assistance to small entrepreneurs at the bottom of the pyramid. This will help in creating more job opportunities and boosting economic growth.
3. Promoting entrepreneurship: the government should implement schemes like the Stand Up India Scheme to promote entrepreneurship among marginalized sections of society such as SC/ST and women. This can be done by providing financial support and professional assistance to entrepreneurs.
4. Improving infrastructure: the government should focus on infrastructure development by implementing projects in sectors like power, coal, highways, sugar, and steel. This will not only create employment opportunities but also attract more investments.
5. Encouraging foreign direct investment (FDI): the government should continue to make reforms in the FDI policy to attract more investments in sectors like insurance, pension, asset reconstruction companies, and stock exchanges. This will help in boosting economic growth and creating employment opportunities.
6. Strengthening dispute resolution mechanisms: the government should develop a comprehensive legal framework for dispute resolution in infrastructure projects, PPP contracts, and public utility contracts. This will help in resolving disputes efficiently and promoting transparency in contract agreements.
7. Promoting skills development and education: the government should allocate more funds for skills development programs and focus on improving the quality of education. This can be done by providing an enabling regulatory framework

[99] Budget Speech Budget 2016–2017—https://www.indiabudget.gov.in/doc/bspeech/bs201617.pdf

for educational institutions and opening new Navodaya Vidyalayas in uncovered districts.

8. Supporting the export sector: the government should continue to take measures to support the export sector by widening and deepening the duty drawback scheme. This will help in promoting exports and boosting economic growth.
9. Addressing the needs of farmers and food processing industry: the government should allow 100% FDI in the marketing of food products produced and manufactured in India. This will benefit farmers, promote the food processing industry, and create employment opportunities.
10. Implementing comprehensive legislation: the government should bring in comprehensive central legislation to deal with illicit deposit-taking schemes and strengthen the Securities Appellate Tribunal by amending the SEBI Act. This will help in protecting the interests of the financially illiterate and ensuring fair practices in the financial sector.

Budget Speech 2022–2023[100] Smt. Nirmala Sitharaman

Based on the provided context, here are some strategies that can be considered:

- Promote mass transit: Encourage the use of public transport in urban areas by investing in the development and expansion of mass transit systems. This can include improving the quality and accessibility of public transportation, implementing clean tech solutions, and creating special mobility zones with zero fossil-fuel policies.
- Support sustainable mobility: encourage the adoption of electric vehicles (EVs) by formulating a battery swapping policy and setting interoperability standards. Encourage the private sector to develop sustainable and innovative business models for "Battery or Energy as a Service" to improve efficiency in the EV ecosystem.
- Enhance land records management: encourage states to adopt Unique Land Parcel Identification Numbers for efficient management of land records. Implement transliteration of land records across different languages, and promote the adoption or linkage with the National Generic Document Registration System for uniform and efficient registration processes.
- Promote clean energy transition: allocate funds for the production of high-efficiency solar modules, and incentivize domestic manufacturing of solar equipment. Emphasize the transition to a circular economy by addressing crosscutting issues such as infrastructure, reverse logistics, technology upgradation, and integration with the informal sector.

[100] Budget Speech 2009–2010—https://www.indiabudget.gov.in/budget_archive/ub2009-10(I)/bs/speecha.htm

- Strengthen digital infrastructure: allocate funds for the development of broadband and mobile services in rural and remote areas. Promote research and development, as well as commercialization of technologies and solutions, to ensure affordable and accessible digital services for all.
- Foster domestic defense industry: increase the allocation of the capital procurement budget for domestic industry in the defense sector. Encourage defense research and development by earmarking a portion of the defense R&D budget for industry, start-ups, and academia. Facilitate collaboration between private industry, DRDO, and other organizations for the design and development of military platforms and equipment.
- Tap into sunrise opportunities: focus on sectors with immense potential for sustainable development and modernization, such as artificial intelligence, geospatial systems and drones, semiconductors, space economy, genomics and pharmaceuticals, green energy, and clean mobility systems. Promote research, development, and collaboration among academia, industry, and public institutions in these sectors.

These strategies aim to promote sustainable development, enhance infrastructure, create employment opportunities, and strengthen domestic capacities in various sectors.

Budget Speech 2022–2024[101] *Smt. Nirmala Sitharaman*

1. Economic empowerment of women
   - Provide financial support and access to credit for women-led self-help groups.
   - Offer advanced skills training and knowledge of modern digital techniques.
   - Facilitate branding and marketing support for their products.
   - Encourage the formation of large producer enterprises or collectives.
   - Enable scaling up of operations to serve larger consumer markets.
2. PM Vishwakarma Kaushal Samman (PM VIKAS)
   - Provide financial assistance to traditional artisans and craftspeople.
   - Offer access to advanced skills training and knowledge of modern digital techniques.
   - Promote integration with the MSME value chain.
   - Facilitate brand promotion and linkage with local and global markets.
   - Ensure social security for artisans belonging to marginalized sections.
3. Tourism

[101] Budget Speech 2009–2010—https://www.indiabudget.gov.in/budget_archive/ub2009-10(I)/bs/speecha.htm

- Promote tourism on a mission mode with active participation of states.
- Encourage convergence of government programs and public-private partnerships.
- Develop and improve urban infrastructure in tourist destinations.
- Enhance the availability and affordability of urban land for tourism development.
- Focus on skills development and entrepreneurship opportunities in the tourism sector.

4. Sustainable cities of tomorrow

- Encourage urban planning reforms for efficient use of land resources.
- Provide adequate resources for urban infrastructure development.
- Promote transit-oriented development for better connectivity.
- Incentivize cities to improve creditworthiness for municipal bonds.
- Focus on urban sanitation and waste management for sustainable cities.

5. Aspirational districts and blocks program

- Ensure saturation of essential government services in 500 blocks.
- Improve socioeconomic conditions of vulnerable tribal groups.
- Provide basic facilities such as housing, clean water, education, and healthcare.
- Enhance road and telecom connectivity in tribal areas.
- Create sustainable livelihood opportunities for tribal communities.

6. Teachers' training

- Implement innovative pedagogy and curriculum transaction methods.
- Provide continuous professional development for teachers.
- Use technology and ICT tools for effective teaching and learning.
- Establish vibrant institutes of excellence for teacher training.
- Conduct regular dipstick surveys to assess the quality of education.

These strategies aim to empower women, promote traditional artisans, boost tourism, develop sustainable cities, uplift vulnerable tribal groups, and enhance the quality of education.

Index

A
Adaptive machine learning, 66–67
Akaike Knowledge Criterion (AIC), 21
Artificial intelligence (AI), 1–6, 32, 36, 50, 57, 71–77, 90, 98, 122
Augmented Dickey-Fuller (ADF), 13–17, 19, 20, 166, 170, 173, 240, 249
Autonomously, 2, 72
Auto-regressive (AR), vii, 21, 77, 80
Autoregressive integrated moving average (ARIMA), 21

B
Bag of words (BOW), 53, 56–66
Balance of payment (BOP), 58, 134, 138, 139, 159, 180, 186, 187, 190, 191, 193, 195, 196, 206, 228–230, 233–235, 344, 366, 376, 383, 422, 423, 428, 432
Bank credits, 146, 149–153, 156, 177, 199, 233, 235, 241, 244, 254, 256, 352–353, 364, 373, 414
Bidirectional encoder representations from transformers (BERT), 79–83
Boosting regressor, 30, 154–155, 219–220
Breadbasket, 276, 285
Budget speeches, 36, 38, 43, 46, 47, 53–65, 67, 68, 78–80, 98, 100, 101, 108, 115, 120–129, 138, 139, 174, 197, 198, 208, 229, 265, 272, 294, 298, 299, 305, 319, 327, 373, 375–455, 457–460

C
Cash reserve ratio (CRR), 144–157, 430
Chain-of-thought, 98, 99, 103, 119
ChatGPT, 77, 83
Churchill, Winston, 269
Climate change, 134, 186, 194, 195, 225, 271, 281, 285, 289, 298, 300, 324, 414
Climate Change Knowledge Portal (CCKP), 286, 287, 291
Consumer price index (CPI), 158, 159, 190, 232, 333, 337, 344, 364
Consumer Pyramid Household Survey (CPHS), 314–317
Contracyclical policies, 206, 453, 454
Cosine similarity, 46–50, 105
COVID 19, 140, 153, 180, 208, 212, 213, 215, 224, 229, 334, 336, 348, 362
Cow Necklace, 175, 185
Credit ratings, 137, 138, 179, 180, 389

D
Deep learning, 3, 5–6, 50–53, 73, 76, 77, 108
Deficit, 21, 57, 124, 136, 228, 342, 375
Democracy, 58, 373
Democracy movement, 98, 373
Digital public infrastructure (DPI), 175, 391, 407
Discrete Fourier Transform, 6
Downgrade, 138, 180

© The Editor(s) (if applicable) and The Author(s), under exclusive license to Springer Nature Switzerland AG 2024
C. Vuppalapati, *Assessing Policy Effectiveness using AI and Language Models*, International Series in Operations Research & Management Science 354, https://doi.org/10.1007/978-3-031-56097-2

E

Economic sustainability, 60–66, 100, 101, 125–127, 134, 187, 203, 225, 227, 265, 272, 300, 306, 375–391
Education, 46, 125, 134, 139, 179, 182, 194, 195, 225, 228, 272, 300, 306, 308, 309, 323, 325, 329, 331–333, 336, 346, 350, 352, 364, 365, 367, 373, 377, 380, 384, 386, 388, 389, 393, 394, 396, 397, 399–403, 408–410, 412, 413, 415–419, 427, 429, 438, 440, 441, 443–445, 447, 448, 451, 456–458, 461
Embedding, 49–53, 55–68, 80, 85–87, 103–106, 122–123
Exchange rate stability, 135, 431, 432
Exploratory data analysis (EDA), 10, 12

F

Fair price shop (FPS), 208–210, 226, 227, 236, 237, 255, 421
Fertiliser bonds, 57
Fertilizers, 55, 266, 268, 270, 271, 273, 286, 293, 294, 299, 320, 321, 332, 384, 392, 407, 410, 417, 420, 429, 439
Fiscal consolidation, 56, 57, 124, 230, 366, 382, 384, 398, 412, 433, 435, 446
Fiscal deficit, 57, 125–127, 139–142, 144, 157, 163, 166, 167, 174, 183, 186, 187, 199, 228, 265, 266, 322, 364, 376, 384, 388, 390, 408, 409, 411, 423–426, 428, 435, 436, 440, 442, 443, 446, 457
Fiscal discipline, 138–141, 299, 380, 381, 419, 423, 425, 427, 432, 442
Fiscal prudence, 135, 136, 267, 387, 430, 431, 436, 448, 457
Food Corporation of India (FCI), 210
Food security, 179, 207–210, 213, 226, 254–256, 266, 271, 275, 283, 285, 289, 294, 309, 319, 336, 351, 364, 404, 409, 412, 416, 417, 444, 452
Foreign direct investment (FDI), 22, 322, 337, 379, 389, 399, 416, 426, 455, 458, 459

G

General Language Understanding Evaluation (GLUE), 79
Generative pre-trained transformer (GPT), 77, 98–99, 101–103, 126
GFS analytics, 48–50
Gini index, 195, 322, 343
Global capital, 265, 367, 387
Global recession, 206, 366, 453, 454
Goddard Earth Observing System (GEOS), 283
Government finance statistics (GFS), 53, 58
Government Finance Statistics Manual (GFSM), 53, 58
Grassroots, 127, 175, 185, 198, 404
Gross domestic product (GDP), 22, 57, 125, 126, 140, 142, 158–161, 163, 166, 167, 170, 174, 177, 192, 195–198, 207, 208, 229, 230, 232, 233, 235, 237, 241, 242, 246, 267, 272, 322–329, 334, 337, 339–340, 342–346, 352, 364, 367, 373, 382, 384, 388, 424, 430, 435, 436

H

Haar wavelet, 9
Hanumayamma, 185
Headcount, 187, 195, 215, 310, 312, 313, 322, 329
Head count ratio (HCR), 310
Heat waves, 259, 275, 278–285, 288–299
Horizon, 7, 8

I

Inequalities, 90, 146, 175–187, 193–195, 206, 307, 308, 314–322, 324, 327, 329, 335–337, 366, 382, 385, 398, 412–414, 419, 432, 446, 448, 452
Inflation, 21, 124, 134, 204, 320, 373, 376
Inflation rates, 124, 158, 159, 161, 163, 166, 167, 170, 177, 232, 235, 236, 242, 246
Inflows, 192, 206, 229, 233, 235, 236, 240, 242, 246, 249, 250, 254, 255, 322, 337, 378
Intergovernmental Panel on Climate Change (IPCC), 280
International Monetary Fund (IMF), 53, 58, 187, 188, 317–319, 367
International payments, 58
Inverse document frequency (IDF), 51
Investment, 22, 134, 205, 307, 373, 375

J

Jawaharlal Nehru, 269

K

Keynes, 144
Kharif, 273
Kisan Credit Card, 177, 255, 395
Kissan Vikas Patra, 197

L
Labour, 91, 198, 341
Lal Bahadur Shastri, 269, 270
Large Language Model Meta AI (Llama), 83, 90–106
Large language models (LLMs), 38, 50, 56, 78, 79, 83, 90–129, 256–265, 333, 373
Legacy, 7, 8, 387
Loan-to-deposit ratio (LDR), 142, 181
Logic based technique, 3, 73
Lorenz curve, 322

M
Macroeconomic stability, 90, 124, 125, 134, 136, 139, 142, 146, 163, 179, 183, 187, 204, 205, 327, 366, 389, 390, 422–437
Main economic indicators (MEI), 188–193
Manufacturing, 1, 71, 90, 124, 125, 181–187, 191–193, 324, 326, 328, 353, 364, 384, 388, 398, 412, 414, 430, 431, 435, 437, 446, 448, 456, 459
Marginalized, 46, 78, 119–129, 138, 139, 157, 176, 180, 185, 205, 206, 215, 223, 225, 227, 250, 254, 256, 294, 298, 300, 306, 307, 333–367, 373, 380, 384, 394–396, 408, 414, 416, 418, 419, 426, 439, 448, 451, 457, 458, 460
Microfinance, 143, 159, 321, 399, 410, 413, 416, 418, 448
Middle-classes, 198, 330, 334
Minimum Support Prices (MSPs), 289, 290, 293, 294, 350, 351, 365
Minister of Finance, 55, 56, 91, 137, 139, 141, 144, 157, 205–207, 265, 266, 306, 325, 366
M1 money supply, 9
Monetary stability, 135, 145, 430, 431
Money Stock, 144, 160, 161, 163, 166, 170, 233, 235

N
Narasimham Committee, 145, 146
The National Food Security Act (NFSA), 210, 319
National Rural Employment Guarantee Scheme (NREGS), 207, 208, 309, 364, 385, 386, 402, 403, 414, 415, 417, 418, 451
National Sample Survey (NSS), 310–312, 314
National Sample Survey Office (NSSO), 310, 311, 329
National Saving certificates, 197
Natural language processing (NLP), vii, 4, 35–45, 49–63, 67, 68, 74, 77, 79, 80, 84, 85, 98, 99, 129
Natural Language Tool Kit (NLTK), 37–39
Nirmala Sitharaman, 389–393, 420–421, 437–439, 459, 460

O
OCED, 188–193
Off-Take, 211–213, 221, 223, 226–256, 300, 348–349, 363, 365
Ordinary Least Squares Regression (OLSR), 22

P
Part of Speech (POS), 39–40
Phased reduction, 143, 144, 157, 199
Pillars, 134, 225, 300, 367, 373
Plunge, 330
Polynomial regression, 30–32
Population growth, 18, 214, 216, 221, 223, 232, 235, 271, 300, 329, 333, 334, 352, 358, 363
Porter stemming, 41
Poverty, 46, 134, 204, 307, 373, 382
Poverty line, 195, 226, 237, 310, 312, 313, 317, 319, 326, 327, 332, 335, 364, 397, 406, 409, 417, 419, 433, 434, 453, 454, 457
Poverty line basket (PLB), 310, 313
Poverty reduction policies, 337
Poverty reduction strategies (PRS), 125, 129, 141, 182, 183, 185, 308, 326, 330, 364, 408–421
Preservation, 306
Price stability, 67, 68, 124, 134, 135, 137, 158, 179, 187, 205, 307, 319, 376, 389, 393, 422, 431, 432
Prime Minister, 43, 44, 269, 400, 404
Procyclical, 227, 231
Producer Price Index (PPI), 7, 190
Professor Madhu Dandavate, 375–376, 422–423, 439–440
Prompt engineering, 99–102, 124–129
PSUs, 137, 381, 436
Public distribution system (PDS), 139, 179, 199, 206–256, 300, 319–320, 336, 348–349, 351, 362–365, 397, 409, 421–422, 444, 452

R
Rabi, 273, 286
Ration cards, 210, 213, 226
Recessions, 227, 231, 334
Reducing poverty, 139, 176, 194, 327, 329, 336, 351, 355, 366, 413, 433, 452
Regional rural banks (RRBs), 159, 204, 308, 321, 394, 409
Restless, 265, 367, 387
Revenue deficit, 57, 139–141, 199, 228, 364, 376, 380, 408, 411, 423, 440
Revival of growth, 326
Rural areas, 36, 139, 144, 159, 185, 205, 210, 266, 299, 305, 306, 309, 312, 320–322, 325–327, 334, 351, 355, 365, 366, 375, 380, 393, 395, 396, 399, 402, 405, 410, 415, 417, 418, 421, 439, 440, 443, 447, 449, 451
Rural communities, 203, 267, 294, 306, 321, 395, 416
Rural credit, 394, 427

S
Sampling, 31, 111, 113–118, 259
Self-help groups (SHGs), 159, 268, 321, 332, 392, 395, 399, 403, 409, 410, 413, 420, 448, 460
Shanmukham Chetty, 67, 305, 306
Shortages, 53, 139, 186, 199, 207, 209–225, 269, 271, 285, 300, 320, 335, 336
Shortfall, 276, 343
Shri Arun Jaitley, 387–389, 405–406, 418–419, 436, 457, 458
Shri Manmohan Singh, 46, 55, 58, 137, 139, 144, 157, 205, 207, 266, 306, 376–380, 393–396, 408, 409, 423–425, 440–443
Shri P. Chidambaram, 59, 383–385, 399–402, 412–416, 421, 422, 428–432, 447–453
Shri Pranab Mukherjee, 56, 206, 265, 385–388, 403–407, 416–418, 433–435, 453–455
Shri Yashwant Sinha, 47, 141, 366, 380–382, 396–398, 409–411, 426–428, 441–446
Slowdown, 175, 206, 215, 335, 453, 454
Small and medium enterprises (SMEs), 53, 54, 401, 430
Social and economic sustainability, 36, 157, 185
Social sustainability, 53–60, 67, 90–129, 134, 137, 139, 158, 179, 185, 186, 199, 225–256, 272, 305–308, 331–333, 373, 392–407, 421
Stabilization, 115, 137–144, 186, 229, 376, 387, 408, 423, 429, 430, 435, 437, 440
Standard & Poor (S&P), 138
Stationarity, 13–14, 166, 167, 170, 171, 173, 240, 241, 246, 247, 249
Statutory liquidity ratio (SLR), 143–157, 159, 199, 265, 266, 430, 443
Sterling balances, 66
Subsidies, 53, 57, 124, 137, 139, 179, 207, 230, 236, 255, 266, 271, 272, 299, 319, 321, 329, 336, 346, 348, 363, 367, 373, 381, 383, 384, 400, 401, 406, 409, 411, 416–418, 423, 425, 427–429, 435, 436, 439, 445, 448, 455, 456
Surplus, 140, 146, 230, 272, 342, 366, 382, 411
Sustainability, 1, 134, 203, 305, 373, 379
Sustainable development, 134, 137, 176, 195, 271, 281, 336, 390, 399, 407, 414, 416, 419, 440, 460
Sustained growth, 67, 68, 134, 136, 139, 205, 265, 307, 376, 381, 382, 387, 393, 433, 436

T
Tariff, 58, 124, 398, 428, 446, 456
Term frequency (TF), 51
Term frequency-inverse document frequency (TF-IDF), 51–53
TFLOPS, 98
Thinking machines, 2, 72
Time-series, 1, 4–21, 32, 74, 75, 146, 160, 166, 168, 171, 172, 209, 234, 240, 244, 247, 248, 252, 254–260, 319, 323, 339–340
Tokens, 38, 39, 41, 72, 79, 92, 94, 98, 99, 101, 103
Top-K, 113–115, 118
Top-P, 115–118

U
Unemployment rates, 136, 187, 190, 232, 236, 241, 246, 250, 431
The United Nations Statistics Department (UNSD), 188–189
Uptick, 250, 300
US Senate, 8

V
Vector autoregression (VAR), 21, 169, 172, 173, 245, 248–250, 252
Vengeance, 336
Veterinary Sensor, 185
Voting rights, 98, 373

W
Wealth, 90, 91, 98, 124, 125, 187, 229, 299, 306–308, 322, 335, 376, 377, 382, 385, 393, 408, 447, 452, 456
Wholesale price index (WPI), 8, 125, 333, 388
World Bank, 187, 188, 193–198, 214, 232, 313, 317, 319, 339, 341–344, 352

SPRINGER NATURE

GPSR Compliance

The European Union's (EU) General Product Safety Regulation (GPSR) is a set of rules that requires consumer products to be safe and our obligations to ensure this.

If you have any concerns about our products, you can contact us on ProductSafety@springernature.com

In case Publisher is established outside the EU, the EU authorized representative is:

Springer Nature Customer Service Center GmbH
Europaplatz 3
69115 Heidelberg, Germany

The manufacturer's authorised representative in the EU is Springer Nature Customer Service Centre GmbH, Europaplatz 3, 69115 Heidelberg, Germany. If you have any concerns regarding our products, please contact ProductSafety@springernature.com

Printed and bound by CPI Group (UK) Ltd, Croydon, CR0 4YY
01/10/2026
02225865-0018